Quantum Theory from First Principles
An Informational Approach

Quantum theory is the soul of theoretical physics. It is not just a theory of specific physical systems, but rather a new framework with universal applicability. This book shows how we can reconstruct the theory from six information-theoretical principles, by rebuilding the quantum rules from the bottom up. Step by step, the reader will learn how to master the counterintuitive aspects of the quantum world, and how to efficiently reconstruct quantum information protocols from first principles. Using intuitive graphical notation to represent equations, and with shorter and more efficient derivations, the theory can be understood and assimilated with exceptional ease. Offering a radically new perspective on the field, the book contains an efficient course of quantum theory and quantum information for undergraduates. The book is aimed at researchers, professionals, and students in physics, computer science, and philosophy, as well as the curious outsider seeking a deeper understanding of the theory.

Giacomo Mauro D'Ariano is a Professor at Pavia University, where he teaches Quantum Mechanics and Foundations of Quantum Theory, and leads the group QUit. He is a Fellow of the American Physical Society and of the Optical Society of America, a member of the Academy Istituto Lombardo of Scienze e Lettere, of the Center for Photonic Communication and Computing at Northwestern IL, and of the Foundational Questions Institute (FQXi).

Giulio Chiribella is Associate Professor and a CIFAR-Azrieli Global Scholar at the Department of Computer Science of The University of Hong Kong. He is a Visiting Fellow of Perimeter Institute for Theoretical Physics, a member of the Standing Committee of the International Colloquia on Group Theoretical Methods in Physics, and a member of the Foundational Questions Institute (FQXi). In 2010, he was awarded the Hermann Weyl Prize for applications of group theory in quantum information science.

Paolo Perinotti is Assistant Professor at Pavia University where he teaches Quantum Information Theory. His research activity is focused on foundations of quantum information, quantum mechanics, and quantum field theory. He is a member of the Foundational Questions Institute (FQXi), and of the International Quantum Structures Association. In 2016 he was awarded the Birkhoff–von Neumann prize for research in quantum foundations.

Quantum Theory from First Principles

An Informational Approach

GIACOMO MAURO D'ARIANO

Università degli Studi di Pavia, Italy

GIULIO CHIRIBELLA

The University of Hong Kong

PAOLO PERINOTTI

Università degli Studi di Pavia, Italy

CAMBRIDGE
UNIVERSITY PRESS

CAMBRIDGE
UNIVERSITY PRESS

University Printing House, Cambridge CB2 8BS, United Kingdom

One Liberty Plaza, 20th Floor, New York, NY 10006, USA

477 Williamstown Road, Port Melbourne, VIC 3207, Australia

314-321, 3rd Floor, Plot 3, Splendor Forum, Jasola District Centre, New Delhi - 110025, India

79 Anson Road, #06-04/06, Singapore 079906

Cambridge University Press is part of the University of Cambridge.

It furthers the University's mission by disseminating knowledge in the pursuit of education, learning and research at the highest international levels of excellence.

www.cambridge.org
Information on this title: www.cambridge.org/9781108714419
10.1017/9781107338340

First published 2017
First paperback edition 2019

A catalogue record for this publication is available from the British Library

Library of Congress Cataloging in Publication data
Names: D'Ariano, G. M. (Giacomo M.), author. | Chiribella, Giulio, author. | Perinotti, Paolo, author.
Title: Quantum theory from first principles : an informational approach / Giacomo Mauro D'Ariano (Università degli Studi di Pavia, Italy), Giulio Chiribella (The University of Hong Kong), Paolo Perinotti (Università degli Studi di Pavia, Italy).
Description: Cambridge, United Kingdom ; New York, NY : Cambridge University Press, 2017. | Includes bibliographical references and index.
Identifiers: LCCN 2016040126 | ISBN 9781107043428 (hardback ; alk. paper) | ISBN 1107043425 (hardback ; alk. paper)
Subjects: LCSH: Quantum theory.
Classification: LCC QC174.12 .C475 2017 | DDC 530.12–dc23 LC record available at https://lccn.loc.gov/2016040126

ISBN 978-1-107-04342-8 Hardback
ISBN 978-1-108-71441-9 Paperback

To our wives and children:

Rosanna and Gilda,
Amy and Francesco,
Silvia and Marco.

Contents

Preface

The book is the result of 20 years of teaching and research by the three authors in the fields of quantum foundations and quantum information, which culminated in two long joint papers [Phys. Rev. A **81** 062348 (2010) and **84** 012311 (2011)] that derive quantum theory from six simple information-theoretical principles. We have now the opportunity of presenting quantum theory in a radically new way, based on a conceptual understanding from the new principles. By "quantum theory" we mean the general theory of physical systems that lies at the core of "quantum mechanics," the latter broadly viewed as the quantum generalization of classical Hamiltonian mechanics. The book will not cover applications to "mechanics," but rather focus on applications to quantum information. For this reason, and with the aim of keeping the center of attention more on conceptual issues, rather than on the mathematical technicalities, we consider finite numbers of finite-dimensional systems, and restrict to probabilities of finite set of events, with some extensions to the infinite/continuous case discussed in the notes at the end of chapters.

The book includes 220 exercises and problems. The exercises are given within the body of each chapter, and selected solutions can be found at the end of that chapter. The exercises represent an integral part of the book and we warmly recommend the reader to solve them (or to check out the solutions), because the results proven therein are often used in our arguments. The problems presented at the end of each chapter build up additional knowledge and problem-solving skills, not strictly needed for the understanding of the arguments, but it is still recommended to solve them (or study the solutions). References, historical comments, and citations are provided in the notes at the end of chapters.

The book can be used for teaching at all levels, ranging from undergraduate, to master, up to PhD, and for pursuing personal research interests. The book is divided into four parts, organized as follows:

Part I *The Status Quo* (Chapter 2) introduces the mathematical structure of quantum theory, deriving it from three simple Hilbert-space postulates (systems, states, and the no-restriction hypothesis), and proving, in the form of theorems, what will later become our six principles for the derivation of quantum theory. The full mathematical structure of quantum theory is derived, including all relevant results in quantum open systems and quantum information. The derivation uses original powerful proving techniques based on tensor operators. In this part, the reader will have the chance to become acquainted with the relevant notions in operational probabilistic theories (OPT) and in convex analysis, and will start using the six principles for deriving results. This entire part can be used for an undergraduate semester course of quantum theory and quantum information, for physicists, mathematicians, and computer scientists.

Part II *The Informational Approach* (Chapters 3–7) presents the framework of operational probabilistic theories and introduces the six principles. Three separate chapters are devoted to the main principles of causality, local discriminability, and purification. Parts I and II together make a complete semester course for a masters-level course.

Part III *Quantum Information Without Hilbert Spaces* (Chapters 8–12) uses all the six principles to derive key results of quantum information theory and general features of quantum theory, including no-go theorems such as the no-cloning and no-bit-commitment theorems. Some parts of these chapters can be incorporated in a masters-level course.

Part IV *Quantum Theory from the Principles* (Chapters 13–17) derives quantum theory from the six principles.

Chapters on causality, local discriminability, and purification are of interest also for philosophers and, more generally, for readers who are seeking for a deeper understanding of these concepts in the light of quantum information.

For possible errors and corrections found after the print of the current edition of the book, the reader is addressed to the webpage: www.qubit.it/errata/homeerrata.html

Acknowledgments

In putting this book together, we have benefitted from inspiring conversations and from the encouragement of many colleagues and friends over many years. In particular, we would like to express our gratitude to Scott Aaronson, Samson Abramsky, Antonio Acín, Gennaro Auletta, Howard Barnum, Jonathan Barrett, Gilles Brassard, Časlav Brukner, Paul Busch, Jeremy Butterfield, Vladimir Bužek, Adàn Cabello, Gianni Cassinelli, Ariel Caticha, Bob Coecke, Roger Colbeck, Maria Luisa Dalla Chiara, Olivier Darrigol, Giancarlo Ghirardi, Nicolas Gisin, Gerald Goldin, Philip Goyal, Alexei Grinbaum, Teiko Heinosaari, Louis Kauffman, Michael Keyl, Gen Kimura, Pekka Lahti, Raymond Lal, Matthew Leifer, Lev Levitin, Seth Lloyd, Norman Margolus, Izumi Ojima, Matthew Pusey, Renato Renner, Alberto Rimini, Valerio Scarani, Dirk Schlingemann, Anthony Short, John Smolin, Robert Spekkens, Reinhard Werner, and Mario Ziman.

GMD wishes to express a very special thanks to Alexander Holevo, Masanao Ozawa, and Horace Yuen for their generous mentoring of the Pavia school in quantum theory and quantum information from the very beginning of the QUit group, and especially to Attilio Rigamonti for his constant enthusiastic support and encouragement from the very beginning of this unconventional and high-risk research line. He is also very grateful to Andrei Khrennikov for hosting the whole evolution of this research in the Växjö conference, and to Gregg Jaeger and Arkady Plotnitsky for very stimulating and intense discussions that lasted from the beginning to the end of the entire project. GMD also owes much to David Finkelstein, for day-long inspiring and enthusiastic private discussions. David unfortunately died just a few days before this book was finished.

Finally, we are all indebted to Chris Fuchs, Lucien Hardy, Robert Spekkens, Bob Coecke, and Alex Wilce for stimulating, through their own work, our thinking on many of the topics presented in this book, as well as for numerous insightful discussions.

A special undeliverable thanks goes to the fond memory of Viacheslav Belavkin, from whom we all learnt a wealth of original concepts.

The joint writing of the book has also been made possible by the financial support of the Foundational Questions Institute (FQXi) (minigrant *Informational Principles for Quantum Theory* and large grant FQXi-RFP3-1325 *The Fundamental Principles of Information Dynamics*). GMD and PP acknowledge financial support from the Templeton Foundation under the project ID#43796 *A Quantum Digital Universe*. GC acknowledges financial support from the 1000 Youth Fellowship Program of China and from the NSFC through grants 11675136, 11450110096 and 11350110207, as well as the hospitality of the Simons Center for the Theory of Computation and of the Perimeter Institute for Theoretical Physics, where part of his contribution was completed. Research at Perimeter is supported in part by the Government of Canada through NSERC and by the Province of Ontario through MRI.

Introduction

Quantum theory is the soul of contemporary physics. It was discovered in an adventurous way, under the urge to solve the puzzles posed by atomic spectra and blackbody radiation. But after its invention, it immediately became clear that it was not just a theory of specific physical systems: it was rather a new language of universal applicability. Already in 1928, the theory had received solid mathematical foundations by Hilbert, von Neumann, and Nordheim,[1] and this work was brought to completion in the monumental work of von Neumann,[2] in the form that we still study nowadays. The theory is extraordinarily successful, and its predictions have been confirmed to an astonishing level of precision in a large spectrum of experiments.

However, almost 90 years after von Neumann's book, quantum theory remains mysterious. Its mathematical formulation – based on Hilbert spaces and self-adjoint operators – is far from having an intuitive interpretation. The association of physical systems to Hilbert spaces whose unit vectors represent pure states, the representation of transformations by unitary operators and of observables by self-adjoint operators – all such postulates look artificial and *ad hoc*. A slightly more operational approach is provided by the C^*-algebraic formulation of quantum theory[3] – still, this formulation relies on the assumption that observables form an *algebra*, where the physical meanings of the multiplication and the sum are far from clear.

In short, the postulates of quantum theory impose mathematical structures without providing any simple reason for this choice: the mathematics of Hilbert spaces is adopted as a magic blackbox that "works well" at producing experimental predictions. However, in a satisfactory axiomatization of a physical theory the mathematical structures should emerge as a consequence of postulates that have a direct physical interpretation. By this we mean postulates referring, e.g., to primitive notions like physical system, measurement, or process, rather than notions like, e.g., Hilbert space, C^*-algebra, unit vector, or self-adjoint operator.

The crucial question thus remains unanswered: why quantum theory? Which are the principles at the basis of the theory? A case that is often invoked in contrast is that of Special Relativity theory, which directly follows from the simple understandable principle of relativity.

---

[1] Hilbert *et al.* (1928).

[2] The book (von Neumann, 1932) has been recently reprinted (von Neumann, 1996).

[3] Haag (1993).

1.1 The Quest for Principles: von Neumann

The need for a deeper understanding of quantum theory in terms of fundamental principles was clear since the very beginning. Von Neumann himself expressed his dissatisfaction with his own mathematical formulation of quantum theory with the surprising words, "I don't believe in Hilbert space anymore."[4] Realizing the physical relevance of the axiomatization problem, Birkhoff and von Neumann made an attempt at understanding quantum theory as a new kind of *propositional calculus*,[5] motivated by the opinion that the main difficulties in accepting the "quantumness" of elementary physical systems stem from the inadequacy of classical logic to encompass the unpredictable nature of quantum measurement outcomes. In their attempt, von Neumann and Birkhoff proposed to treat the propositions about the physical world in a suitable logical framework, different from classical logic, where the operations AND and OR are no longer distributive. The lack of interpretation of the observables algebra led Jordan, von Neumann, and Wigner to consider the possibility of a commutative algebra of observables, with a product that only requires the definition of squares and sums of observables – the so-called *Jordan product*.[6] These works inaugurated the tradition of quantum logics, which led to several attempts at an axiomatization of quantum theory, most notably by Mackey,[7] Varadarajan,[8] and Jauch and Piron,[9, 10] with ramifications still the object of active research.[11]

Researchers in quantum logic managed to derive a significant part of the quantum framework from logical axioms. In general, a certain degree of technicality (mainly related to the emphasis on infinite-dimensional systems) makes these results far from providing a clear-cut description of quantum theory in terms of fundamental principles. Even among the experts there is a general consensus that the axioms are not as insightful as one would have hoped. For both experts and non-experts, it is hard to figure out what is the moral of the quantum logic axiomatizations: what is special about quantum theory after all? Why should quantum theory be preferred to alternative theories?

A notable alternative axiomatization program was that of Ludwig,[12] who adopted an operational approach, where the basic notions are those of preparation and measuring devices, and the postulates specify how preparations and measurements combine to give the probabilities of experimental outcomes. However, even Ludwig's program never succeed in deriving Hilbert spaces from operational principles, as some of the postulates still contained mathematical notions with no operational interpretation.

---

[4] This was reported by Birkhoff (1984).

[5] Birkhoff and von Neumann (1936).

[6] Jordan *et al.* (1934). See the recent encyclopedic books of Alfsen and Shultz (2001, 2003).

[7] Mackey (1963).

[8] Varadarajan (1962).

[9] Jauch and Piron (1963); Piron (1964, 1976). Foulis and Randall developed an empirical counterpart of Piron's approach (Foulis *et al.*, 1983; Foulis and Randall, 1984).

[10] For a thorough textbook see Beltrametti *et al.* (2010).

[11] For a review on the more recent progresses of quantum logics see Coecke *et al.* (2000).

[12] Ludwig (1983).

1.2   Quantum Information Resurrects the Quest

The ambition to find a more insightful axiomatization re-emerged with the rise of quantum information. The new field showed that the mathematical axioms of quantum theory imply striking operational consequences, such as quantum key distribution,[13] quantum algorithms,[14] no cloning,[15] quantum teleportation,[16] and dense coding.[17] A natural question is then: can we reverse the implication and *derive* the mathematics of quantum theory from some of its operational features? This question lies at the core of a research program launched by Fuchs[18] and Brassard,[19] which can be synthesized by the motto "quantum foundations in the light of quantum information."[20] The ultimate goal of the program is to reconstruct the whole structure of quantum theory from a few simple principles of information-theoretic nature.

One may wonder why quantum information theorists should be more successful than their predecessors in the axiomatic endeavor. A good reason is the following. In the pre-quantum information era, quantum theory was viewed like an impoverished version of classical theory, lacking the ability to make deterministic predictions about the outcomes of experiments. Clearly, this perspective offered no vantage point for explaining why the world should be quantum. Contrarily, quantum information provided plenty of positive reasons for preferring quantum theory to its classical counterpart – as many good reasons as the number of useful quantum information and computation protocols. Turning some of these reasons into axioms then appeared as a promising route towards a compelling axiomatization.[21]

The quantum information approach can also be regarded as an evolution of the quantum logic program, where quantum theory – rather than being considered as an alternate logical system – is regarded as an alternate theory of information processing, namely describing information sources and information-processing channels. Indeed, in classical probability theory, logic can be regarded as the special case of information-processing theory where the probabilities of events are bound to the truth values $\{0, 1\}$. In non-deterministic theories like quantum theory, however, there are events whose truth value cannot be assessed, and one must concede that all we know about them is their occurrence probability.

Another new feature of the quantum information approach has been to shift the emphasis to finite-dimensional systems, which allow for a simpler treatment but still possess all the

---

[13] Wiesner (1983); Bennett *et al.* (1984); Ekert (1991).

[14] Grover (1996); Shor (1997).

[15] Dieks (1982); Wootters and Zurek (1982).

[16] Bennett *et al.* (1993).

[17] Bennett and Wiesner (1992).

[18] Fuchs (2002, 2003).

[19] Brassard (2005).

[20] Fuchs *et al.* (2001). This was also the title of one influential conference, held in May 2000 at the Université de Montréal, which kickstarted the new wave of quantum axiomatizations.

[21] See Clifton *et al.* (2003). This work, however, assumed a C^*-algebra framework, and used informational-theoretical constraints for selecting the algebra, in particular for adopting the quantum versus the classical algebra.

relevant quantum features. In a sense, the study of finite-dimensional systems allows one to decouple the conceptual difficulties in our understanding of quantum theory from the technical difficulties of infinite-dimensional systems.

In this scenario, Hardy in 2001[22] reopened the debate about axiomatization with fresh ideas resting on the quantum information experience. Some of his axioms, however, contained mathematical notions with no interpretation, e.g. statements about the dimensionality of the state space, or the continuity of the set of pure states. Stimulated by Hardy's and Fuchs's works one of the authors of this book addressed a new axiomatization approach[23] based on operational principles about tomography, calibration and composition of transformations, and generally on the reduction of experimental complexity, such as the existence of a pure faithful state, a property that allows for tomography of transformations preparing a single input pure state. However, a thorough derivation of the theory was still missing, and also in this case there remained mathematical postulates with no interpretation. Later, building on Hardy's work the program flourished, leading to an explosion of new axiomatizations based on a variety of conceptual and mathematical frameworks,[24,25] including the framework and axiomatization contained in the present book.[26] These works realized the old dream of Wheeler's program "it from bit," for which he argued that "all things physical are information-theoretic in origin."[27]

1.3   Quantum Theory as an OPT

A lesson that we learned from the experience of quantum information is to regard quantum theory as a theory of information processing in the first place. We thus realized the crucial role played by the description of processes in the form of quantum circuits. This has led us to consider quantum theory as an extension of probability theory, to which we add the crucial ingredient of *connectivity* among events. This means that to the joint events we associate not only their joint probability, but also a circuit that connects them. When the events in the circuit have a well-defined order, the circuit is mathematically described by a *directed acyclic graph* (a graph with directed edges and without loops). Therefore, if we want to predict a joint probability, the varibles to be specified are not only the events but also the circuit connecting them.

A theory for making predictions about joint events depending on their reciprocal connections is what we call an *operational probabilistic theory* (OPT). We see that OPT is a non-trivial extension of probability theory, which, according to Jaynes and Cox,[28] in turn

---

[22] Hardy (2001).

[23] D'Ariano (2006a,b, 2007a,b, 2010); D'Ariano and Tosini (2010).

[24] Goyal *et al.* (2010); Dakic and Brukner (2011); Hardy (2011); Masanes and Müller (2011); Masanes *et al.* (2013); Wilce (2012); Barnum *et al.* (2014).

[25] For a comprehensive collection of papers see the book by Chiribella and Spekkens (2015).

[26] Chiribella *et al.* (2010a, 2011).

[27] Wheeler (1990).

[28] Jaynes (2003); Cox (1961).

is an extension of logic.[29] We now realize how, in the previous axiomatization attempts, only one facet of quantum theory was considered, consisting of propositional calculus and probability, whereas the connectivity facet was missing.

From what we have said, we now understand how the basic element of an OPT – the notion of *event* – gets dressed with *wires* that allow us to connect it with other events. Such wires are the *systems* of the theory. In agreement with the directed nature of the graph, there are *input* and *output* systems. The events are the *transformations*, whereas the transformations with no input system are the *states* (corresponding to preparations of systems), and those with no output system are the *effects* (corresponding to observations of systems). Since the purpose of a single event is to describe a process connecting an input with an output, the full circuit associated to a probability is a closed one, namely a circuit with no input and no output.

The circuit framework is mathematically formalized in the language of *category theory*.[30] In this language, an OPT is a category, whose systems and events are *objects* and *arrows*, respectively. Every arrow has an input and an output object, and arrows can be sequentially composed. The associativity, existence of a trivial system, and commutativity of the parallel composition of systems of quantum theory technically correspond to having a *strict symmetric monoidal category*.[31] Although the OPT language can be rephrased in purely category theoretical terms, its original version[32] is more physicist-friendly, and it will be adopted in the present book. Expressions in such a language have an immediate meaning as the description of elementary physical processes and their relations within an experimental setting – for example, specifying whether two events occur in sequence or in parallel. However, we note the indispensable role of the probabilistic structure in promoting the OPT language from a merely descriptive tool to a framework for prediction, which is the crucial feature of a scientific theory. Two OPTs will then be different if they have different rules for assigning probabilities to the circuits.

1.4 The Principles

OPTs provide a general unified framework to formalize theories of information, including classical information theory and quantum information theory. In this framework, we characterize quantum theory *as a theory of information*. In short, quantum theory is the theory which allows for the optimal validation of randomness: all the six principles of the theory come together in such respect from complementary standpoints. Five of the six principles – causality, local discriminability, perfect discriminability, ideal compression, and atomicity of composition – express ordinary properties that are shared by quantum and

---

[29] We would like to mention the famous quote of J. C. Maxwell: "the true logic for this world is the calculus of probabilities." See also Keynes (2004).

[30] Mac Lane (1978).

[31] For an introduction to the graphical language of monoidal catagories we recommend the beautiful surveys by Selinger (n.d.) and Coecke (2008).

[32] The language of OPTs was introduced in Chiribella *et al.* (2010a).

classical information theory. The sixth principle – purification – identifies quantum theory uniquely.

In non-technical words, the six principles are the following:

- **Causality**. Measurement results cannot depend on what is done on the system at the output of the measurement. Equivalently: no signal can be sent from the future to the past.
- **Local discriminability**. We can reconstruct the joint state of multiple systems by performing only local measurements on each system.
- **Perfect discriminability**. Every state that is not completely mixed can be perfectly distinguished from some other state.
- **Ideal compression**. Every source of information[33] can be encoded in a lossless and maximally efficient fashion (*lossless* means perfectly decodable, *maximally efficient* means that every state of the encoding system represents a state of the source).
- **Atomicity of composition**. No side information can hide in the composition of two atomic transformations. Equivalently: the sequential composition of two precisely known transformations is precisely known.
- **Purification**. Every random preparation of a system can be achieved by a pure preparation of the system with an environment, in a way that is essentially unique.

The first five principles of the list are satisfied by classical information theory. Hence, in our axiomatization, the purification principle is highlighted as the distinctive axiom of quantum theory. All the six principles have an epistemological nature. Causality is necessary for control of observations, shielding them from the influence of external agents acting in the future or from far apart. Local discriminability allows for the local accessibility of information. Perfect discriminability allows for falsifiability of propositions of the theory. Atomicity of composition allows for control in composing transformations and observations. Purification allows for validation of randomness, by leaving to an agent access to both system and environment.

It is important to remark here the value of the six principles for philosophy of science. For example the local discriminability principle reconciles the holism of a theory with the reductionistic approach, as explained in Chapter 6. Paradigmatic is the principle of causality, which would be matter for a treatise, in consideration of the wealth of literature on the subject in philosophy and physics. To realize the subtlety of the notion one can just consider the simple fact that causality has never been formally stated as a principle in physics.[34] Mostly the causality notion has been misunderstood due to a spurious connection with the independent notion of determinism.[35] The causality principle for quantum theory is the logical quintessence of the meaningful notions debated within the

---

[33] An information source technically is a set of states of a fixed system.

[34] Only very recently it has been explicitly remarked by some authors that causality is built in quantum theory (Ellis, 2008).

[35] The logical independence between the notion of causality and that of determinism is proved by the existence of causal OPTs that are not deterministic, e.g. quantum theory, and vice versa of deterministic theories that are not causal, as those constructed in D'Ariano *et al.* (2014a).

specialized literature since Hume, and ranging to modern and contemporary authors.[36] The language of OPT provides the right framework for formalizing the notion of causality in a theory-independent manner, offering a rigorous notion for philosophical analysis. Such notion also corresponds to the standard use of causality in inference and scientific modeling,[37] and coincides with the Einstenian causality, as explained in Chapter 5.

The purification principle is also of great relevance for philosophy of science. It is the axiom that selects quantum theory, thus containing its essence. Its conceptual content is the expression of a law of *conservation of information,* stating that irreversibility is in principle reducible to a lack of control over an environment. More precisely, the principle is equivalent to stating that every irreversible process can be simulated in an essentially unique way by a reversible interaction of the system with an environment, initially prepared in a pure state.[38] This statement includes the case of measurement processes, and in that case it implies the possibility of arbitrarily shifting the cut between the observer and the observed system. The arbitrariness of such a shift was considered by von Neumann as a "fundamental requirement of the scientific viewpoint,"[39] and his discussion of the measurement process was exactly aimed at showing that quantum theory fulfills it. Finally, the principle of purification is of great relevance for philosophy of probability,[40] since it provides the existence of random sources that can be validated by a measurement performed jointly on the source and on the purifying system.

[36] Salmon (1967); Dowe (2007).
[37] Pearl (2012).
[38] Chiribella *et al.* (2010a).
[39] See p. 418 of von Neumann (1996).
[40] Gillies (2000).

PART I

THE STATUS QUO

I think I can safely say that nobody understands quantum mechanics.
Richard Feynman, *The Character of Physical Law (1965)*

Quantum Theory from Hilbert Spaces 2

> It is not surprising that our language should be incapable of describing processes occurring within atoms, for, as has been remarked, it was invented to describe the experiences of daily life, ... Fortunately, mathematics is not subject to this limitation, and it has been possible to invent a mathematical scheme–the quantum theory–which seems entirely adequate for the treatment of atomic processes.
>
> Werner Heisenberg, *The Physical Principles of the Quantum Theory* (1930)

In this chapter we introduce the mathematical structure of quantum theory, for a finite number of systems and finite dimension, starting from three simple Hilbert-space postulates, and based on the intuitive notions of state, transformation, and effect, within a causal context. Such notions will be formalized in a general non-causal context in Chapter 3, which is devoted to the informational framework. We will assume the reader has a basic knowledge of matrix analysis and of operators over Hilbert spaces. Comments about the extension to infinite dimensions, continuous outcome-spaces, and infinitely many systems can be found in the notes at the end of the chapter.

From the three simple Hilbert-space postulates, the present chapter extracts six principles, which will become the postulates for quantum theory in the axiomatic derivation presented in Part III of the book. In addition, we will take the opportunity to present and explore many relevant features of the theory – including its convex and causal structure, the Choi–Jamiołkowski isomorphism, and relevant notions, such as those of POVM, complete positivity of maps, quantum operation and channels, entangled states, state discrimination, tomography of states, and transformations. The purification principle will open the stage to the dilation theorems of the theory of quantum open systems, along with other consequences, such as the existence of faithful and steering states, the theorem of no information without disturbance, quantum teleportation, and more. We will review the quantum no-cloning and no-programming theorems, and finally provide a thorough derivation of the theory of quantum channel inversion, which is the backbone of the general theory of quantum error correction in quantum information. Additional interesting topics will be classical information theory as a restriction of quantum information theory, and a critical reconsideration of the von Neumann postulate and the notion of observable.

A few historical records can be found within the notes at the end of the chapter. Many results are derived in exercises proposed in the main body of the chapter, with most solutions provided at the end, where additional problems can be found.

2.1 Primitive Notions

Quantum theory is a theory of abstract systems. Depending on the context, the *system* can have different physical realizations, ranging from the spin of a particle, or one of its orbital degrees of freedom, or a mode of the electromagnetic field. In all these contexts, quantum theory provides a high-level language that can be used to describe physical processes and to make predictions about the outcome probabilities in the possible experiments.

A theory of abstract systems is based on a number of primitive notions, such as those of state, transformation, and observation. Before discussing how these notions are embodied in the specific mathematical framework of quantum theory, it is good to present them in a general, theory-independent manner. The intuitive notions presented in this paragraph will be the starting point for our presentation of quantum theory in the rest of the chapter, and will also serve as a first introduction to the broader framework of operational probabilistic theories, later formalized in Chapter 3.

Let us denote abstract systems by capital Roman letters, such as A, B, ..., Z. A *state* of system A – denoted by $\rho \in$ St(A) – describes a *preparation* of A. The preparation procedure can be implemented either by an experimenter (like e.g. the preparation of two carts at certain distance and with a certain velocity in a collision experiment) or by a spontaneous physical process (like e.g. the preparation of solar neutrinos by nuclear fusion). Analogously to what we have in classical mechanics, where the knowledge of the state allows us to predict the evolution of the system, here the state allows us to predict the probabilities for all transformations and observations undergone by the system. A *transformation* $\mathcal{A}$ modifies the state of the system, and, more generally, can produce an output system different from the input system (think e.g. of a chemical reaction, or photon absorption by an atom). A transformation with input A and output B will be denoted by $\mathcal{A} \in$ Transf(A→ B). In general, transformations can occur probabilistically. Hence, we should regard a transformation $\mathcal{A}$ as an element of a complete *test* $\{\mathcal{A}_j\}_{j \in Y}$, where j (belonging to the set Y) is the *outcome* of the test, and labels all the alternative transformations that can take place in a given process. Typically the test describes a measuring apparatus which randomly performs transformations, signaling which transformation occurred through the corresponding outcome. A state $\rho \in$ St(A) itself can be regarded as a special type of transformation, with no input system and with output system A. Generally, state preparations are achieved probabilistically, which means that we should also regard $\rho \in$ St(A) as an element of a complete test $\{\rho_i\}_{i \in X}$ performed on A. Such a test will be called a *preparation test*. Also, a transformation $\mathcal{A} \in$ Transf(A → B) occurring on system A prepared in state ρ results in the preparation of the state $\mathcal{A}\rho \in$ St(B) of system B.

In addition to preparation tests, it is convenient to single out another special class of tests, which have no output system and only provide an outcome.[1] Such tests will be called *observation tests*, and each of the alternative transformations in an observation test will be

---

[1] These tests do not necessarily destroy the system, but just represent a process in which the output system is neglected.

called *effect*. In general, an observation test will be denoted as $\{a_k\}_{k \in Z}$, with the effect a_k corresponding to the outcome $k \in Z$. The set of effects on system A will be denoted by $\mathsf{Eff}(A)$.

Suppose now that a system, initially prepared according to the preparation test $\{\rho_i\}_{i \in X}$, undergoes a test $\{\mathcal{A}_j\}_{j \in Y}$, and then an observation test $\{a_k\}_{k \in Z}$. Then, the purpose of the theory is to predict the *probability* of observing the preparation ρ_i, followed by transformation $\mathcal{A}_j$ and by the effect a_k. We denote such a probability by

$$p(\rho_i, \mathcal{A}_j, a_k) = (a_k | \, \mathcal{A}_j \, | \rho_i) \,, \tag{2.1}$$

or, in the absence of the transformation $\mathcal{A}_j$, $p(\rho_i, a_k) = (a_k \, | \, \rho_i)$. Note that, since at least one sequence of outcomes must occur, we have the normalization condition

$$\sum_{i \in X} \sum_{j \in Y} \sum_{k \in Z} p(\rho_i, \mathcal{A}_j, a_k) = 1 \,.$$

Since the aim of the theory is to predict the probabilities of outcomes, two transformations that give the same probabilities in all possible experiments will be identified.

An important notion in every probabilistic theory is the notion of *coarse-graining*. Given a test $\{\mathcal{A}_j\}_{j \in Y}$, one can consider a coarse-grained event, such as the event that the outcome j belongs to a subset $V \subseteq Y$ of the outcome set Y. The probability of such an event is then given by the sum of probabilities of all outcomes in V, e.g. for a state ρ_i and an effect a_k the probability of the coarse-grained event V is given by

$$p(\rho_i, \mathcal{A}_V, a_k) := \sum_{j \in V} p(\rho_i, \mathcal{A}_j, a_k) \,. \tag{2.2}$$

Since all transformations are completely specified by their joint probabilities, we use the above equation to define the *coarse-grained transformation* $\mathcal{A}_V := \sum_{j \in V} \mathcal{A}_j$, the sum referring to the fact that the probability of the coarse-grained transformation is the sum of the probabilities of the corresponding transformations, for every state and for every effect.

A special kind of coarse-graining arises when we choose randomly between two different tests and ignore the information of which test has been performed. Then, according to the previous definition of coarse-grained transformation and linearity of joint probabilities, the following defines the convex combination of two transformations:

$$\mathcal{T} = q\mathcal{A} + (1 - q)\mathcal{B}, \tag{2.3}$$

with $q \in [0, 1]$. In the same fashion, the transformation $q\mathcal{A}$ will denote the transformation $\mathcal{A}$ with all joint probabilities rescaled by q. Assuming that all possible randomizations can be implemented, it follows that the set of transformations $\mathsf{Transf}(A \to B)$ is closed under convex combinations, and the same holds for states $\mathsf{St}(A)$ and effects $\mathsf{Eff}(A)$ as special cases. We conclude that $\mathsf{Transf}(A \to B)$, $\mathsf{St}(A)$, and $\mathsf{Eff}(A)$ are all convex sets.

Multiple systems, say A and B, can be jointly prepared through mutual interaction, and it is convenient to treat them as a single *composite* (also called *multipartite*) system AB. A transformation $\mathcal{A} \in \mathsf{Transf}(A \to B)$ acting on a single system should be more generally regarded as acting on system A in the presence of other systems, which will be jointly denoted as R. Therefore, any transformation $\mathcal{A} \in \mathsf{Transf}(A \to B)$ is actually a transformation $\mathcal{A} \in \mathsf{Transf}(AR \to BR)$ that acts on input system AR and results in the output system BR. It is however convenient to keep the notation $\mathcal{A} \in \mathsf{Transf}(A \to B)$ to emphasize that $\mathcal{A}$ acts non-trivially only on system A, whereas, in order to emphasize the extension we will also write $\mathcal{A}^{ext}_R \in \mathsf{Transf}(AR \to BR)$, or simply $\mathcal{A}^{ext}$. Also states $\rho \in \mathsf{St}(A)$ and effects $a \in \mathsf{Eff}(A)$ should be considered as transformations $\rho^{ext}_R \in \mathsf{Transf}(R \to AR)$ and $a^{ext}_R \in \mathsf{Transf}(AR \to R)$, respectively.

Finally, we introduce a special requirement, which expresses the notion of *causality*, discussed in depth in Chapter 5. The requirement is the following: for every preparation test $\{\rho_i\}_{i \in X}$ we require that the marginal probability $p(\rho_i)$ obtained from Eq. (2.1) depends only on the specific preparation ρ_i, and not on the other tests performed thereafter. We call $p(\rho_i)$ the *probability of the preparation ρ_i*.

The requirement that the probability of preparations be well-defined is equivalent to a normalization condition for observation tests. Indeed, well-definiteness of the preparation probability implies the relation

$$\sum_{j \in Y} (a_j \mid \rho) = p(\rho),$$

for every state ρ and for every observation test $\{a_j\}_{j \in Y}$. This relation is equivalent to the normalization condition

$$\sum_{j \in Y} a_j = e_A, \qquad (2.4)$$

where $e_A \in \mathsf{Eff}(A)$ is a fixed effect, independent of the test $\{a_j\}_{j \in Y}$. We call e_A the *deterministic effect* of system A.[2] Using the deterministic effect, we can compute the probability of a preparation as

$$p(\rho) = (e|\rho). \qquad (2.5)$$

If the probability is equal to 1, we say that the state ρ is *deterministic*. The set of deterministic states of system A will be denoted by $\mathsf{St}_1(A) \subseteq \mathsf{St}(A)$ (this notation will be extended to deterministic transformations from A to B, which will be denoted by $\mathsf{Transf}_1(A \to B)$).

2.2 Hilbert-space Postulates for Quantum Theory

In the following $\mathcal{H}$ denotes a finite-dimensional Hilbert space. For $s > 0$, $\mathsf{B}_s(\mathcal{H}) := \{|\lambda\rangle \in \mathcal{H} : \|\lambda\| \leq s\}$ denotes the s-radius ball in $\mathcal{H}$ centered at the 0 vector, and $\mathsf{S}_s(\mathcal{H}) := \{|\lambda\rangle \in \mathcal{H} : \|\lambda\| = s\}$ denotes the corresponding sphere. We denote by $\mathsf{Lin}(\mathcal{H})$ the set of linear operators over $\mathcal{H}$, and recall that an operator $A \in \mathsf{Lin}(\mathcal{H})$ is completely specified by its expectations $\langle \lambda|A|\lambda \rangle$ on $\mathsf{S}_s(\mathcal{H})$ for any fixed $s > 0$. We also denote by $\mathsf{Lin}_+(\mathcal{H}) \subseteq \mathsf{Lin}(\mathcal{H})$ the convex cone of positive operators, recalling that an operator $A \in \mathsf{Lin}(\mathcal{H})$ is *positive* – denoted as $A \geq 0$ – when $\langle \lambda|A|\lambda \rangle \geq 0 \; \forall |\lambda\rangle \in \mathcal{H}$. The notion of positivity

---

[2] The assumption of causality is then equivalent to the *uniqueness* of the deterministic effect.

establishes the partial ordering "$\geq$" over $\mathsf{Lin}(\mathcal{H})$, corresponding to writing $A \geq B$ whenever $A - B \geq 0$.

Exercise 2.1 [Polarization identity] Using the *polarization identity* for any two vectors $|x\rangle, |y\rangle \in \mathcal{H}$

$$|x\rangle\langle y| = \frac{1}{4} \sum_{k=0}^{3} i^k \left(|x\rangle + i^k |y\rangle \right) \left(\langle x| + (-i)^k \langle y| \right), \qquad (2.6)$$

prove that an operator $A \in \mathsf{Lin}(\mathcal{H})$ is completely specified by its expectations $\langle \lambda | A | \lambda \rangle$ on $\mathsf{S}_s(\mathcal{H})$ for any fixed $s > 0$, i.e. $\langle \lambda | A | \lambda \rangle = 0$ for every $\lambda \in \mathsf{S}_s(\mathcal{H})$ iff $A = 0$.

Exercise 2.2 Show that iteration of Eq. (2.3) for a set of states $\{\rho_i\}_{i=1,...N} \in \mathsf{St}(A)$ with probability distribution $\mathbf{p} = \{p_i\}_{i=1,...,N}$ leads to the general form for convex combinations

$$\rho_{\mathbf{p}} := \sum_{i=1}^{N} p_i \rho_i,$$

and if $\rho_i \in \mathsf{St}_1(A) \; \forall i = 1, \ldots, N$, then also $\rho_{\mathbf{p}} \in \mathsf{St}_1(A)$.

Exercise 2.3 [Partial ordering] A convex cone C is a set closed under addition and multiplication by positive real numbers. We can always regard the cone as embedded in the real vector space $\mathsf{C}_{\mathbb{R}} := \mathsf{Span}_{\mathbb{R}}(\mathsf{C})$, corresponding to the extension of linear positive combinations to real ones. The cone introduces a partial ordering "$\geq$" in $\mathsf{C}_{\mathbb{R}}$, namely $a \geq b$ for $a, b \in \mathsf{C}_{\mathbb{R}}$ when $a - b \in \mathsf{C}$. Show that the set of positive operators $\mathsf{Lin}_+(\mathcal{H})$ is a convex cone, and thus it establishes a partial ordering in $\mathsf{Lin}(\mathcal{H})$.

Exercise 2.4 [Pauli matrices] Show that the Pauli matrices

$$\sigma_t = \begin{pmatrix} 1 & 0 \\ 0 & 1 \end{pmatrix}, \quad \sigma_x = \begin{pmatrix} 0 & 1 \\ 1 & 0 \end{pmatrix}, \quad \sigma_y = \begin{pmatrix} 0 & -i \\ i & 0 \end{pmatrix}, \quad \sigma_z = \begin{pmatrix} 1 & 0 \\ 0 & -1 \end{pmatrix}, \qquad (2.7)$$

(also denoted as $\sigma_0 = \sigma_t$, $\sigma_1 = \sigma_x$, $\sigma_2 = \sigma_y$, and $\sigma_3 = \sigma_z$) when multiplied by $\frac{1}{\sqrt{2}}$ form an orthonormal basis for the Hilbert space of operators $\mathsf{Lin}(\mathbb{C}^2)$ with the Hilbert–Schmidt scalar product $(A, B) := \mathrm{Tr}[A^\dagger B]$, for $A, B \in \mathsf{Lin}(\mathbb{C}^2)$. Show that the Pauli matrices are traceless (apart from σ_t), unitary, involutive, and self-adjoint. Show that the expansion of $X \in \mathsf{Lin}(\mathbb{C}^2)$ can be written as

$$X \in \mathsf{Lin}(\mathbb{C}^2), \quad X = \tfrac{1}{2} \sum_{j=0}^{3} n^j(X)\sigma_j, \quad n^j(X) = \mathrm{Tr}[X\sigma_j].$$

2.2.1 The Postulates

Postulate 1 (Systems and their Composition)   To each system A we associate a complex Hilbert space $\mathcal{H}_A$. To the composition AB of systems A and B we associate the tensor product $\mathcal{H}_{AB} = \mathcal{H}_A \otimes \mathcal{H}_B$.

Postulate 2 (States)   To each state $\omega \in \mathsf{St}(A)$ of system A corresponds a positive operator ρ_ω on $\mathcal{H}_A$ with $\mathrm{Tr}\,\rho_\omega \leq 1$. Vice versa, to every such operator on $\mathcal{H}_A$ corresponds a state in $\mathsf{St}(A)$. The correspondence preserves convex combinations.

The positive operator ρ_ω representing the state ω in Postulate 2 is called the *density operator* or *density matrix*. With a little abuse of notation we conveniently write $\rho \in \mathsf{St}(\mathrm{A})$ to denote both the state and the corresponding density operator of system A, and call ρ the *quantum state*.

Postulate 3 (No-restriction Hypothesis)   All maps that satisfy all mathematical requirements for representing a transformation within the theory will be actual transformations of the theory.

The mathematical requirements for representing a transformation are left implicit in Postulate 3. This will be clarified in the following sections, where we will derive the complete mathematical framework of quantum theory from Postulates 1–3.

Exercise 2.5   [Convexity of $\mathsf{St}(\mathrm{A})$] Show that the set of quantum states of system A is convex.

Exercise 2.6   Show that for any $|\lambda\rangle \in \mathsf{B}_1(\mathcal{H}_\mathrm{A})$ the rank-one positive operator $|\lambda\rangle\langle\lambda|$ represents a state of A.

2.3   Density Matrices and POVMs

Lemma 2.1 (Effects)   *To each effect $a \in \mathsf{Eff}(\mathrm{A})$ corresponds a positive operator E_a on $\mathcal{H}_\mathrm{A}$ with $E_a \leq I_\mathrm{A}$, where I_A denotes the identity on $\mathcal{H}_\mathrm{A}$ and represents the unique deterministic effect of system A. Vice versa, each operator $E \leq I_\mathrm{A}$ on $\mathcal{H}_\mathrm{A}$ describes an effect. The joint probability of state and effect is given by the Born rule*

$$p(\rho, a) := (a|\rho) = \mathrm{Tr}(E_a \rho), \tag{2.8}$$

where the trace is performed over the Hilbert space $\mathcal{H}_\mathrm{A}$ of system A.

Proof   According to Postulate 3 each probability functional[3] over states of a system A is an effect for A. States are in correspondence with positive operators on $\mathcal{H}_\mathrm{A}$, and these span the whole $\mathsf{Lin}(\mathcal{H}_\mathrm{A})$. Thus, by the linearity of coarse-graining, effects can be uniquely extended to linear functionals over operators on $\mathcal{H}_\mathrm{A}$. By the Riesz–Fréchet representation theorem it immediately follows that such functionals are of the form $\mathrm{Tr}(\cdot\, E)$ with $E \in \mathsf{Lin}(\mathcal{H}_\mathrm{A})$,[4] the correspondence between operators and functionals being one-to-one.[5] The functional $\mathrm{Tr}(\cdot\, E)$ is positive and bounded from above by 1 if and only if $0 \leq E \leq I_\mathrm{A}$. Indeed, $|\lambda\rangle\langle\lambda|$ is a quantum state for any $|\lambda\rangle \in \mathsf{B}_1(\mathcal{H}_\mathrm{A})$, hence $\langle\lambda|E|\lambda\rangle = \mathrm{Tr}[|\lambda\rangle\langle\lambda|E] \geq 0 \; \forall|\lambda\rangle \in \mathcal{H}_\mathrm{A}$, which implies positivity of E. On the other hand, $\langle\lambda|E|\lambda\rangle \leq 1$ for $|\lambda\rangle \in \mathsf{B}_1(\mathcal{H}_\mathrm{A})$ implies $\langle\lambda|E|\lambda\rangle \leq \langle\lambda|\lambda\rangle$ for $|\lambda\rangle \in \mathsf{S}_1(\mathcal{H}_\mathrm{A})$, namely $\langle\lambda|I_\mathrm{A} - E|\lambda\rangle \geq 0 \; \forall|\lambda\rangle \in \mathcal{H}_\mathrm{A}$, i.e. $E \leq I_\mathrm{A}$. The operator $E = I_\mathrm{A}$ is the only one achieving the upper bound, since $\langle\lambda|E|\lambda\rangle = 1$ for

---

[3] A probability functional is a positive functional bounded by 1.

[4] We will often use the functional notation $f(\cdot)$ denoting the variable with a central dot.

[5] See e.g. Holevo (1982).

$|\lambda\rangle \in \mathsf{S}_1(\mathcal{H}_A)$ implies $E = I_A$. Finally, every operator E such that $0 \leq E \leq I_A$ satisfies $0 \leq \langle\lambda|E|\lambda\rangle \leq 1 \ \forall|\lambda\rangle \in \mathsf{B}_1(\mathcal{H}_A)$, hence it satisfies $0 \leq \mathrm{Tr}[E\rho] \leq 1$ for any $\rho \geq 0$ with $\mathrm{Tr}\rho \leq 1$. Then by Postulate 3 E represents an effect.   □

Hence, a quantum effect for system A is a probability functional of the form $\mathrm{Tr}[E\cdot]$, with $E \in \mathsf{Lin}(\mathcal{H}_A)$ and $0 \leq E \leq I_A$, and the deterministic effect is given by the trace. Therefore, according to Eq. (2.5), the state-preparation probability is

$$p(\rho) = \mathrm{Tr}\,\rho. \qquad (2.9)$$

As for states, with a little abuse of notation we call the *quantum effect* for system A any operator $E \in \mathsf{Lin}(\mathcal{H}_A)$ satisfying $0 \leq E \leq I_A$, and also write $E \in \mathsf{Eff}(A)$.

As emphasized in Section 2.1, the effect should be regarded as a transformation in $\mathsf{Transf}(AR \rightarrow R)$. We therefore need to check that the result of Lemma 2.1 can be extended to any additional system R. Indeed, the quantum effect $a \in \mathsf{Eff}(A)$ can be extended to a transformation from AR to R, namely

$$(a|_A: \qquad |\sigma)_{AR} \in \mathsf{St}(AR) \mapsto (a|_A|\sigma)_{AR} \in \mathsf{St}(R), \qquad (2.10)$$

by extending $E_a \mapsto E_a \otimes I_R$ and performing the partial trace Tr_A instead of the full trace Tr. One has

$$(a|_A = \mathrm{Tr}_A[\cdot\,(E_a \otimes I_R)], \quad E_a \in \mathsf{Lin}_+(A),\ 0 \leq E_a \leq I_A.$$

In particular, the deterministic effect of system A is

$$(e|_A = \mathrm{Tr}_A. \qquad (2.11)$$

In Section 2.3.2 we will see that Eq. (2.10) corresponds to the notion of conditional state.

Ensembles of Quantum States   A preparation test $\mathcal{R}$ for system A prepares an *ensemble of states* for A. According to Postulate 2 the test is described by a set of density operators $\{\rho_i\}_{i \in X}$ on $\mathcal{H}_A$ satisfying the normalization

$$\sum_{i \in X} \mathrm{Tr}\,\rho_i = 1,$$

due to Eq. (2.9), since the sum of their preparation probabilities must be 1. The deterministic state

$$\rho_X := \sum_{i \in X} \rho_i,$$

is called the *prior state*, and represents the state that is prepared in average, namely the expected state when one does not read the outcome $i \in X$ of the test.

It is customary in the literature to present the ensemble of states as the collection $\{\hat{\rho}_i, p_i\}_{i \in X}$ of deterministic states $\hat{\rho}_i$ along with the corresponding probabilities of preparation $p_i > 0$. In our case simply one has $p_i = \mathrm{Tr}\,\rho_i$ and $\hat{\rho}_i = \rho_i/p_i$, $\forall i \in X$. The reader will soon appreciate the simplification of using non-deterministic states, instead of always resorting to normalized ones.

Exercise 2.7 [Convexity of $\mathsf{Eff}(A)$] Show that the set of quantum effects for given system A is convex.

Exercise 2.8 [The trivial quantum system] Show that a system I with $\mathcal{H}_I = \mathbb{C}$ has states and effects given by probability values, and that the system composition is trivially AI $= A$.

Exercise 2.9 [The qubit and the Bloch ball] The *qubit* A is the quantum system with $\dim \mathcal{H}_A = 2$. It is the lowest dimensional non-trivial quantum system. Show that its convex set of deterministic states $\mathsf{St}_1(A)$ is the unit ball B^3 in $\mathbb{R}^3$. Precisely, one has that any deterministic state of a qubit is represented by a density operator on $\mathbb{C}^2$ written in a unique way as follows:

$$\rho_{\mathbf{n}} = \tfrac{1}{2}(I_2 + \mathbf{n} \cdot \boldsymbol{\sigma}), \quad \mathbf{n} \in B^3. \tag{2.12}$$

The ball of states in Eq. (2.12) is called the *Bloch ball*. Extremal states (usually called *pure*) correspond to the points of the surface, which is the *Bloch sphere*.

Exercise 2.10 Provide the analytical expression of the full set of probabilistic states $\mathsf{St}(A)$ of a qubit.

Exercise 2.11 Show that the convex set of effects $\mathsf{Eff}(A)$ of a qubit A is given by the *spindle-shaped* convex set

$$E = xI_2 + \mathbf{m} \cdot \boldsymbol{\sigma}, \quad \|\mathbf{m}\| \leq x \leq 1 - \|\mathbf{m}\|.$$

2.3.1 Observation Tests

An observation test $\mathcal{O}$ on system A is a collection of effects for A that sum to the deterministic effect. From Lemma 2.1 this kind of test is described by a collection $\mathcal{O} = \{E_i\}_{i \in \mathsf{X}}$ of positive operators satisfying the identity

$$\sum_{i \in \mathsf{X}} E_i = I_A. \tag{2.13}$$

Such a collection of positive operators is called *POVM* (*probability-operator-valued measure* or *positive-operator-valued measure*: see notes at the end of the chapter).

Quantum Mantra 1 (Observables) A special case of observation test is the so-called *observable*, corresponding to a resolution of the identity made of orthogonal projectors

$$\forall i, j \in \mathsf{X}: \quad P_i P_j = \delta_{ij} P_i, \quad \sum_{i \in \mathsf{X}} P_i = I_A. \tag{2.14}$$

The observable is associated to a random variable through the connection of each outcome $i \in \mathsf{X}$ with a real value x_i, occurring with probability $p_i = \mathrm{Tr}[P_i \rho]$. The expectation of the random variable with the system in the state ρ is then given by

$$\langle X \rangle_\rho = \mathrm{Tr}[X\rho], \quad X = \sum_{i \in \mathsf{X}} x_i P_i,$$

where X is an Hermitian operator with spectrum $\mathsf{Sp}(X) = \{x_i\}_{i \in \mathsf{X}}$. Notice that the notation $\langle X \rangle_\rho$ matches the usual notation for expectation of random variables (for this reason in the literature the same capital letter X often denotes both the Hermitian operator and the random variable), with the additional specification that the expectation depends on state ρ.

Von Neumann Observable A special case of observable X is the *von Neumann observable*, that is an observable with non-degenerate spectrum $\mathsf{Sp}\,(X)$. In this case, the Born rule simplifies as $p(x) = \langle x|\rho|x\rangle$, with $x \in \mathsf{Sp}\,(X)$ the measured value.

Property Another special case of observable is that corresponding to a *property* of system A, described by an orthogonal projector P, namely an observable with binary eigenvalue 0 and 1 corresponding to the orthogonal projectors $I_A - P$ and P. We say that system A has the property P when the outcome 1 occurs with certainty. In this case its state ρ gives $\mathrm{Tr}[\rho P] = 1$, namely $\mathsf{Supp}\,\rho \subseteq \mathsf{Supp}\,P$.[6]

In the literature the identity-resolution with orthogonal operators $\{P_i\}_{i\in X}$ is also called *PVM* ("projector-valued measure": see also the notes at the end of the chapter). Apart from the specific values, the observable is just a resolution of the identity made with orthogonal projectors on $\mathcal{H}_A$, which is a special case of observation test. The "measured values" are just a relabeling of the outcomes with a function $f : X \to \mathbb{R}$ (called *post-processing*). If the function is not one-to-one, the observable is degenerate, namely it has eigen-spaces with dimension greater than one.

It is clearly possible to associate a random variable to a generally non-orthogonal observation test in the same way as for the observable. However, if the POVM $\{P_i\}_{i\in X}$ is not orthogonal, the expectation value of the function of the random variable will not be equal to the expectation value of the operator $f(X)$. Instead, we have

$$\langle f(X) \rangle_\rho = \sum_{i\in X} f(x_i)p_i \neq \mathrm{Tr}[f(X)\rho], \qquad p_i = \mathrm{Tr}[P_i\rho] \qquad (2.15)$$

whereas equality holds only for observables.

2.3.2 Conditional and Marginal States

The action of effect $(a|_A$ on a state of AB produces a state of B, namely

$$\forall|\sigma)_{AB} \in \mathsf{St}(AB), \quad (a|_A|\sigma)_{AB} = \mathrm{Tr}_A[\sigma\,(E_a \otimes I_B)] := |\sigma_a)_B \in \mathsf{St}(B). \qquad (2.16)$$

The state $|\sigma_a)_B \in \mathsf{St}(B)$ is called the *conditional state*. Specifically, $|\sigma_a)$ in Eq. (2.16) is the state conditioned by the observation event a of the performed test. We can represent the application of the effect $a \in \mathsf{Eff}(A)$ to a state in $\mathsf{St}(AB)$ using the diagrammatic equation

where the left-to-right direction goes from input to output. The left-rounded boxes with no input represent states, and the right-rounded boxes with no output represent effects.

---

[6] The notation $\mathsf{Supp}\,X$ denotes the support of the operator X, namely the orthogonal complement of its kernel $\mathsf{Ker}\,X$.

We will formalize such diagrammatic representation in Chapter 3, which is devoted to the informational framework. For the moment we will use a few simple diagrams as an intuitive representation, just to become familiar with them.

Exercise 2.12   Show that σ_a in Eq. (2.16) is a quantum state for B.

The state conditioned by the deterministic effect is called the *marginal state* $|\sigma)_B$ of system B in the joint state $|\sigma)_{AB}$. According to Eq. (2.11), it is given by the density operator

$$\sigma^B = \mathrm{Tr}_A[\sigma^{AB}].$$

The marginal state $\sigma^B \in \mathsf{St}(A)$ allows us to evaluate marginal probabilities of observations on system B only, as follows:

$$p(b_j) = \mathrm{Tr}[\sigma^B E_{b_j}]. \tag{2.17}$$

Indeed, upon marginalizing the observation test $\{a_i\}_{i \in X}$ on system A, one has

$$p(b_j) = \sum_{i \in X} p(a_i, b_j) = \sum_{i \in X} (a_i \otimes b_j | \sigma) = (e_A \otimes b_j | \sigma)$$

corresponding to

$$\sum_{i \in X} p(a_i, b_j) = \sum_{i \in X} \mathrm{Tr}[\sigma^{AB}(E_{a_i} \otimes E_{b_j})] = \mathrm{Tr}[\sigma^{AB}(I_A \otimes E_{b_j})]$$

$$= \mathrm{Tr}[\mathrm{Tr}_A(\sigma^{AB}) \otimes E_{b_j}] = \mathrm{Tr}[\sigma^B E_{b_j}],$$

where we used Eq. (2.4). Diagrammatically, Eq. (2.17) becomes

The marginal state of system B provides all expectations of local observations on B, in particular expectations of its observables, e.g. for any observable X of B one has

$$\langle X \rangle = \mathrm{Tr}[(I_A \otimes X)\sigma^{AB}] = \mathrm{Tr}[X\sigma^B].$$

2.4 Causality, Convex Structure, Discriminability

2.4.1 Causality

The reader may have already noticed the asymmetry between marginalizing over effects and over states. Consider a preparation test $\{\rho_i\}_{i \in X} \subseteq \mathsf{St}(A)$ followed by an observation test $\{a_j\}_{j \in Y} \subseteq \mathsf{Eff}(A)$. The Born rule (2.8) gives the joint probability

$$p(i, j) = (a_j | \rho_i) = \mathrm{Tr}[\rho_i E_{a_j}].$$

The marginal probability in which we sum up states is given by

$$p(j) = \sum_{i \in X} p(i, j) = \sum_{i \in X} \mathrm{Tr}[\rho_i E_{a_j}] = \mathrm{Tr}[\rho_X E_{a_j}] \equiv (a_j | \rho_X),$$

where $\rho_X := \sum_{i \in X} \rho_i$ is the prior state of the preparation test. Therefore, the marginal probability of the effect depends on the specific preparation test performed. On the contrary, the marginal probability of preparation of the i-th state of the ensemble does not depend on the chosen observation test $\{a_j\}_{j \in Y}$, since one has

$$p(i) = \sum_{j \in Y} p(i,j) = \sum_{j \in Y} \mathrm{Tr}[\rho_i E_{a_j}] = \mathrm{Tr}\,\rho_i \equiv (e|\rho_i).$$

Therefore, quantum theory satisfies the following relevant principle.

Principle 1 (Causality)   The probability of a preparation is independent of the choice of the observation.

No wonder that Principle 1 holds, as we assumed it in Section 2.1 when we defined the preparation probability $p(\rho)$ as a function of the state $\rho \in \mathrm{St}(\rho)$ alone. This is actually in agreement with our intuition of what a "preparation" is. We will, however, consider a more general probabilistic context in Parts II–IV of the book, and formalize the general framework of operational probabilistic theories in Chapter 3, in a way independent of the causality assumption. We will also devote the whole of Chapter 5 to the causality principle within the general operational context. There, we will also re-derive the result that causality is equivalent to the uniqueness of the deterministic effect, and we will also see that it is necessary for the possibility of achieving any probabilistic state deterministically. We have already checked that both assertions are true in quantum theory: on one hand the unique deterministic effect e_A is given by Tr_A, on the other hand corresponding to the quantum state $\rho \neq 0$ there exists a unique deterministic state given by $\hat{\rho} = \rho/\mathrm{Tr}\,\rho \in \mathrm{St}_1(A)$. A procedure for preparing $\hat{\rho}$ consists in repeating the preparation test for ρ, and selecting the state a posteriori upon occurrence of the corresponding outcome. Such a procedure – called *post-selection* – is generally not available for an operational probabilistic theory that doesn't satisfy causality, since the preparation probability may depend on the choice of the observation, and not only on the state ρ.

2.4.2 A Convex Interlude

Cone Structure of Quantum Theory   In Exercises 2.5 and 2.7 we have seen that both sets $\mathrm{St}(A)$ and $\mathrm{Eff}(A)$ are convex. They both span the respective convex cones $\mathrm{St}_+(A)$ and $\mathrm{Eff}_+(A)$.[7] The cones $\mathrm{St}_+(A)$ and $\mathrm{Eff}_+(A)$ span the two real linear spaces $\mathrm{St}_{\mathbb{R}}(A)$ and $\mathrm{Eff}_{\mathbb{R}}(A)$. The two cones $\mathrm{St}_+(A)^{\vee} = \mathrm{Eff}_+(A)$ and $\mathrm{St}_+(A) = \mathrm{Eff}_+(A)^{\vee}$ are reciprocally dual under the state-effect pairing $(\cdot|\cdot) : \mathrm{Eff}_+(A) \times \mathrm{St}_+(A) \to \mathbb{R}$, since every positive linear functional over $\mathrm{Lin}_+(\mathcal{H}_A)$ is the trace with a positive operator on $\mathcal{H}_A$ (see Lemma 2.1). Correspondingly $\mathrm{St}_{\mathbb{R}}(A)$ and $\mathrm{Eff}_{\mathbb{R}}(A)$ are also reciprocally dual as linear spaces. In quantum theory both $\mathrm{St}_{\mathbb{R}}(A)$ and $\mathrm{Eff}_{\mathbb{R}}(A)$ coincide with the real linear space $\mathrm{Herm}(\mathcal{H}_A)$ of Hermitian operators on $\mathcal{H}_A$, whereas both cones $\mathrm{St}_+(A)$ and $\mathrm{Eff}_+(A)$ coincide with the cone $\mathrm{Lin}_+(\mathcal{H}_A)$ of positive operators on $\mathcal{H}_A$.

---

[7] We recall that a convex cone C is a set closed under conic combinations, namely for $x_1, x_2 \in C$ and $a, b \geq 0$ one has $ax_1 + bx_2 \in C$.

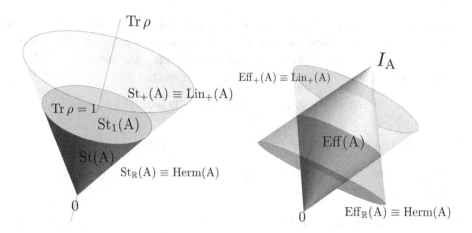

Fig. 2.1 Convex structure of the states and effects of a quantum system. This is the typical convex structure of a causal operational probabilistic theory, as will be shown in Chapter 5 (see Fig. 5.2).

We can now easily recover the "shape" of the two convex sets $\mathsf{St}(A)$ and $\mathsf{Eff}(A)$. The "normalization" condition $\mathrm{Tr}[\rho] = q > 0$ of the positive operator $\rho \in \mathsf{Lin}_+(\mathcal{H}_A)$ corresponds to an hyperplane H_q normal to the axis of the cone $\mathsf{Lin}_+(\mathcal{H}_A)$ at coordinate q. The set of deterministic states $\mathsf{St}_1(A)$ is thus the intersection $\mathsf{Lin}_+(\mathcal{H}_A) \cap \mathsf{H}_1$ of the cone of positive operators with the hyperplane at $q = 1$, resulting itself in a convex set. The set $\mathsf{St}(A)$ is thus the truncated cone contained between the hyperplane H_1 at $q = 1$ and the zero operator, which is the vertex of the cone.

As regards the convex set of effects, it is the intersection of the convex cone of positive operators $\mathsf{Lin}_+(\mathcal{H}_A)$, with the domination cone $\{X \in \mathsf{Herm}(\mathcal{H}_A) : X \leq I_A\}$, resulting in a spindle-shaped convex set. The two cone structures are depicted in Fig. 2.1. As we will see in Chapter 5, the present convex structure for states and effects is common to all causal operational probabilistic theories with no restriction.

Extremal Points and Rays   In the following we will call *extremal* any point of a convex set that cannot be written as a convex combination of different points of the set. A *ray* of the convex cone C is any set of the form $\mathsf{R}_p := \{ap, \forall a > 0\}$ for $p \in \mathsf{C}$. We also call this set the *ray through point* $p \in \mathsf{C}$. We call a ray *extremal* if none of its points can be written as a convex combination of two points not belonging to the same ray. Both definitions of ray and extremal ray correspond to the common geometrical intuition. In a convex cone the only extremal point is its tip, namely the 0 point. In a convex cone of events the sum corresponds to *coarse-graining*. For example, given two events x_1 and x_2 in a given test, corresponding to the outcomes $j = 1$ and $j = 2$, respectively, one can form the event corresponding to the union-outcome "$j = 1$ or $j = 2$" by taking $x = x_1 + x_2$.

Refinement Sets   The reverse notion of coarse-graining is that of *refinement*. For a convex cone C and $x, x_1, x_2 \in \mathsf{C}$ not belonging to the same ray (i.e. not pairwise proportional) satisfying $x = x_1 + x_2$, we say that both x_1 and x_2 *refine* x. On the other hand, we say that $x \in \mathsf{C}$ is *atomic* if it is not refinable, namely it cannot be written as the sum of two points that do not belong to the same ray.

For a generic convex set C we have also the notion of *convex refinement*, defined in terms of convex combinations instead of sums. Precisely, for $x, x_1, x_2 \in \mathsf{C}$ with $x_1 \neq x_2$ satisfying $x = ax_1 + (1-a)x_2$ for some $0 < a < 1$ we say that x_1 (x_2) *convexly refines* x. Clearly, an *extremal* point $x \in \mathsf{C}$ is not convexly refinable (its convex refinement is the point itself). In the following we will also use the word "refinement" in place of "convex refinement" when the meaning will be clear from the context.

Summarizing:

$x, x_1, x_2 \in \mathsf{C}, \mathsf{C}_+, x \neq x_1, x \neq x_2, p \in (0, 1)$:

C_+ convex cone:	$x = x_1 + x_2, \ x \not\propto x_{1,2} \Rightarrow x_{1,2}$ *refines* x,
C convex set:	$x = px_1 + (1-p)x_2 \Rightarrow x_{1,2}$ *convexly refines* x

and correspondingly:

$x \in \mathsf{C}_+$ *atomic*:	$x = x_1 + x_2 \Rightarrow x_1 \propto x_2$,
$x \in \mathsf{C}$ *extremal*:	$\exists p \in (0, 1) : x = px_1 + (1-p)x_2 \Rightarrow x_1 = x_2$.

Both notions of refinement correspond to a *partial ordering* within the pertaining set. Upon denoting the ordering by the symbol $\leq$ (which ordering is used will be clear from the context), we write $y \leq x$ to denote that y (convexly) refines x. Precisely, one has

$x, y \in \mathsf{C}$ convex set,	$y \leq x$ if $\ \exists p \in (0, 1)$ such that	$x = py + (1-p)z, \ z \in \mathsf{C}$,
$x, y \in \mathsf{C}_+$ convex cone,	$y \leq x$ if	$x = y + z, \ z \in \mathsf{C}_+$.

Notice that in the conic case the ordering $\leq$ coincides with the operator ordering.[8] The reader can easily check that both relations "$\leq$" are partial orderings over their respective convex sets. Again, the convex cone definition applies also to the case of a truncated convex cone.

In the following we will call the *refinement set* of $x \in \mathsf{C}_+$ – denoted as $\mathsf{RefSet}\, x$ – the set of points that refine x, and we will call the *convex refinement set* of $x \in \mathsf{C}$ – denoted as $\mathsf{RefSet}_1 x$ – the set of points that convexly refine x. When it is clear from the context, we will also simply say "refinement set" for both cases. For an illustration of these notions see Fig. 2.2.

Exercise 2.13 Show that the convex refinement set $\mathsf{RefSet}_1 x$ of a point x in a convex set is also a convex set. Show that the refinement set $\mathsf{RefSet}\, x$ of a point x in a convex cone is a spindle.

When a point $x \in \mathsf{C}$ is *internal* in a convex set C (namely it does not belong to its border), one has $\mathsf{RefSet}_1 x \equiv \mathsf{C}$, i.e. the refinement set is the full convex set. On the other hand, when the point x belongs to the border of C its refinement set is a *face* of convex set, with dimensionality strictly smaller than that of the original convex set. Faces are convex sets themselves. A face of a convex set is always the convex refinement set of any of

---

[8] In the literature the symbol $\preceq$ is often used to denote convex refinement.

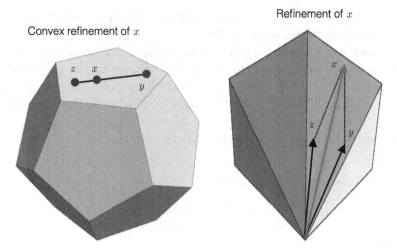

Fig. 2.2   Illustration of the notion of *refinement* and *convex refinement*. Left: the convex set is a dodecahedron, and the refinement set of the point x lying on the border is a pentagon-shaped face, which is itself a convex set. Right: the cone is a polygon-base pyramid. For a point x lying on the border of the cone, i.e. belonging to a face, the refinement set is a convex subset of such a face.

its interior points.[9] For a convex cone a face is also the conic hull of the refinement set RefSet x of a point x on the border.

Exercise 2.14   Show that a face of a convex set is always the convex refinement set of any of its interior points.

Extremal, Atomic, Pure, ...   We will call a state, effect, or transformation *extremal* or *atomic* according to our previous definitions. For continuity with the past literature the extremal states will also be called *pure states*, and we can extend such term for duality to the extremal effects, which may be called pure. Notice that corresponding to an extremal state $\rho \in \mathsf{St}_1(A)$, the state $p\rho \in \mathsf{St}(A)$ $(0 < p < 1)$ is atomic, since it belongs to an extremal ray of the cone $\mathsf{St}_+(A)$. Since, however, it is just proportional to an extremal normalized state, by abuse of nomenclature thay can also be called pure.[10] Since any positive operator can be always decomposed as the sum of rank-one operators, it is immediate to see that a pure state $\rho \in \mathsf{St}(A)$ is of the form $\rho = |\lambda\rangle\langle\lambda|$ with $|\lambda\rangle \in \mathsf{B}_1(\mathcal{H}_A)$ ($|\lambda\rangle \in \mathsf{S}_1$, i.e. $\|\lambda\| = 1$, for a deterministic state). Indeed, the set of operators $\mathsf{R}_\lambda := \{q|\lambda\rangle\langle\lambda|, q \in [0,+\infty)\}$ describes an extremal ray of the cone of positive operators. The deterministic effect provides an example of a point which is extremal in $\mathsf{Eff}(A)$, but it is not atomic. Also the effects at the intersection ring between the two cones in Fig. 2.1 are extremal.

Exercise 2.15   Show that an extremal ray of $\mathsf{Lin}_+(\mathcal{H}_A)$ is of the form $\mathsf{R}_\lambda = \{q|\lambda\rangle\langle\lambda|, q \in [0,+\infty)\}$.

---

[9] To be more precise these are the points of the relative interior of the face (Rockafellar, 2015).

[10] Extending the nomenclature "pure" also to effects and transformations to unify the notion of atomicity and extremality will generate a confusion between two different notions. In the case of states, instead, there would be no confusion, since it is only a matter of normalization.

2.5 Quantum States

We can now go back to the old notions that we learned in elementary courses of quantum mechanics:

Quantum Mantra 2 (The Superposition Principle)   Pure deterministic quantum states are in correspondence with equivalence classes of unit vectors in $\mathcal{H}_A$ modulo a phase factor, then also called *state vectors*. The following *superposition principle* holds: unit-normalized complex linear combination of state vectors are themselves state vectors.

Remark (Alternate Decompositions Into Pure States)   The decomposition of a density matrix into pure states is not unique: all decompositions are connected to each other via isometries. The connecting isometry is unique when it connects to a linearly independent decomposition $\rho = \sum_i |\lambda_i\rangle\langle\lambda_i|$, e.g. the spectral decomposition. One can then easily check that any other decomposition must be of the form

$$\rho = \sum_j |\psi_j\rangle\langle\psi_j|, \quad |\psi_i\rangle = \sum_j V_{ij}|\lambda_j\rangle, \quad (V^\dagger V)_{ij} = \delta_{ij}, \tag{2.18}$$

namely it is connected to the minimal decomposition via a matrix V with orthogonal columns (namely an isometry).

Exercise 2.16   Check Eq. (2.18).

2.5.1 States Separate Effects and Vice Versa

Given two states $\rho_0 \neq \rho_1 \in \mathsf{St}(A)$, is there always an observation test $\{a_i\}_{i\in X} \subset \mathsf{Eff}(A)$ that allows us to discriminate between them? If such an observation test exists, it must contain at least an effect, say a, that occurs with different probabilities over ρ_0, ρ_1.[11] Then, we will say that the effect a *separates* the states ρ_0 and ρ_1 (or, the effect a is *separating* for states ρ_0, ρ_1) when $(a|\rho_1) \neq (a|\rho_0)$, namely when the effect occurs with different joint probabilities over the two states. Thanks to the polarization identity, we can answer our initial question for the positive, namely given any two different states $\rho_0 \neq \rho_1 \in \mathsf{St}(A)$, there is always an effect $a \in \mathsf{Eff}(A)$ that separates them. Indeed, if this were not the case, one would have

$$0 = (a|\rho_0 - \rho_1) = \mathrm{Tr}[E_a(\rho_0 - \rho_1)], \quad \forall a \in \mathsf{Eff}(A), \tag{2.19}$$

and since $|\lambda\rangle\langle\lambda|$ belongs to $\mathsf{Eff}(A)$ for every $|\lambda\rangle \in \mathsf{B}_1(\mathcal{H}_A)$, taking $E_a = |\lambda\rangle\langle\lambda|$ in Eq. (2.19) would imply $\rho_0 = \rho_1$ (see Exercise 2.1).

---

[11] If we can rely on the hypothesis that always the same state is prepared in a sequence of repeated preparations, repeating the same measurement many times we can practically assess which of the two states was prepared, from the frequency of the outcome. Nevertheless, this would require a procedure that always selects the same state for all repetitions. In the absence of such a procedure the two probabilistic states are prepared randomly. However, even in this case it is possible to prove that, by suitably coarse-graining the test $\{a, e-a\}$, it is always possible to build a binary observation test that discriminates the states with error probability smaller than $\frac{1}{2}$ (see Lemma 3.1 in Chapter 3).

We also have the symmetrical situation in which we exchange states with effects, namely we use a state σ to discriminate between two effects a_0 and a_1. In this case we will say that a state σ separates the two effects a_0 and a_1 when $(a_0|\sigma) \neq (a_1|\sigma)$, namely when the state as a functional over effects has different values over the two effects. Also in the case of effects, for every $a_0 \neq a_1 \in \mathsf{Eff}(A)$ there exists a separating state $\sigma \in \mathsf{St}(A)$.

The notion of "separating" is actually a general mathematical notion, which applies to functionals over any kind of set. Consider a set B, and denote by $\mathsf{B}^\vee$ the space of real functionals over B. We say that the functional $a \in \mathsf{B}^\vee$ is separating for $b_1 \neq b_2 \in \mathsf{B}$ iff $(a|b_1) \neq (a|b_2)$. We then say that a subset $\mathsf{A} \subseteq \mathsf{B}^\vee$ is separating for B when

$$\forall b_1, b_2 \in \mathsf{B}, \quad b_1 \neq b_2 \Longleftrightarrow \exists a \in \mathsf{A} : \ (a|b_1) \neq (a|b_2). \tag{2.20}$$

This is equivalent to saying that a subset $\mathsf{A} \subseteq \mathsf{B}^\vee$ is separating for B when

$$\forall b_1, b_2 \in \mathsf{B}, \quad b_1 = b_2 \Longleftrightarrow \forall a \in \mathsf{A} \ (a|b_1) = (a|b_2). \tag{2.21}$$

In the case when B is vector space, the condition (2.21) can be rewritten as

$$\mathsf{B} \ni b = 0 \Longleftrightarrow \forall a \in \mathsf{A} : \ (a|b) = 0. \tag{2.22}$$

In the following for a given subset $\mathsf{A} \subseteq \mathsf{B}$ of a linear space B we will define $\mathsf{A}^\perp \subseteq \mathsf{B}^\vee$ as $\mathsf{A}^\perp := \{a \in \mathsf{B}^\vee | (a|\alpha) = 0, \forall \alpha \in \mathsf{A}\}$. By Eq. (2.22) we have that $\mathsf{A} \subseteq \mathsf{B}^\vee$ is separating for B iff $\mathsf{A}^\perp = \{0\} \subseteq \mathsf{B}$.

We use the two following simple lemmas.

Lemma 2.2 *Let $\gamma \in \mathsf{B}$, with X subspace of the linear space B. If $(y|\gamma) = 0$ for all $y \in \mathsf{X}^\perp$, then $\gamma \in \mathsf{X}$.*

Lemma 2.3 *For a given subset $\mathsf{A} \subseteq \mathsf{B}$ of a linear space B, $\mathsf{A}^\perp$ is a vector space coinciding with $\mathsf{Span}(\mathsf{A})^\perp$. Moreover, $(\mathsf{A}^\perp)^\perp \subseteq \mathsf{B}$ is a subspace coinciding with $\mathsf{Span}(\mathsf{A})$.*

Exercise 2.17   Prove Lemmas 2.2 and 2.3.

Using Lemmas 2.2 and 2.3 we can now prove the following relevant one.

Lemma 2.4 *For B a vector space and $\mathsf{B}^\vee$ the space of linear functionals on B, a set $\mathsf{A} \subseteq \mathsf{B}^\vee$ is separating for B if and only if A spans $\mathsf{B}^\vee$.*

Proof   Consider $\beta \in (\mathsf{A}^\perp)^\perp$, namely

$$(a|\beta) = 0, \quad \forall a \in \mathsf{A}^\perp = \mathsf{Span}(\mathsf{A})^\perp.$$

Lemma 2.2 implies that $\beta \in \mathsf{Span}(\mathsf{A})$, hence we have $(\mathsf{A}^\perp)^\perp \subseteq \mathsf{Span}(\mathsf{A})$, and Lemma 2.3 implies that

$$(\mathsf{A}^\perp)^\perp = \mathsf{Span}(\mathsf{A}).$$

Since $\mathsf{A} \subseteq \mathsf{B}^\vee$ is separating for B iff $\mathsf{A}^\perp = \{0\} \subseteq \mathsf{B}$, and since $\{0\}^\perp = \mathsf{B}^\vee$, we conclude that $\mathsf{Span}(\mathsf{A}) = (\mathsf{A}^\perp)^\perp = \{0\}^\perp = \mathsf{B}^\vee$.   □

From Lemma 2.4 it follows also that A separates B iff Span A does, and this holds for both the linear span $\mathsf{Span}_\mathbb{R}$ and for the conic span Span_+. Also A separates B iff A separates Span B, and again this holds for both $\mathsf{Span}_\mathbb{R}$ and Span_+. Finally, all the above statements hold for $\mathsf{B}^\vee$ the cone of positive linear functionals.

Since for dual cones C and $\mathsf{C}^\vee$ the only element of one cone which is orthogonal to all elements of the other is the null vector (the tip of the cone), the cones C and $\mathsf{C}^\vee$ separate each other. Therefore, since, as seen in Section 2.4, the cones of states and of effects are dual each other, and $\mathsf{Span}_+\mathsf{St}_1(A) = \mathsf{Span}_+\mathsf{St}(A) = \mathsf{St}_+(A)$ and $\mathsf{Span}_+\mathsf{Eff}(A) = \mathsf{Eff}_+(A)$, we conclude that: *states separate effects and effects separate states.* This means that there is always an observation test discriminating two states or a preparation test discriminating two effects.

Exercise 2.18 Show that pure states separate effects, and atomic effects separate states.

We will call *local effects* those effects that are factorized into single-system effects, e.g.

$$(a|_A(b|_B = \mathrm{Tr}[\cdot (E_a \otimes E_b)].$$

Analogously, we will call *local states* those states that are factorized into single-system states, e.g.

$$|\rho)_A|\sigma)_B = \rho \otimes \sigma.$$

A relevant property of quantum theory which originates from the tensor product structure of composition of systems in Postulate 1 is the following lemma.

Lemma 2.5 *Local effects separate joint multipartite states, and local states separate joint multipartite effects.*

Proof The proof is an immediate consequence of Lemma 2.4, since local effects are a spanning set for all effects. More precisely, one has

$$\mathsf{Span}_\mathbb{R}[\mathsf{Eff}(A) \otimes \mathsf{Eff}(A)] = \mathsf{Span}_\mathbb{R}[\mathsf{Eff}_\mathbb{R}(A) \otimes \mathsf{Eff}_\mathbb{R}(A)]$$
$$= \mathsf{Span}_\mathbb{R}[\mathrm{Herm}(\mathcal{H}_A) \otimes \mathrm{Herm}(\mathcal{H}_B)]$$
$$= \mathrm{Herm}(\mathcal{H}_A \otimes \mathcal{H}_B) = \mathsf{Eff}_\mathbb{R}(AB) = \mathsf{Span}_\mathbb{R}[\mathsf{Eff}(AB)],$$

namely local effects are separating for $\mathsf{Span}_\mathbb{R}\mathsf{St}(AB)$, hence, in particular for the subset $\mathsf{St}(AB)$. Similarly one proves that local states are a spanning set for all joint states.   □

From the above lemma, one has the following relevant principle that holds for quantum theory.

Principle 2 (Local Discriminability)   It is possible to discriminate any pair of joint states of multipartite systems by using only local measurements.

This principle has a deep conceptual value, and is a crucial property of quantum theory, bringing dramatic simplifications to the structure of the sets of states and transformations, and reconciling the holism of entangled states (see Section 2.6) with the reductionist experimental approach based on local observations. The principle will be the subject of Chapter 6.

2.5.2 Tomography of States and Effects

In Section 2.5.1 we proved that a set $A \subseteq B^\vee$ is separating for B iff $\mathsf{Span}\, A$ coincides with $B^\vee$, and this holds for both the linear span $\mathsf{Span}_\mathbb{R}$ and the conic span Span_+. As a consequence, if we consider a test $\{l_i\}_{i\in X} \subseteq \mathsf{Eff}(A)$ made with a set of effects that is separating for states, we can expand any effect a as follows:

$$\forall a \in \mathsf{Eff}(A), \ a = \sum_{i\in X} c_i(a) l_i.$$

The coefficients $c_i(a)$ can always be chosen to be linear in a, namely $c_i \in \mathsf{St}_\mathbb{R}(A)$. The set $\{c_i\}_{i\in X}$ is a *dual set* for $\{l_i\}_{i\in X}$, and it is unique when the set $\{l_i\}_{i\in X}$ is linearly independent. Upon pairing the effect a with a state $\rho \in \mathsf{St}(A)$ one has

$$\forall a \in \mathsf{Eff}_\mathbb{R}(A), \ (a|\rho) = \sum_{i\in X} c_i(a)(l_i|\rho).$$

Therefore, if we know the probabilities $(l_i|\rho)$ we know the probability $(a|\rho)$ of any effect a on the state ρ, namely we know the state ρ. Indeed, we can expand the state over the dual set $\{c_i\}$ as $\rho = \sum_{i\in X}(l_i|\rho)c_i$. Therefore, upon performing the observation test $\{l_i\}_{i\in X}$ we can reconstruct the state ρ. We say that we are performing a *state tomography* of ρ, and call the observation test $\{l_i\}_{i\in X}$ *informationally complete*. In principle the state tomography can be performed also with a set of effects that don't constitute a complete test, and this would correspond to running different tests. We can therefore consider also a generic set of effects as informationally complete. In such case we can also conclude that:

Proposition 2.6 *A set of effects is separating for states iff it is informationally complete for states.*

In practice we need to repeat the experiment many times in order to recover the probabilities $(l_i|\rho)$ as frequencies of the outcomes, or use other statistical estimation methods, such as the maximum likelihood. The feasibility of the tomography with a generic set of effects is granted by the procedure consisting in the randomization of different tests and of suitable coarse-graining (an example of such a procedure is illustrated in Problem 2.21). The test resulting from such procedure would then be an informationally complete observation test.

Quorum of Observables A common case of informationally complete test is that consisting of a set of observables that span the set of all Hermitian operators. Such a set is called a *quorum of observables*. The simplest example of quorum is the set of Pauli matrices for $\dim \mathcal{H}_A = 2$. Indeed, as shown in Exercise 2.4 the three Pauli matrices plus the identity I_A make an orthonormal basis for $\mathsf{Lin}(\mathbb{C}^2)$. The notion of quorum of observables corresponds to randomly measured observables of the quorum in repeated measurements, and can also be mathematically represented by a POVM.

Exercise 2.19   Upon writing the general expansion of any operator over the set of Pauli matrices plus the identity, derive the general matrix element $\langle u|\rho|v\rangle$ of the state as a function of the expectations of the Pauli matrices.

SIC POVMs   The randomized Pauli observable of Exercise 2.19 is the six-element POVM $\{|s\rangle_\alpha \langle s|_\alpha\}_{\alpha=x,y,z,\,s=\pm}$ where $\sigma_\alpha|s\rangle_\alpha = s|s\rangle_\alpha$. Such POVM is not minimal (a minimal POVM would contain four elements). A minimal informationally complete POVM which is also rank-one is provided by a *SIC POVM* (SIC stands for *symmetric informationally complete*). This is made of four rank-one projectors made with Bloch vectors of the form Eq. (2.12) at the vertices of a tetrahedron.[12]

Tomography of Multipartite States   What about the possibility of experimentally determining the joint state of a composite system $\sigma \in \mathrm{St}_1(AB\ldots Z)$? Do we need an informationally complete observation test made of joint effects of $AB\ldots Z$? Here the local discriminability principle (Principle 2) comes to help: since local effects are separating for joint states, we can use local informationally complete observation tests for each single system A, B, ..., Z. Local discriminability guarantees us that the test $\{a_i \otimes b_j \otimes \ldots \otimes z_k\}_{i\in X_A, j\in X_B,\ldots k\in X_Z}$ is separating for all states $\mathrm{St}_1(AB\ldots Z)$, namely we can achieve the tomography of any unknown joint state (including the entangled states that we will see in Section 2.6) as follows:

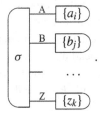

For example, for testing a state $\sigma \in \mathrm{St}_1(AA)$ we just need two copies of the same informationally complete observation test for system A; we don't need to build up a new informationally complete observation test, e.g. $\{w_{i,j}\}_{i,j\in X}$ for AA, as in

$$\left(\boxed{\sigma}\!-\!\!\begin{array}{c}A\\ \\A\end{array}\!\!-\!\boxed{\{w_{i,j}\}}\right).$$

This is the power of local discriminability! As an example, for N qubits, the expectation value of a joint operator J can be expressed in terms of the quorum of local Pauli observables as follows:

$$\langle J\rangle = 2^{-N} \sum_{\{l_j\}} \mathrm{Tr}[J(\otimes_j \sigma_{l_j})]\langle \otimes_j \sigma_{l_j}\rangle.$$

---

[12] At the moment the existence of SIC POVMs for any dimension is still an open mathematical problem. Analytical solutions are known only up to dimension 13, and for dimensions 15 and 19, whereas numerically they have been found up to dimension 67 (Scott and Grassl, 2009). The existence of SIC POVMs for any dimension is relevant as a foundation of the quantum-Bayesian interpretation of quantum theory (QBism) (Fuchs and Schack, 2013).

Informationally Complete Preparation test   Dually to the notion of informationally complete observable we have the notion of *informationally complete preparation test*. Similarly to the case of informationally complete observables, an informationally complete preparation test $\{\omega_i\}_{i \in X}$ is simply a separating set of states, in terms of which one can write any state as a real linear combination, namely $\mathsf{Span}_{\mathbb{R}}\{\omega_i\}_{i \in X} = \mathsf{St}_{\mathbb{R}}(A)$. Thus, by an informationally complete preparation test we can achieve the tomography of effects and of observation tests. It is clear that the same arguments expressed for the observation tests hold dually here for the preparation tests (methods for recovering the effect, minimality, uniqueness of expansion coefficients, etc.), and as a consequence of Principle 2, also for informationally complete preparation tests we can use local preparations in order to achieve the tomography of any joint effect.

2.6   Entangled Quantum States and Effects

By definition, for any two density operators $\rho^A \in \mathsf{St}(A)$ and $\rho^B \in \mathsf{St}(B)$ the operator $\rho^{AB} = \rho^A \otimes \rho^B \in \mathsf{St}(AB)$ is a quantum state for the composite system AB. Analogously for effects one has that $E^{AB} = E^A \otimes E^B \in \mathsf{Eff}(AB)$ is an effect for the composite system AB if $E^A \in \mathsf{Eff}(A)$ and $E^B \in \mathsf{Eff}(B)$. States and effects of this form are called *factorized*. According to the Born rule they produce factorized probabilities, namely of the form $\mathrm{Tr}(\rho^A E^A)\mathrm{Tr}(\rho^B E^B)$. They represent preparations and observations independently performed over the two systems. We can prepare mixtures of factorized states by randomly choosing between factorized states, leading to states of the form $\rho^{AB} = \sum_i \rho_i^A \otimes \rho_i^B$ with $\mathrm{Tr}\,\rho^{AB} \leq 1$.[13] States of this form are called *separable*. Similarly, an effect of the form $E^{AB} = \sum_i E_i^A \otimes E_i^B \leq I_{AB}$ with $E_i^A \in \mathsf{Eff}(A)$ and $E_i^B \in \mathsf{Eff}(B)$ is also called separable.[14]

It is now natural to ask if there exist states and effects that are not separable. The answer is positive. Indeed, it is easy to see that a rank-one positive operator of the form

$$R = |\Lambda\rangle\langle\Lambda|, \quad |\Lambda\rangle = \sum_i |\lambda_i\rangle_A \otimes |\lambda_i\rangle_B \in \mathcal{H}_A \otimes \mathcal{H}_B, \qquad \{|\lambda_i\rangle\} \text{ non-collinear} \quad (2.23)$$

is not separable. States and effects that are not separable are called *entangled*.

Exercise 2.20   Show that any positive operator R of the form in Eq. (2.23) is not separable.

Exercise 2.21   Show that for any entangled pure state of the form $|R\rangle\langle R|$ with

$$|R\rangle = \sum_{i=1}^{s} |i\rangle \otimes |v_i\rangle \in \mathcal{H}_A \otimes \mathcal{H}_B, \quad \{|i\rangle\}_{i=1,\ldots,s} \text{ orthonormal set for } \mathcal{H}_A,$$

---

[13] Such preparations can be performed by local preparations and classical communication between the two systems.

[14] It is still an open problem what is the most general procedure to obtain separable effects. Although a random choice of local measurement gives rise to a separable measurement, there exist separable measurements that are non-local, for example a POVM for system AB of the form $\{A_i \otimes B_i\}_{i \in X}$, with $A_i \geq 0$ and $B_i \geq 0$ $\forall i \in X$ and $\sum_{i \in X} A_i \otimes B_i = I_A \otimes I_B$ (whereas, generally $\sum_{i \in X} A_i \neq I_A$ and $\sum_{i \in X} B_i \neq I_B$).

the marginal state for system B is given by the density matrix

$$\rho = \sum_{i=1}^{s} |v_i\rangle\langle v_i|.$$

The notion of entangled state is easily extended to n-partite states with $n > 2$ for pure states. A popular example is that of the GHZ state, with wavevector

$$|\Psi_{GHZ}\rangle = \tfrac{1}{\sqrt{2}}(|0\rangle \otimes |0\rangle \otimes |0\rangle - |1\rangle \otimes |1\rangle \otimes |1\rangle),$$

which generalizes the *singlet state* for two systems with wavevector

$$|\Psi^-\rangle = \tfrac{1}{\sqrt{2}}(|0\rangle \otimes |1\rangle - |1\rangle \otimes |0\rangle).$$

Clearly there are also mixed states that are not separable. A relevant example is given by the Werner states in Exercise 2.28.

How Can I Know that a State is Entangled?　For a pure bipartite state with wavevector $|\Psi\rangle \in \mathcal{H}\otimes\mathcal{K}$ it is very simple to establish if the state is entangled: the state is not entangled iff it can be written as the tensor product of two pure states. How can we establish if this is the case, independently of the chosen basis? It is sufficient to write the vector $|\Psi\rangle$ in the form

$$|\Psi\rangle = \sum_{n=1}^{\dim\mathcal{H}} \sum_{m=1}^{\dim\mathcal{K}} \Psi_{nm}|h_n\rangle \otimes |k_m\rangle,$$

for any couple of orthonormal bases $\{|h_n\rangle\}_{n=1,\dots\dim\mathcal{H}}$ $\{|k_m\rangle\}_{m=1,\dots\dim\mathcal{K}}$. Then the state is factorized iff $\mathrm{rank}(\Psi) = 1$, where $\mathrm{rank}(A)$ denotes the rank of A, and Ψ is the matrix of coefficients Ψ_{nm}.

Exercise 2.22　Prove that a pure bipartite state is entangled iff the rank of its matrix of coefficients is greater than one.

For non-pure bipartite states there is no known general criterion to assess if a density operator is a separable state or not. For $d = 2$ a criterion is available only for bipartite states. The criterion is the following. For $A \in \mathrm{Lin}(\mathcal{H}\to\mathcal{K})$ (the symbol $\mathrm{Lin}(\mathcal{H}\to\mathcal{K})$ denotes operators from Hilbert space $\mathcal{H}$ to Hilbert space $\mathcal{K}$) define the transposed $A^{\mathsf{T}} \in \mathrm{Lin}(\mathcal{K}\to\mathcal{H})$ with respect to the orthonormal basis $\{|h_n\rangle\}_{n=1,\dim\mathcal{H}}$ and $\{|k_m\rangle\}_{m=1,\dim\mathcal{K}}$ as

$$A = \sum_{n=1}^{\dim\mathcal{H}} \sum_{m=1}^{\dim\mathcal{K}} A_{nm}|k_m\rangle\langle h_n| \mapsto A^{\mathsf{T}} = \sum_{n=1}^{\dim\mathcal{H}} \sum_{m=1}^{\dim\mathcal{K}} A_{nm}|h_n\rangle\langle k_m|, \qquad (2.24)$$

which corresponds to taking the complex conjugate of the adjoint and keeping each operator $|k_m\rangle\langle h_n|$ in the sum in Eq. (2.24) as real. Clearly, $(A^{\mathsf{T}})^{\mathsf{T}} = A$. The transposition can be extended to the partial transposition defined for operators on tensor product spaces through the relation $X = A \otimes B \mapsto X^\theta = A \otimes B^{\mathsf{T}}$, and then extended by linearity. Thus, the criterion for assessing the entanglement of a pure state $\rho \in \mathrm{Lin}_+(\mathbb{C}^2 \otimes \mathbb{C}^2)$ is the so-called *PPT criterion*, namely: the state ρ is separable iff the density operator has positive partial transpose (PPT) $\rho \mapsto \rho^\theta$.[15] For $d > 2$ the PPT criterion is only a necessary condition for separability, namely if a state is separable then it has PPT but the converse is generally not true. The Werner states also provide counter-examples for $d > 2$ (see Exercise 2.29).

Exercise 2.23   Consider the *swap operator E*, which exchanges the two spaces of the tensor product $\mathcal{H} \otimes \mathcal{H}$, namely $E|\psi\rangle \otimes |\varphi\rangle = |\varphi\rangle \otimes |\psi\rangle$. Show that it can be written

$$E = \sum_{n,m=0}^{d-1} |n\rangle\langle m| \otimes |m\rangle\langle n|. \qquad (2.25)$$

Show that $E^\theta = |I\rangle\rangle\langle\langle I|$, where $|I\rangle\rangle := \sum_{n=1}^{d} |n\rangle \otimes |n\rangle$, where $d = \dim \mathcal{H}$.

Exercise 2.24   Using the above exercise, show that for $R = \frac{1}{d}|I\rangle\rangle\langle\langle I|$, with $d = \dim \mathcal{H}$, R^θ is not positive.

Exercise 2.25   Show that the PPT criterion does not depend on the chosen orthonormal bases.

Exercise 2.26   Show that if a state is separable then it has PPT.

Exercise 2.27   Show that for $d = 2$ the following Werner states are separable for $p \le \frac{1}{2}$ and entangled for $p > \frac{1}{2}$

$$\rho = p|\Psi^-\rangle\langle\Psi^-| + \tfrac{1}{4}(1-p)I_2 \otimes I_2, \quad p \in [0,1], \qquad (2.26)$$

where $|\Psi^-\rangle\langle\Psi^-|$ is the singlet state. Notice that ρ is the convex combination of the maximally mixed state $\frac{1}{4}I_2 \otimes I_2$ with the state $|\Psi^-\rangle\langle\Psi^-|$ which is maximally entangled (for quantification of entanglement see the following).

Exercise 2.28   [Werner states] The Werner states in $\mathcal{H} = \mathbb{C}^d$ are given by

$$\rho = p\frac{2}{d^2+d}P_+ + (1-p)\frac{2}{d^2-d}P_-, \quad p \in [0,1], \qquad (2.27)$$

where P_+ and P_- are the orthogonal projectors over the symmetric and antisymmetric subspaces of $\mathcal{H} \otimes \mathcal{H}$

$$P_\pm := \tfrac{1}{2}(I_2 \pm E),$$

and E is the swap operator in Eq. (2.25). Show that the states (2.27) can be equivalently written as

$$\rho = \frac{1}{d^2 - d\alpha}(I_2 - \alpha E), \qquad \alpha = \frac{(1-2p)d+1}{1-2p+d}.$$

Exercise 2.29[16]   Show that the Werner states are separable for $p \ge \frac{1}{2}$ and entangled for $p < \frac{1}{2}$, and thus they violate the PPT criterion for $d > 2$.

Between separable states and entangled states there is a full gradation of entanglement. There are many ways to quantify the entanglement of a state, motivated by different operational protocols.[17] A common quantity for pure bipartite states is the *entropy of entanglement*,[18] which is given by the von Neumann entropy of the marginal state, namely

---

[15] Horodecki *et al.* (1996).

[16] Werner (1989).

[17] For a review see Plenio and Virmani (2007).

[18] Bennett *et al.* (1996).

$$\text{Ent}(|\Psi\rangle) = S(\text{Tr}_2|\Psi\rangle\langle\Psi|), \tag{2.28}$$

where the von Neumann entropy S is given by

$$S(\rho) := -\text{Tr}\,\rho\log\rho \le \log d.$$

The upper bound is achieved only when the state ρ is maximally mixed, namely when $\rho = \frac{1}{d}I_d$. In such a case the state $|\Psi\rangle\langle\Psi|$ is said to be *maximally entangled*.

Since the property of separability is shared by both states and effects, the notion of entanglement can be extended to effects. Indeed, any pure entangled state is also an atomic entangled effect, and maximally entangled effects correspond to maximally entangled states. A rank-one PVM consisting of maximally entangled projectors is called the *Bell measurement*. We will see examples of such measurements in the continuation of this chapter.

2.7 Compression

Using the definitions of refinement sets given in Section 2.4.2, for the special case of a state $\omega \in \text{St}_1(A)$ we have

$$\text{RefSet}_1\,\omega \quad := \quad \{\rho \in \text{St}_1(A)\,|\,\exists p \in [0,1]\,|\,\omega = p\rho + (1-p)\sigma,\ \sigma \in \text{St}_1(A)\}, \tag{2.29}$$

$$\text{RefSet}\,\omega \quad := \quad \{\rho \in \text{St}(A)\,|\,\omega = \rho + \sigma,\ \sigma \in \text{St}(A)\}. \tag{2.30}$$

The use of the convex refinement set $\text{RefSet}_1\,\omega$ will be more common for $\omega \in \text{St}_1(A)$, but makes sense also for $\omega \in \text{St}(A)$, in which case it is a face of the convex set of states $\text{St}_q(A)$ having probability $q = \text{Tr}\,\omega$.

We have seen that any face of a convex set of $\text{St}_1(A)$ is a convex set itself. It is natural to ask if it is also the convex set of states for some other system. For quantum theory the answer is positive: any face of the convex set of states $\text{St}_1(A)$ coincides with the convex set of states $\text{St}_1(B)$ of another quantum system B. More precisely, for a quantum state ρ the quantum system corresponding to the face $\text{RefSet}_1\,\rho$ is the one having the support of ρ as Hilbert space. In equations

$$\rho \in \text{St}_1(A) \implies \text{RefSet}_1\,\rho = \text{St}_1(B) \text{ with } \mathcal{H}_B = \text{Supp}\,\rho. \tag{2.31}$$

One also has that

$$\rho \in \text{St}(A), \quad \text{RefSet}\,\rho \subseteq \text{St}(B) \text{ with } \mathcal{H}_B = \text{Supp}\,\rho. \tag{2.32}$$

The above result follows from the statements of the following exercises.

Exercise 2.30 Prove that for any two positive operators ρ, σ on $\mathcal{H}$ one has

$$\rho - \sigma \ge 0 \implies \text{Supp}\,\sigma \subseteq \text{Supp}\,\rho, \tag{2.33}$$

$$\text{Supp}\,\sigma \subseteq \text{Supp}\,\rho \implies \exists p > 0 \text{ such that } \rho - p\sigma \ge 0. \tag{2.34}$$

Exercise 2.31 [*Refinement set of ρ*] Prove Eq. (2.31).

We conclude that a state ρ on the border of $\mathsf{St}_1(A)$ can be encoded on a *smaller* system B with Hilbert space having $\dim \mathcal{H}_B < \dim \mathcal{H}_A$. In simple words, this means that the state can be *compressed*. Such a compression can be done in a perfectly invertible way via a partial isometry V embedding $\mathsf{Supp}\,\rho$ in $\mathcal{H}_B$, as follows:

$$\rho \in \mathsf{St}_1(A), \qquad \forall \sigma \in \mathsf{RefSet}_1\,\rho : \quad \sigma_{compressed} = V\sigma^A V^\dagger$$

where

$$V \in \mathsf{Lin}(\mathcal{H}_A \to \mathcal{H}_B), \quad V^\dagger V = \mathsf{Proj}\,(\mathsf{Supp}\,\rho).$$

The compression scheme is also *maximally efficient*: the encoding system C has the smallest possible size, namely it has no more states than exactly those needed to compress ρ, e.g. for a state on a face of $\mathsf{St}_1(A)$ the dimension of the Hilbert space of encoding system C will be strictly smaller than that of $\mathcal{H}_A$. We will call such a lossless and maximally efficient compression scheme *ideal*. The existence of an ideal compression scheme is a relevant feature of quantum theory, which we will state in simple words in terms of the following principle.

Principle 3 (Ideal Compression)   For every state there exists an ideal compression scheme.

Exercise 2.32   Show that the set of states $\mathsf{St}_1(A)$ of the qutrit is an eight-dimensional convex set whose border is made of a continuum of faces shaped as three-balls.

For an *internal state* $\rho \in \mathsf{St}_1(A)$ one has $\mathsf{Supp}\,\rho \equiv \mathcal{H}_A$, namely the support of its density matrix is the full Hilbert space of system A, and the state cannot be compressed. An internal state is also named *completely mixed*, since it can be obtained with a convex combination including any state in $\mathsf{St}_1(A)$. On the contrary, a state on the border of the convex set $\mathsf{St}_1(A)$ is also called *not completely mixed*, to emphasize that it cannot be obtained as a mixture of some states. Such a state corresponds to assess that the system A has a definite *property*. For example, when measuring the observable $\{P, I - P\}$ with $P = \mathsf{Proj}\,\mathsf{Supp}\,\rho$ one will always obtain the same outcome, corresponding to the projector P (see Quantum Mantra 1). Clearly such a state can be perfectly discriminated by another state, namely any state $\rho^\perp$ with $\mathsf{Supp}\,(\rho^\perp) \subseteq \mathsf{Ker}\,\rho$. The property P itself will provide a binary test that perfectly discriminates between the two states, with the outcome 1 corresponding to state ρ, and the outcome 0 to state $\rho^\perp$. This leads us to the following principle for quantum theory.

Principle 4 (Perfect Discriminability)   Every state in $\mathsf{St}_1(A)$ that is not completely mixed can be perfectly discriminated from some other state in $\mathsf{St}_1(A)$.

In simple words, Principle 4 states that if we have some definite knowledge about system A, then the state of the system can be perfectly discriminated from another state of the same system A. This means that there is an observation test with probabilities 0 and 1 that is capable of ascertaining a non-trivial property of the system.

Principle 4 is crucial for the falsifiability of propositions. Consider the proposition "the system is in the state ρ": unless ρ is completely mixed, Principle (4) guarantees that the proposition is logically falsifiable. Indeed, performing the binary test with POVM $\{P, I-P\}$ on the state $\rho^\perp$ one would deterministically obtain the outcome indicating that the system is *not* in the state ρ.

2.8 Quantum Transformations

In this section we derive the mathematical representation of general probabilistic transformations in quantum theory, known as *quantum operations* in the literature. In the following section we will then derive the complete convex structure of transformations via the powerful Choi–Jamiołkowski isomorphism.

2.8.1 Completely Positive Maps

Which kind of mapping of states is mathematically admissible in quantum theory? According to the no-restriction hypothesis (Postulate 3), if a map is a mathematically admissible transformation, then it will represent an actual transformation allowed by the theory. In the following we will denote transformation and map with the same symbol $\mathcal{A}$. The transformation $\mathcal{A}$ sends states of system A to states of system B, namely

$$\mathcal{A} : \mathsf{St(A)} \ni |\rho)_A \mapsto \mathcal{A}|\rho)_A = |\mathcal{A}\rho)_B \in \mathsf{St(B)}.$$

The new state $|\mathcal{A}\rho)_B \in \mathsf{St(B)}$ will lead to the new probabilities

$$\forall a \in \mathsf{Eff(B)} : \quad (a|\mathcal{A}\rho)_B = (a|_B \mathcal{A}|\rho)_A =: (\mathcal{A}^\dagger a|\rho)_A, \tag{2.35}$$

$\mathcal{A}^\dagger$ denoting the *adjoint map* of $\mathcal{A}$ under the pairing $(\cdot|\cdot)$.[19] Since states are separating for effects, the above identity uniquely defines the action of the transformation over effects

$$\mathcal{A} : \mathsf{Eff(B)} \ni (a|_B \mapsto (a|_B \mathcal{A} = (\mathcal{A}^\dagger a|_A \in \mathsf{Eff(A)}.$$

We will denote by $\mathcal{A} \in \mathsf{Transf(A \to B)}$ the fact that $\mathcal{A}$ transforms states of A into states of B, or, vice versa, effects of B into effects of A. The corresponding map in quantum theory must send state operators to state operators and effect operators to effect operators. We will use both mapping notations $\mathcal{A}(\rho)$ and $\mathcal{A}\rho$ when acting on density operator ρ and $\mathcal{A}^\dagger(E)$ and $\mathcal{A}^\dagger E$ when acting as the adjoint on an operator effect E. We now analyze separately the admissibility properties of the map $\mathcal{A}$ in order to be a transformation of quantum theory.

(1) Linearity Since the transformation $\mathcal{A}$ must preserve convex combinations of states (or effects), namely

$$\forall p \in [0, 1] : \quad \mathcal{A}(p\rho_1 + (1-p)\rho_2) = p\mathcal{A}\rho_1 + (1-p)\mathcal{A}\rho_2,$$

the maps $\mathcal{A}$ must be linear.

---

[19] The standard notation for the adjoint map in the literature is $\mathcal{A}^*$; in this book, however, we reserve the $*$ symbol to complex conjugation. Notice that being the pairing defining the adjoint a real one, the adjoint would indeed correspond to a matrix transposition.

(2) Complete positivity   Since in quantum theory states are positive operators, the map $\mathcal{A}$ must send positive operators to positive operators. We say that the map is *positive* (also positive-preserving). In equations: $\forall Q \geq 0, \mathcal{A}(Q) \geq 0$. We must however consider that the physical transformation can act in the presence of other physical systems, and the specifying system A as the definition domain for $\mathcal{A}$ must be regarded as a locality assertion about the action of the map, meaning that $\mathcal{A}$ acts non-trivially on states of system A only, whereas it leaves the states of other systems invariant. Upon denoting jointly by R any additional system different from A, we represent diagrammatically the locality of transformation $\mathcal{A}$ as follows:

$$\begin{array}{c} \omega \end{array} \begin{array}{c} \text{A} \quad \boxed{\mathcal{A}} \quad \text{B} \\ \text{R} \end{array} \,. \tag{2.36}$$

The assertion that $\mathcal{A}$ acts non-trivially only on states of A leaving the states of R invariant makes sense only on factorized states $\rho \otimes \nu \in \mathsf{Lin}_+(\mathcal{H}_A \otimes \mathcal{H}_R)$, along with their linear combinations, via the linearity of the map. Let's denote temporarily by $\mathcal{A}_R^{ext}$ the map extended to R, to distinguish it from the map $\mathcal{A}$ acting on system A only. One has

$$\mathcal{A}_R^{ext}\left(\sum_i \rho_i \otimes \nu_i\right) = \sum_i \mathcal{A}(\rho_i) \otimes \nu_i.$$

Thanks to the local discriminability Principle 2, in quantum theory factorized states span the whole set of joint states, and we can therefore define the extended map $\mathcal{A}_R^{ext} \in \mathsf{Lin}[\mathsf{Lin}(\mathcal{H}_{AR}) \to \mathsf{Lin}(\mathcal{H}_{BR})]$ by linearity and consistently denote the extension as

$$\mathcal{A}_R^{ext} = \mathcal{A} \otimes \mathcal{I}_R.$$

For example, over the entangled state (2.23) the map acts as follows:

$$\mathcal{A} \otimes \mathcal{I}_R(|\Lambda\rangle\langle\Lambda|) = \sum_{ij} \mathcal{A}(|\lambda_i\rangle\langle\lambda_j|) \otimes |\lambda_i\rangle\langle\lambda_j|. \tag{2.37}$$

In order to be an admissible transformation, $\mathcal{A}_R^{ext} \in \mathsf{Transf}(AR \to BR)$, also the extended map $\mathcal{A}_R^{ext}$ must be positive preserving, and this must be true for any possible additional system R. We call such a property of the original map $\mathcal{A}$ *complete positivity* (CP), namely: the map $\mathcal{A} \otimes \mathcal{I}_R$ must preserve positivity when applied to operators on the extended Hilbert space $\mathcal{H}_A \otimes \mathcal{H}_R$, for any system R. The notion of complete positivity is non-trivial, since there are maps that preserve positivity over operators on Hilbert space $\mathcal{H}$, but not when extended to operators over $\mathcal{H} \otimes \mathcal{H}'$. An example of such a positive non-CP map is the transposition map, as shown in Exercise 2.34.

(3) Trace-non-increasing   The last property that the map $\mathcal{A}$ must satisfy in order to be mathematically admissible is that the joint probability $p(\mathcal{A}, \rho) = (e|\mathcal{A}\rho)$ of input preparation ρ and transformation $\mathcal{A}$ must be smaller than the marginal probability of the input state preparation alone $p(\rho) = (e|\rho)$, namely

$$\forall \rho \in \mathsf{St}(A) : \quad (e|\mathcal{A}\rho) \leq (e|\rho). \tag{2.38}$$

Being the probability of preparation of ρ given by $\mathrm{Tr}\,\rho$, Eq. (2.38) rewrites

$$\forall \rho \in \mathsf{St}(A): \quad \mathrm{Tr}\,\mathcal{A}\rho \leq \mathrm{Tr}\,\rho. \tag{2.39}$$

A map with the property in Eq. (2.39) is called *trace-non-increasing* (often imprecisely called "trace-decreasing" for short).

Quantum Operation   A linear map $\mathcal{A} \in \mathsf{Lin}(\mathsf{Lin}(\mathcal{H}_A) \rightarrow \mathsf{Lin}(\mathcal{H}_B))$ that is linear CP and trace-non-increasing represents a transformation in quantum theory, and is called a *quantum operation*. In the special case of a CP trace-preserving map (equals sign in Eq. (2.39)), one has a deterministic transformation, also called a *quantum channel*.

Dual Map   The adjoint transformation, acting on effects instead of states, is

$$\forall b \in \mathsf{Eff}(B): \quad (b|_B \mathcal{A} \mapsto (\mathcal{A}^\dagger b|_A.$$

This is obtained by duality, considering the action of $\mathcal{A}$ over a separating set of states of A, e.g. the whole set $\mathsf{St}(A)$, as in Eq. (2.35). In operator form, this becomes

$$\forall \rho \in \mathsf{St}(A), \forall b \in \mathsf{Eff}(B): \quad \mathrm{Tr}[E_b\,\mathcal{A}(\rho)] = \mathrm{Tr}[\rho\,\mathcal{A}^\dagger(E_b)], \quad E_b \in \mathsf{Eff}(B). \tag{2.40}$$

Identity (2.40) defines univocally the dual linear map $\mathcal{A}^\dagger$ (see Exercise 2.33). It is easy to show that the dual map $\mathcal{A}^\dagger$ is itself linear CP, whereas the trace-non-increasing condition (2.39) translates to the condition for $\mathcal{A}^\dagger$ to be *sub-unital*

$$\mathcal{A}^\dagger(I_B) \leq I_A. \tag{2.41}$$

with equality holding iff the map $\mathcal{A}$ is a *quantum channel*. In this case the map $\mathcal{A}^\dagger$ is *unital*.

Exercise 2.33   Provide an explicit expression of $\mathcal{A}^\dagger$ in terms of $\mathcal{A}$.

Exercise 2.34   [A positive map that is not CP] Show that the *transposition map* defined in Eq. (2.24)

$$\mathcal{T}: \rho \in \mathsf{Lin}(\mathcal{H}) \mapsto \mathcal{T}(\rho) := \rho^\mathsf{T} \in \mathsf{Lin}(\mathcal{H}), \tag{2.42}$$

is positive preserving, but not CP.

Exercise 2.35   Show that a linear map $\mathcal{A}$ is CP iff $\mathcal{A}^\dagger$ is CP. Derive Eq. (2.41)

Exercise 2.36   Show that a linear map of the following form:

$$\mathcal{A}\rho = \sum_i A_i \rho A_i^\dagger,$$

is completely positive.

Quantum "Pictures"   In quantum theory the evolution of states and effects are called the *Schrödinger picture* and *Heisenberg picture*, respectively. Historically, these have been the two different ways of evolving expected values of observables, namely: (1) by evolving the state vector keeping the observable constant (Schrödinger picture), (2) by evolving the observable while keeping the state constant (Heisenberg picture).

Causality and Transformations One can regard the action of the transformation $\mathcal{A} \in$ Transf(A→ B) on the state $\omega \in$ St(AR) as producing the new state $|\mathcal{A}\omega)_{BR}$. The causality principle (Principle 1) guarantees that the marginal probability of the transformation $\mathcal{A}$ occurring on state $|\omega)_{AB}$ is independent of the choice of observations of the output, hence it is equal to the probability

$$(e|_{BR}\mathcal{A}|\omega)_{AR} = \text{Tr}[(\mathcal{A} \otimes \mathcal{I}_R)\omega] = \text{Tr}[\mathcal{A}(\rho)] = \text{Tr}[\rho\mathcal{A}^\dagger(I_B)] = (a_\mathcal{A}|\rho)_A,$$

where $\rho \in$ St(A) is the marginal state of $\omega \in$ St(AR). We have therefore associated the transformation $\mathcal{A} \in$ Transf(A→ B) to an effect $a_\mathcal{A} \in$ Eff(A) which provides the joint probability $p(\mathcal{A}, \rho) = (a_\mathcal{A}|\rho)$ of occurrence of transformation $\mathcal{A}$ on any state ρ. This is described by the positive operator

$$0 \le P_\mathcal{A} = \mathcal{A}^\dagger(I_B) \le I_A.$$

Thanks to causality, we can also define the *conditional state*

$$\rho_\mathcal{A} := \frac{\mathcal{A}\rho}{(a_\mathcal{A}|\rho)} = \frac{\mathcal{A}\rho}{\text{Tr}(\mathcal{A}\rho)} = \frac{\mathcal{A}\rho}{\text{Tr}(P_\mathcal{A}\rho)},$$

describing the state provided that the state was ρ and the transformation $\mathcal{A}$ occurred.

Summarizing, we have:

Theorem 2.7 (Quantum Operations) *The transformation $\mathcal{A} \in$ Transf(A→ B)*

$$\mathcal{A}|\rho)_A \mapsto |\mathcal{A}\rho)_B, \quad or \quad (a|_B\mathcal{A} \mapsto (\mathcal{A}^\dagger a|_A$$

is represented in quantum theory by a CP trace-non-increasing linear map

$$\mathcal{A} : \text{Lin}(\mathcal{H}_A) \to \text{Lin}(\mathcal{H}_B)$$

acting on quantum states $\rho \in$ St(A) as follows:

$$\rho \mapsto \mathcal{A}\rho, \quad \rho \in \text{Lin}(\mathcal{H}_A), \quad \text{(Schrödinger picture)},$$

and acting on quantum effects $E \in$ Eff(B) as the dual sub-unital map $\mathcal{A}^\dagger$

$$E \mapsto \mathcal{A}^\dagger(E), \quad E \in \text{Lin}(\mathcal{H}_B), \quad \text{(Heisenberg picture)}.$$

The effect $P_\mathcal{A}$ corresponding to the transformation $\mathcal{A}$ is the positive operator

$$P_\mathcal{A} = \mathcal{A}^\dagger(I_B) \le I_A. \tag{2.43}$$

The transformation is deterministic (called a quantum channel) when we have equality in Eq. (2.43) (i.e. the map $\mathcal{A}$ is trace-preserving), whereas it is probabilistic (called a quantum operation), when it is trace-decreasing.

Exercise 2.37 Evaluate the effect corresponding to each of the transformations: $A \cdot A^\dagger$, $\sum_{i \in X} A_i \cdot A_i^\dagger$, $\text{Tr}[E\cdot]$ for $0 \le E \le I_A$.

2.8.2 The *Double-ket* Notation for Bipartite States

In the following we will make extensive use of a tensor notation that uses bras, kets, and operators. For example, for $|v\rangle \in \mathcal{H}$ we regard the bra $\langle v|$ as the operator in $\mathsf{Lin}(\mathcal{H}, \mathbb{C})$ $\langle v| : |\psi\rangle \in \mathcal{H} \mapsto \langle v|\psi\rangle \in \mathbb{C}$. Similarly, for $X \in \mathsf{Lin}(\mathcal{K})$ one has the tensors, $|v\rangle \otimes X \in$ $\mathsf{Lin}(\mathcal{K} \to \mathcal{H} \otimes \mathcal{K})$, $\langle v| \otimes X \in \mathsf{Lin}(\mathcal{H} \otimes \mathcal{K}, \mathcal{K})$, etc. For $\{|h_i\rangle\}_1^{\dim \mathcal{H}}$ and $\{|k_i\rangle\}_1^{\dim \mathcal{K}}$ orthonormal basis of $\mathcal{H}$ and $\mathcal{K}$, respectively, with $\dim \mathcal{H} = \dim \mathcal{K} = d$, the tensor $\sum_1^d \langle h_i| \otimes |k_i\rangle \equiv$ $\sum_1^d |k_i\rangle \otimes \langle h_i| \in \mathsf{Lin}(\mathcal{H} \to \mathcal{K})$ will represent an isomorphism between the two spaces.

We now introduce a convenient matrix notation that exploits the Hilbert-space isomorphism between $\mathsf{Lin}(\mathcal{H} \to \mathcal{K})$ and $\mathcal{K} \otimes \mathcal{H}$. For infinite-dimensional Hilbert spaces the reader can find suggestions for further studies in the notes at the end of the chapter.

We define the complex conjugate $|v\rangle^*$ of a vector $|v\rangle \in \mathcal{H}$ with respect to a fixed orthonormal basis $\{|h_n\rangle\}$ for $\mathcal{H}$ as follows:

$$|v\rangle = \sum_n c_n |h_n\rangle \implies |v\rangle^* = \sum_n c_n^* |h_n\rangle. \tag{2.44}$$

The map $\langle v| \leftrightarrow |v\rangle$ between $\mathcal{H}$ and its dual $\mathcal{H}^\vee \simeq \mathcal{H}$ is antilinear. Consider then the linear map $\mathbf{Vec} : \mathcal{H}^\vee \to \mathcal{H}$ given by

$$\langle v| \in \mathcal{H}^\vee \mapsto \mathbf{Vec}\,\langle v| = |v\rangle^*,$$

which is an isomorphism between Hilbert spaces. We can regard $\mathbf{Vec}$ as the application to $\langle v|$ extended as $\langle v| \otimes I$ of the following vector regarded as a linear map:

$$|I\rangle\!\rangle_{\mathcal{H} \otimes \mathcal{H}} = \sum_n |h_n\rangle \otimes |h_n\rangle, \tag{2.45}$$

giving

$$\left(\langle v| \otimes I \right) |I\rangle\!\rangle_{\mathcal{H} \otimes \mathcal{H}} = \sum_n \langle h_n|v\rangle^* |h_n\rangle = |v\rangle^*. \tag{2.46}$$

As we will see soon, the vector $|I\rangle\!\rangle_{\mathcal{K} \otimes \mathcal{K}}$ is a convenient notational tool for expressing the isomorphism $\mathbf{Vec}$. The map $\mathbf{Vec}$ naturally extends to the map $\mathbf{Vec} : \mathsf{Lin}(\mathcal{H} \to \mathcal{K}) \to \mathcal{K} \otimes \mathcal{H}$ for any pair of Hilbert spaces $\mathcal{H}$ and $\mathcal{K}$, as follows:

$$|u\rangle \in \mathcal{K}, \quad |v\rangle \in \mathcal{H}, \quad \mathbf{Vec}\,|u\rangle\langle v| = |u\rangle \otimes |v\rangle^*, \tag{2.47}$$

which, by linearity, is extended to any operator $M \in \mathsf{Lin}(\mathcal{H} \to \mathcal{K})$ as follows:

$$M = \sum_{ij} M_{ij} |k_i\rangle\langle h_j| \mapsto \mathbf{Vec}\,M = \sum_{ij} M_{ij} |k_i\rangle \otimes |h_j\rangle, \tag{2.48}$$

where $\{k_j\}$ is an orthonormal basis for $\mathcal{K}$. We introduce the convenient notation

$$|M\rangle\!\rangle_{\mathcal{K} \otimes \mathcal{H}} := \sum_{ij} M_{ij} |k_i\rangle \otimes |h_j\rangle,$$

with the "double-ket" symbol reminding us of the correspondence between vectors $|M\rangle\!\rangle_{\mathcal{K} \otimes \mathcal{H}} \in \mathcal{K} \otimes \mathcal{H}$ and operators $M \in \mathsf{Lin}(\mathcal{H} \to \mathcal{K})$. Notice that the notation is consistent with that of Eq. (2.45). In the following we will drop the Hilbert-space label from the double-ket vectors when no confusion can arise. The scalar product induced by **Vec** on operators is the *Frobenius product* of Hilbert–Schmidt operators

$$P, Q \in \mathsf{Lin}(\mathcal{H} \to \mathcal{K}), \qquad (Q, P) := \langle\!\langle Q | P \rangle\!\rangle = \mathrm{Tr}[Q^\dagger P], \qquad (2.49)$$

defining the norm $\|P\|_2 := \sqrt{\mathrm{Tr}\, P^2}$. In addition, we have the rules for partial traces

$$\mathrm{Tr}_{\mathcal{H}}[|P\rangle\!\rangle\langle\!\langle Q|] = PQ^\dagger \in \mathsf{Lin}(\mathcal{K}), \quad \mathrm{Tr}_{\mathcal{K}}[|P\rangle\!\rangle\langle\!\langle Q|] = P^\mathsf{T} Q^* \in \mathsf{Lin}(\mathcal{H}), \qquad (2.50)$$

where the transposed M^T of an operator M has been defined in Eq. (2.24). The action of elementary tensor products $A \otimes B$ $[A \in \mathsf{Lin}(\mathcal{H} \to \mathcal{H}')$ and $B \in \mathsf{Lin}(\mathcal{K} \to \mathcal{K}')]$ on $|C\rangle\!\rangle \in \mathcal{H} \otimes \mathcal{K}$ is given by the following simple rule:

$$A \otimes B|C\rangle\!\rangle = |ACB^\mathsf{T}\rangle\!\rangle \in \mathcal{H}' \otimes \mathcal{K}'. \qquad (2.51)$$

Care should be taken in the bookkeeping of the operator domains in Eq. (2.51), keeping in mind that in the notation $|M\rangle\!\rangle$ it is implicit that $|M\rangle\!\rangle \in \mathcal{K} \otimes \mathcal{H}$ whenever $M \in \mathsf{Lin}(\mathcal{H} \to \mathcal{K})$. The isomorphism **Vec** and its inverse map write formally as follows:

$$|M\rangle\!\rangle = \mathbf{Vec}(M) = (M \otimes I_{\mathcal{H}})|I\rangle\!\rangle_{\mathcal{H} \otimes \mathcal{H}}, \qquad M = \mathbf{Vec}^{-1}|M\rangle\!\rangle = \mathrm{Tr}_{\mathcal{H}}[|M\rangle\!\rangle\langle\!\langle I|]. \quad (2.52)$$

Exercise 2.38   Check identities (2.49), (2.50), (2.51), and (2.52).

Exercise 2.39   [Schmidt form of bipartite vectors] Using the isomorphism in Eq. (2.48) and the *singular value decomposition* of the operator $M \in \mathsf{Lin}(\mathcal{K} \to \mathcal{H})$ (see Appendix 2.1)

$$M = \sum_{l=1}^{k} \sigma_l(M)|v_l\rangle\langle w_l|,$$

show that the following general expansion holds for any bipartite vector in $\mathcal{H} \otimes \mathcal{K}$

$$|M\rangle\!\rangle = \sum_{l=1}^{k} \sigma_l(M)|v_l\rangle \otimes |w_l\rangle, \qquad (2.53)$$

with $\{|v_l\rangle\}$ and $\{|w_l\rangle\}$ orthonormal set of vectors. The expansion (2.53) is called the *Schmidt form* of the vector $|M\rangle\!\rangle$. Show that the number k of non-vanishing terms in Eq. (2.53) is equal to rank(M). Such number is called the *Schmidt number* of the vector.

Exercise 2.40   Show that the vector $|I\rangle\!\rangle_{\mathcal{H} \otimes \mathcal{H}} = \sum_{i=1}^{\dim \mathcal{H}} |h_i\rangle \otimes |h_i\rangle$ defined in terms of the canonical orthonormal basis $\{|h_i\rangle\}_{i=1}^{\dim \mathcal{H}}$ for both copies of $\mathcal{H}$ is invariant under the operator $U \otimes U^*$, with $U \in \mathsf{Lin}(\mathcal{H})$ unitary. Therefore, for any chosen orthonormal basis $\{|v_i\rangle\}_{i=1}^{\dim \mathcal{H}}$ we can write

$$|I\rangle\!\rangle_{\mathcal{H} \otimes \mathcal{H}} = \sum_{i=1}^{\dim \mathcal{H}} |v_i\rangle \otimes |v_i\rangle^*.$$

where the complex conjugate is defined with respect to the canonical basis (2.44).

Exercise 2.41 [Maximally entangled state] According to the quantification of entanglement given in Eq. (2.28), a pure bipartite state is maximally entangled when the local state is maximally mixed. Using the **Vec** isomorphism, show that a state on $\mathbb{C}^d \otimes \mathbb{C}^d$ is maximally entangled iff it is of the form $\frac{1}{d}|U\rangle\!\rangle\langle\!\langle U|$, with U unitary operator.

Some useful identities about unitary operators in dimension $d = 2$ can be found in Problems 2.13–2.16 at the end of this chapter.

Exercise 2.42 [Bell measurement] Consider a set $\{U_i\}_{i=0}^{d-1}$ of unitary operators on $\mathbb{C}^d$, orthogonal in the Hilbert–Schmidt scalar product. Using the **Vec** isomorphism, show that the following is a *Bell measurement*, namely a rank-one PVM made with maximally entangled effects

$$P_i = \tfrac{1}{d}|U_i\rangle\!\rangle\langle\!\langle U_i|, \quad i = 0, \ldots, d-1.$$

Exercise 2.43 [The shift-and-multiply basis] Show that the following unitary operators on $\mathbb{C}^d$ are an orthogonal basis for $\mathsf{Lin}(\mathbb{C}^d)$

$$U_j = Z^p W^q, \ j \equiv (p,q) \in \mathbf{Z}_d \times \mathbf{Z}_d,$$

$$Z = \sum_{k=0}^{d-1} |k\rangle\langle k| \exp\left(\tfrac{2\pi i k}{d}\right), \qquad W = \sum_{k=0}^{d-1} |k \oplus 1\rangle\langle k|, \tag{2.54}$$

where $\oplus$ denotes the sum modulo d, and the orthogonality relation is given by

$$\mathrm{Tr}[U_j^\dagger U_l] = d\delta_{jl}.$$

Notice that all unitary operators are traceless, apart from $U_{(0,0)} = I_d$.

In Eq. (2.24) we defined the transposition of an operator M with respect to two orthonormal bases considered as real, as e.g. the sets $\{|h_i\rangle\}$ and $\{|k_j\rangle\}$ in Eq. (2.48). Clearly, there are infinitely many bases that correspond to the same transposition map. Using identity (2.51) one can see that the choice of vector $|I\rangle\!\rangle_{\mathcal{H}\otimes\mathcal{H}}$ for every Hilbert space $\mathcal{H}$ uniquely defines the transposition of an operator $A \in \mathsf{Lin}(\mathcal{H} \to \mathcal{K})$ as follows

$$(A \otimes I)|I\rangle\!\rangle_{\mathcal{H}\otimes\mathcal{H}} = (I \otimes A^\mathsf{T})|I\rangle\!\rangle_{\mathcal{K}\otimes\mathcal{K}}.$$

2.8.3 The Choi–Jamiołkowski Isomorphism

Using the isomorphism Iso we can build up the following linear bijective map between the map-space $\mathsf{Lin}(\mathsf{Lin}(\mathcal{H}_A) \to \mathsf{Lin}(\mathcal{H}_B))$ and the operator space $\mathsf{Lin}(\mathcal{H}_B \otimes \mathcal{H}_A)$:

$$\mathbf{Iso} : \mathsf{Lin}(\mathsf{Lin}(\mathcal{H}_A) \to \mathsf{Lin}(\mathcal{H}_B)) \ni \mathcal{A} \mapsto R_{\mathcal{A}} \in \mathsf{Lin}(\mathcal{H}_B \otimes \mathcal{H}_A),$$
$$\mathcal{A} \mapsto R_{\mathcal{A}} := \mathcal{A} \otimes \mathcal{I}_A(|I\rangle\!\rangle\langle\!\langle I|), \quad \mathcal{A}(\rho) = \mathrm{Tr}_2[(I_B \otimes \rho^\mathsf{T})R_{\mathcal{A}}]. \tag{2.55}$$

Linearity and invertibility of the **Iso** map are evident from the identities above.

Exercise 2.44 Prove that the two maps in Eq. (2.55) are inverse of each other.

The most relevant feature of map **Iso** is that it provides a *cone isomorphism between the cone of positive operators* $\mathsf{Lin}_+(\mathcal{H}_B \otimes \mathcal{H}_A)$ *and the cone of CP maps* $\mathsf{CP}(\mathsf{Lin}(\mathcal{H}_A) \to \mathsf{Lin}(\mathcal{H}_B))$. Precisely, this means that **Iso** puts CP maps and positive operators in one-to-one correspondence, preserving their cone structure, namely sending extremal rays to extremal rays and conic combinations to conic combinations. This is helpful in analyzing the cone structure of CP maps, since this reflects the cone structure of positive operators, with which we are already familiar. In the literature, the isomorphism **Iso** is called the "Choi–Jamiołkowski isomorphism," and the operator $R_{\mathcal{A}} := \mathbf{Iso}\,\mathcal{A}$ is named the "Choi–Jamiołkowski operator." [20]

In order to show that **Iso** is a cone isomorphism it is sufficient to prove that it sends positive operators to CP maps and vice versa. This is evident, since **Iso** is linear, and if the map $\mathcal{A}$ is CP, then by definition $R_{\mathcal{A}}$ is positive (since $|I\rangle\!\rangle\langle\!\langle I| \geq 0$), whereas the fact that the inverse map $R_{\mathcal{A}} \mapsto \mathcal{A}$ always gives a CP map for a positive $R_{\mathcal{A}}$ is easily proved by decomposing $R_{\mathcal{A}}$ into extremal rays $R_{\mathcal{A}} = \sum_i |A_i\rangle\!\rangle\langle\!\langle A_i|$, which using identities (2.55) gives the map

$$\mathcal{A}\rho = \sum_i A_i \rho A_i^\dagger, \tag{2.56}$$

which we showed is CP in Exercise 2.36. Finally notice that since **Iso** is linear, it preserves conic combinations.

From the Choi–Jamiołkowski cone isomorphism it follows that:

Corollary 2.8   *The cone structure of CP maps is given by the following statements:*

1. *The extremal-ray maps in $\mathcal{A} \in \mathsf{Lin}(\mathsf{Lin}(\mathcal{H}_A) \to \mathsf{Lin}(\mathcal{H}_B))$ are of the form $\mathcal{A}\rho = A\rho A^\dagger$, with $A \in \mathsf{Lin}(\mathcal{H}_A \to \mathcal{H}_B)$.*
2. *Any CP map admits the conic decomposition (called Kraus form)*

$$\mathcal{A}\rho = \sum_i A_i \rho A_i^\dagger, \quad \text{Kraus form of the CP map } \mathcal{A}. \tag{2.57}$$

In the following, we will call a transformation $\mathcal{V}$ *isometric* if it has a Kraus form $\mathcal{V}(\rho) = V\rho V^\dagger$ with V isometric, and *unitary* if V unitary. A special instance of Kraus form is that of the *canonical Kraus form*, in which the vectors $|A_i\rangle\!\rangle$ are the eigenvectors of $R_{\mathcal{A}}$, and therefore are orthogonal, namely $\mathrm{Tr}(A_i^\dagger A_j) = \delta_{ij}\|A_i\|_2^2$. This is only a special case of *minimal Kraus* form, for which the number of terms in the sum in Eq. (2.57) is equal to $\mathrm{rank}(R_{\mathcal{A}})$, also called *rank of the map*. The different Kraus forms are classified by the following simple lemma.

Lemma 2.9[21]   *The quantum operation $\mathcal{A}' = \sum_{i \in \mathsf{X}} A_i' \cdot A_i'^\dagger$ is the same as the quantum operation $\mathcal{A} = \sum_{j \in \mathsf{Y}} A_j \cdot A_j^\dagger$ if the respective Kraus operators are related as follows:*

$$A_i' = \sum_{j \in \mathsf{Y}} U_{ij} A_j,$$

---

[20] (Choi, 1972, 1975; Jamiolkowski, 1972).
[21] Chuang and Nielsen (2000).

with U a (rectangular) matrix with orthonormal columns, i.e. $(U^\dagger U)_{nm} = \delta_{nm}$. Moreover, the if condition is an iff when the Kraus operators $\{A_i\}$ are linearly independent, namely the Kraus form is minimal.

Exercise 2.45 Prove Lemma 2.9.

Exercise 2.46 Take a basis F_k for $\mathsf{Lin}(\mathcal{H}_A)$. Show that $\mathcal{C} := \sum_{kl} c_{kl} F_k \cdot F_l^\dagger$ is CP iff the matrix $\{c_{kl}\}$ is a positive matrix.

Corollary 2.8 allows us to classify all quantum transformations. The classification is given by the following.

Quantum Mantra 3 (Quantum Operations in Kraus Form) A quantum transformation in $\mathsf{Transf}(A \to B)$ is represented by a linear map of the form

$$\mathcal{A}\rho = \sum_i A_i \rho A_i^\dagger, \qquad \text{(Schroedinger picture)}$$

$$\mathcal{A}^\dagger X = \sum_i A_i^\dagger X A_i, \qquad \text{(Heisenberg picture)},$$

with

$$\sum_i A_i^\dagger A_i \leq I_A.$$

Moreover, the operator

$$P_{\mathcal{A}} = \sum_i A_i^\dagger A_i, \tag{2.58}$$

represents the effect associated to the transformation, hence $P_{\mathcal{A}} = I_A$ corresponds to a quantum channel. The *atomic* transformations, apart from a phase factor, are in one-to-one correspondence with contractive operators $A \in \mathsf{Lin}(\mathcal{H}_A \to \mathcal{H}_B)$ (i.e. $\|A\| \leq 1$), acting as follows:

$$\mathcal{A}\rho = A\rho A^\dagger.$$

For pure states $|\psi\rangle\langle\psi|$, apart from an arbitrary phase, one has

$$|\psi\rangle \mapsto A|\psi\rangle,$$

hence *contractive operators describe atomic transformations in quantum theory.*

Quantum Mantra 4 (Quantum Conditional State) The *conditional state* $\rho_{\mathcal{A}} \in \mathsf{St}_1(B)$ for initial state $\rho \in \mathsf{St}(A)$ given the occurrence of the quantum operation $\mathcal{A} \in \mathsf{CP}(\mathcal{H}_A, \mathcal{H}_B)$, is the deterministic post-selected state

$$\rho \mapsto \rho_{\mathcal{A}} = \frac{\mathcal{A}\rho}{\mathrm{Tr}(\mathcal{A}\rho)}, \tag{2.59}$$

and occurs with overall joint probability

$$p(\mathcal{A}, \rho) = \mathrm{Tr}[\mathcal{A}\rho] = \mathrm{Tr}[P_{\mathcal{A}}\rho], \ P_{\mathcal{A}} = \mathcal{A}^\dagger(I), \qquad \text{(Born rule)}. \tag{2.60}$$

For a pure state $|\psi\rangle\langle\psi|$ and atomic quantum operation $\mathcal{A} = A \cdot A^\dagger$ one has

$$|\psi\rangle \mapsto |\psi\rangle_{\mathcal{A}} = \frac{A|\psi\rangle}{\|A\psi\|},$$

apart from an arbitrary phase factor. Therefore multiplication of a Hilbert-space vector by a *contractive operator* A (i.e. with $\|A\| \leq 1$) describes a state-conditioning, and we can regard any quantum evolution as a state-conditioning.

As is shown in Exercise 2.47, states and effects can be considered themselves as special types of transformations, with

$$\mathsf{St}(A) = \mathsf{Transf}(I \to A), \quad \mathsf{Eff}(A) = \mathsf{Transf}(A \to I), \qquad \mathcal{H}_I = \mathbb{C}$$

Exercise 2.47 [States and effects as transformations] Check that a state $\rho \in \mathsf{St}(A)$ can be regarded as a transformation $\rho \in \mathsf{Transf}(I \to A)$ with input trivial system I. Similarly an effect $a \in \mathsf{Eff}(A)$ can be regarded as a transformation $a \in \mathsf{Transf}(A \to I)$ with output trivial system. Write the Choi–Jamiołkowski operator for both cases.

As a corollary of what has been said, we can derive the following principle for quantum theory.

Principle 5 (Atomicity of Composition)   The composition of two atomic transformations is atomic.

Proof   Atomic transformations are of the form $\mathcal{A} = A \cdot A^\dagger$, and their composition e.g. $\mathcal{B}\mathcal{A} = BA \cdot A^\dagger B^\dagger = (BA) \cdot (BA)^\dagger$ is still of the same form.   □

Notice that Principle 5 holds for all transformations, including states and effects (see Exercise 2.47).

Corollary 2.10 (Isometric Transformations)   *An atomic quantum channel $\mathcal{A} \in \mathsf{Transf}_1(A \to B)$ must be necessarily isometric.*

Proof   An atomic quantum channel is of the form $\mathcal{A}(\rho) = A\rho A^\dagger$ with $A \in \mathsf{Lin}(A \to B)$. Being the operation deterministic, one has $A^\dagger A = I_A$, where A is the input system.   □

Corollary 2.11 (Unitary Transformations)   *A quantum operation $\mathcal{A} \in \mathsf{Transf}(A)$ can be inverted by another transformation if and only if it is unitary.*

Proof   Sufficiency is trivial, since unitary transformations are invertible. Let us now suppose that the transformation $\mathcal{A}$ is invertible with inverse $\mathcal{A}^{-1}$. In order to be invertible the transformation $\mathcal{A}$ and its inverse $\mathcal{A}^{-1}$ must be deterministic, since $\mathcal{A}^{-1}\mathcal{A} = \mathcal{I}_A$ is trace-preserving, and thus neither $\mathcal{A}$ nor $\mathcal{A}^{-1}$ can be trace-decreasing. Moreover, $\mathcal{A}$ is atomic. Indeed, suppose $\mathcal{A} = \sum_n \mathcal{A}_n$, then $\sum_n \mathcal{A}^{-1}\mathcal{A}_n = \mathcal{I}_A$. Since $\mathcal{I}_A$ is atomic it must be $\mathcal{A}^{-1}\mathcal{A}_n = p_n\mathcal{I}_A$, and applying $\mathcal{A}$ on the left on both sides one has $\mathcal{A}_n = p_n\mathcal{A}$, namely any refinement for $\mathcal{A}$ is trivial. Being $\mathcal{A}$ deterministic and atomic, it is isometric, namely its Kraus operator is an isometry on a finite-dimensional Hilbert space $\mathcal{H}_A$. Finally, isometries on finite-dimensional spaces are unitary.   □

A transformation is *reversible* if it can be inverted both on the left and on the right by another transformation. Therefore, it follows that:

Corollary 2.12 (Reversible Transformations) *The only reversible transformations are the unitary ones.*

Exercise 2.48 Show that the convex set of states St(BA) and the convex set of transformations Transf(A→ B) are not isomorphic, even though the two respective cones are, according to the Choi–Jamiołkowski isomorphism.

Exercise 2.49 Show that a quantum operation in the refinement set of $\mathcal{A} \in$ Transf(A→ B) has Kraus operators that are linear combinations of those of $\mathcal{A}$.

Solution

Let's denote by $\mathsf{Span}_+ \mathsf{S}$ the conic hull of the set S, and by $\mathsf{Span}_+ x$ the conic hull of the refinement set $\mathsf{RefSet}\, x$ of $x \in \mathsf{C}_+$, with C_+ convex cone. Thanks to the Choi–Jamiołkowski isomorphism, the conic hull $\mathsf{Span}_+ \mathcal{A}$ of the refinement set of $\mathcal{A} \in$ Transf(A→ B) is isomorphic with the conic hull $\mathsf{Span}_+ R_\mathcal{A}$ of the refinement set of the corresponding Choi operator $R_\mathcal{A} \in \mathsf{Lin}_+(\mathcal{H}_B \otimes \mathcal{H}_A)$, and similarly to quantum states, we have

$$\mathsf{Span}_+ R_\mathcal{A} = \mathsf{Lin}_+[\mathsf{Supp}\,(R_\mathcal{A})],$$

with

$$\mathsf{Supp}\,(R_\mathcal{A}) = \mathsf{Span}\{|A\rangle\!\rangle, A \in \mathsf{Lin}(\mathcal{H}_A \to \mathcal{H}_B)\ \text{Kraus operator for}\ \mathcal{A}\}.$$

Being **Iso** a cone isomorphism, it follows that the conic hull of the refinement set of $\mathcal{A}$ is given by

$$\mathsf{Span}_+ \mathcal{A} = \mathbf{Iso}^{-1}\mathsf{Span}_+ R_\mathcal{A} = \mathbf{Iso}^{-1}\mathsf{Lin}_+[\mathsf{Supp}\,(R_\mathcal{A})]$$
$$= \mathsf{Span}_+[B \cdot B^\dagger, B \in \mathsf{Span}\{A \in \mathsf{Lin}(\mathcal{H}_A \to \mathcal{H}_B), A\ \text{Kraus for}\ \mathcal{A}\}].$$

2.8.4 Tomography of Transformations

Can we extend the tomographic method to transformations? Yes, of course! We just need two informationally complete tests: a preparation test $\{\rho_i\}_{i\in X}$ and an observation test $\{l_j\}_{j\in Y}$. Then, by acquiring experimentally the joint probabilities $\{(l_j|\mathcal{A}|\rho_i)\}_{i\in X, j\in Y}$ we can determine the action of the transformation $\mathcal{A}$ on any state and effect via the double expansion

$$(a|\mathcal{A}|\rho) = \sum_{i\in X, j\in Y} c_j(a)v_i(\rho)(l_j|\mathcal{A}|\rho_i),$$

where $\{v_i\}_{i\in X}$ and $\{c_j\}_{j\in Y}$ are dual sets of $\{\rho_i\}_{i\in X}$ and $\{l_j\}_{j\in Y}$, respectively. Then, the local discriminability principle guarantees that we can obtain also the local action of the transformation over any multipartite system. We will later see (Section 2.10.2) that entanglement offers an alternative and efficient way to achieve tomography of a transformation, by applying it to a *single* entangled state.

2.8.5 Quantum Instruments

Transformations generally occur probabilistically. A complete set of probabilistic transformations $\{\mathcal{T}_i\}_{i\in X} \subseteq$ Transf(A→ B) (i.e. one transformation of the set will occur with certainty) is called a *test*. We have already seen two different kinds of test: the preparation test, which has no input (the input is the trivial system), and the observation test, which has no output (the output is the trivial system). The quantum test with non-trivial input and output systems is a complete collection of quantum operations, and is usually called *quantum instrument*. One has

$$\text{quantum instrument } \{\mathcal{T}_i\}_{i \in X} : \quad \mathcal{T}_i \in \mathsf{Transf}(A \to B), \ \forall i \in X,$$

$$\sum_{i \in X} \mathcal{T}_i =: \mathcal{T}_X \ \text{ quantum channel.}$$

Also for tests we have the notion of coarse-graining, namely

$$Z \subseteq X : \qquad \mathcal{T}_Z := \sum_{i \in Z} \mathcal{T}_i.$$

The quantum instrument is also associated with an observation test, namely a POVM, which gives the probability distribution of each outcome as follows:

$$\forall i \in X : \quad P_i = \mathcal{T}_i^{\dagger}(I_B).$$

The output state after the measurement on input state $\rho \in \mathsf{St}(A)$ is given by $\mathcal{T}_Z(\rho)$ for outcome i belonging to the set Z.

A simple example of test is the *measure-and-prepare* test, corresponding to performing an observation test and then preparing a state depending on the outcome. In the quantum case, the test will be assigned by a POVM $\{R_i\}_{i \in X}$ and a set of quantum states $\{\rho_i\}_{i \in X}$. One has

$$\forall i \in X : \quad \mathcal{T}_i(\rho) = \mathrm{Tr}[R_i \rho] \rho_i.$$

We call a quantum instrument $\{\mathcal{T}_i\}_{i \in X} \subseteq \mathsf{Transf}(A)$ *atomic* if the transformation for each outcome is atomic. We will call the instrument *repeatable* if in a cascade of applications to the same system we always get the same outcome.

Exercise 2.50   When is a measure-and-prepare instrument atomic?

Exercise 2.51   [Repeatable instrument] Show that the following are both necessary and sufficient conditions for a quantum instrument $\{\mathcal{T}_j\}$ to be repeatable

$$\mathcal{T}_{i_1}^{\dagger} \mathcal{T}_{i_2}^{\dagger}(I) = \delta_{i_1 i_2} \mathcal{T}_{i_1}^{\dagger}(I),$$

$$\forall i, j \ i \neq j \qquad \forall n, m \quad A_n^{(i)} A_m^{(j)} = 0,$$

where $\mathcal{T}_j = \sum_n A_n^{(i)} \cdot A_n^{(i)\dagger}$.

A relevant example of atomic and repeatable instrument is the von Neumann–Lüders instrument, defined as

$$\forall i \in X : \quad \mathcal{T}_i(\rho) = Z_i \rho Z_i, \qquad Z_i Z_j = \delta_{ij} Z_i,$$

where $\{Z_i\}_{i \in X}$ is an orthogonal-projector resolution of the identity – usually called a projection-valued measure (PVM).

Exercise 2.52   What is the POVM of the von Neumann–Lüders instrument?

Exercise 2.53 Consider the instrument $T_i(\rho) = V_i Z_i \rho Z_i V_i^\dagger$, where V_i is partial isometry with support equal to that of Z_i, and $\{Z_i\}_{i\in X}$ a PVM. Is the instrument repeatable?

A special case of Lüders' instrument is the von Neumann, in which all projectors $Z_i \; \forall i \in X$ are rank-one. This is the case of the well-known collapse-postulate of von Neumann:

Quantum Mantra 5 (von Neumann Postulate) The measurement of the observable X over system A leaves the system in the eigenstate $|x\rangle\langle x|$ of X corresponding to the measured value x.

Further discussion about von Neumann's postulate can be found in Section 2.12.

Exercise 2.54 [von Neumann–Lüders instrument] Show that a von Neumann–Lüders instrument $\{T_i\}_{i\in X}$ perfectly discriminates the quantum states $\{\rho_i\}_{i\in X}$ iff $\mathsf{Supp}\,\rho_n \subseteq \mathsf{Supp}\,Z_n$ (see Exercise 2.55).

A POVM is not required to be orthogonal in order to be able to discriminate a set of discriminable quantum states $\{\rho_i\}_{i\in X}$: it is only required that it satisfies $\mathsf{Supp}\,P_i \supseteq \mathsf{Supp}\,\rho_i$ $\forall i \in X$, with ρ_i having orthogonal supports. It is necessarily orthogonal only if the set of states satisfies the completeness property $\oplus_{i\in X}\mathsf{Supp}\,\rho_i = \mathcal{H}_A$ (see Exercise 2.55). On the other hand, obviously there exist instruments that are discriminating but not repeatable, e.g. $T_n(\rho) = V_n Z_n \rho Z_n V_n^\dagger$, with V_n isometry embedding $\mathsf{Supp}\,Z_n$ into $\mathsf{Supp}\,Z_m$ with $m \neq n$. Vice versa, in order to be repeatable, orthogonality of the POVM is required only for atomic instruments in finite dimensions, as shown in Exercise 2.56, whereas generally non-atomic repeatable instruments can have non-orthogonal POVM. In infinite dimensions there even exist atomic repeatable intruments with non-orthogonal POVM, as shown in Problem 2.29 at the end of the chapter. In summary, repeatability and discriminability are independent properties of the instrument, and there exist repeatable instruments that are non-discriminating and discriminating instruments that are not repeatable. Moreover, there are non-orthogonal measurements that are discriminating, whereas every repeatable instrument must discriminate some set of states.

Exercise 2.55 [Discriminable states and orthogonal POVM] Show that a set of states $\{\rho_i\}_{i\in X}$ is discriminable iff they have orthogonal support, and a POVM $\{P_i\}_{i\in X}$ discriminating such states must have $\mathsf{Supp}\,P_i \supseteq \mathsf{Supp}\,\rho_i \; \forall i \in X$. Show also that if the set satisfies the completeness property $\oplus_{i\in X}\mathsf{Supp}\,\rho_i = \mathcal{H}_A$, then the POVM must necessarily be a PVM.

Exercise 2.56 Show that if an atomic instrument is repeatable, then it necessarily has orthogonal POVM (valid only for $\dim \mathcal{H}_A < \infty$).

For a non-atomic instrument (or for infinite dimensions) the result in Exercise 2.56 does not necessarily hold. An example of a non-atomic repeatable instrument with generally non-orthogonal POVM is the measure-and-prepare instrument

$$\mathcal{A}_i = \mathrm{Tr}[Q_i \cdot] \nu_i,$$

with Q_i POVM discriminating the states ν_i.

Notice that for an atomic instrument $\{\mathcal{T}_i\}_{i \in X}$ with $\mathcal{T}_i = A_i \cdot A_i^\dagger$ the repeatability condition implies that $A_i A_j = 0$. This doesn't imply that $A_i A_j^\dagger = 0$, which is the condition corresponding to orthogonality of the POVM $\{P_i\}_{i \in X}$, since $P_i = A_i^\dagger A_i$.

Moreover, we remark that a discriminating instrument must have cardinality $|X| \leq \dim \mathcal{H}_A$ due to orthogonality of supports of discriminated states. The same bound must also be satisfied by a repeatable instrument, since repeatability is equivalent to discriminability of the output states of the instrument. As we have seen, the von Neumann–Lüders instrument is both discriminating and repeatable, and, among the discriminating measurements it is the one producing the *least disturbance*, since it leaves the states in the supports of its orthogonal projectors invariant. The von-Neumann measurement is the one of the von Neumann–Lüders class which discriminates the maximum number of states, which is equal to $\dim \mathcal{H}_A$. Such states must be necessarily pure and orthogonal – i.e. corresponding to orthogonal vectors.

2.8.6 Probabilistic Exact State Discrimination

From what we have seen above, we deduce that it is impossible to perfectly discriminate between two non-orthogonal states. A direct proof of this fact for pure states is provided by the following exercise.

Exercise 2.57 [Impossibility of perfect discrimination between two non-orthogonal states] Prove that it is impossible to discriminate perfectly between two pure non-orthogonal states.[22]

Even though it is impossible to perfectly discriminate non-orthogonal states, when the states correspond to linearly independent vectors an error-free discrimination is possible with non-unit probability. More precisely, for linearly independent vectors it is possible to have a measurement that discriminates between the corresponding pure states with no errors, however, giving sometimes an inconclusive result. This is shown in Exercise 2.58.

Exercise 2.58 [Probabilistic exact discrimination of non-orthogonal pure states] Consider N linearly independent non-orthogonal normalized vectors $\{|\psi_i\rangle\}_{i \in X} \in \mathcal{H}$. Show that a POVM of the form

$$i \in X : \quad P_i := \alpha |\varphi_i\rangle \langle \varphi_i|, \quad P_? = I - \sum_{i \in X} P_i,$$

for suitable $\alpha > 0$ and $\{|\varphi_i\rangle\}_{i \in X} \in \mathcal{H}$ dual set of unit vectors $\{|\psi_i\rangle\}_{i \in X}$ (namely satisfying $\langle \psi_i | \varphi_j \rangle = 0$ for $i \neq j$) discriminates the states $\{|\psi_i\rangle \langle \psi_i|\}_{i \in X}$ with no error. Find the allowed values of α, and give its optimal value, minimizing the probability of the inconclusive event "?" equally probable.

Exercise 2.59 As in Exercise 2.58, consider the discrimination between the two pure states corresponding to the non-orthogonal vectors $|\psi_1\rangle \equiv |0\rangle$ and $|\psi_2\rangle \equiv \frac{1}{\sqrt{2}}(|0\rangle + |1\rangle)$ in $\mathbb{C}^2$. Show that $\alpha_{\text{opt}} = \sqrt{2}/(1 + \sqrt{2})$.

---

[22] Chuang and Nielsen (2000).

Exercise 2.60 Explain why we cannot discriminate pure states corresponding to a set of linearly dependent vectors?

Exercise 2.61 Show that we can make a set of pure states corresponding to linearly dependent vectors exactly discriminable if we have sufficiently many copies of them.

2.9 Classical Theory as a Restriction of Quantum

In this section we will show that the Classical Theory is a restriction of quantum theory, where the set of states of the classical system is the convex hull of a maximal set of orthogonal pure states of the corresponding quantum system A, and we assume the no-restriction hypothesis.

We will use the vector notation $\mathbf{x} = (x_1, x_2, \ldots, x_n) \in \mathbb{R}^n$, with $\mathbf{x} \geq 0$ denoting that $x_i \geq 0, \forall i = 1, \ldots, n$, $\mathbf{x} \leq 1$ denoting that $x_i \leq 1, \forall i = 1, \ldots, n$, and $\|\mathbf{x}\|_1 := \sum_{i=1}^n |x_i|$. We will denote by S_+^n the convex cone of n-dimensional vectors with $\mathbf{x} \geq 0$, with S^n the convex set of n-dimensional *sub-stochastic vectors* $\mathbf{x} \geq 0$ with $\|\mathbf{x}\|_1 \leq 1$, and with S_1^n the convex set of n-dimensional *stochastic vectors* $\mathbf{x} \geq 0$ with $\|\mathbf{x}\|_1 = 1$.

Exercise 2.62 Show that $\mathsf{S}_1^n \simeq \mathsf{S}^{n-1}$ are isomorphic convex sets, and that S^n is an n-dimensional *simplex*, namely the convex hull of $n+1$ points in $\mathbb{R}^{n-1}$ that are affinely independent.

From Exercise 2.62 we see that $\mathsf{S}_1^1 \simeq \mathsf{S}^0$ is a single point, $\mathsf{S}_1^2 \simeq \mathsf{S}^1$ is a segment, $\mathsf{S}_1^3 \simeq \mathsf{S}^2$ is a triangle, $\mathsf{S}_1^4 \simeq \mathsf{S}^3$ a tetrahedron, etc.

Finite-dimensional classical theory every system A of the theory is associated with an n_A-dimensional linear space $\mathbb{R}^{n_A}$, with a set of states given by S^{n_A} the set of n_A-dimensional sub-stochastic vectors. The composite system AB has $n_{AB} = n_A \times n_B$. Effects and transformations are determined assuming the no-restriction hypothesis.

Since the convex set of states of a system is simplex, we say that the classical theory is *simplicial*. The classical system is called *bit* for $n = 2$, *trit* for $n = 3$, and *n-it* for generic n. It is easy to see that, under the no-restriction hypothesis (Postulate 3), for the *n-it* one has:

1. the convex set of normalized states is $\mathsf{S}_1^n \simeq \mathsf{S}^{n-1}$;
2. the convex set of effects is the set E^n of unit-dominated positive vectors $0 \leq \mathbf{x} \leq 1$;
3. the convex set of transformations from the n-dimensional to the m-dimensional system is the set $\mathsf{T}(n, m)$ of Markov transformations represented by $m \times n$ matrices $\mathbf{M}$ that are sub-stochastic, namely with each column vector $\mathbf{M}(i) \in \mathsf{S}^n$ a sub-stochastic vector $\forall i = 1, \ldots, n$. The transformation is then represented as $\mathbf{x} \mapsto \mathbf{y} = \mathbf{M}\mathbf{x}$ with $\mathbf{x} \in \mathsf{S}_n, \mathbf{y} \in \mathsf{S}_m$.

Exercise 2.63 Find the atomic effects of the n-dimensional classical system.

Exercise 2.64 Find the atomic transformations in $\mathsf{T}(n, m)$.

From what we have said, we can regard classical theory as a restriction of quantum theory, by identifying the classical system A with the quantum system, and restricting the

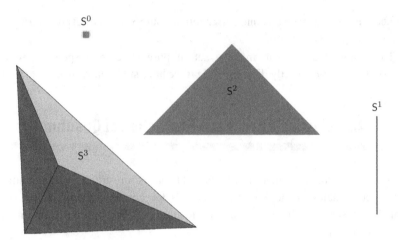

Fig. 2.3   The convex sets of states of Classical Theory.

states to the convex combinations of a fixed maximal set of orthogonal pure states, namely
of the form

$$\mathbf{p} \in \mathsf{S}^d, \quad \rho_\mathbf{p} = \sum_{l=1}^{d} p_l |l\rangle \langle l|, \tag{2.61}$$

with $\{|l\rangle\}_{d=1}^{d}$ a fixed orthonormal basis in $\mathcal{H}_A$, and assuming the no-restriction hypothesis.

Exercise 2.65   Show that the Kraus operators for $\mathsf{T}(n, m)$ are restricted to be proportional to
the dyads $|k\rangle \langle l|$, with $1 \le k \le m$, and $1 \le l \le n$.

Exercise 2.66   Characterize the reversible transformations of classical theory.

Exercise 2.67   Find the deterministic transformations of classical theory.

Exercise 2.68   Show that the identity transformation in classical theory is not atomic.

Exercise 2.69   Show that Principles 1–5 also hold for classical theory.

Exercise 2.70   Show that there are no entangled states in classical theory, hence the
purification principle cannot hold.

Exercise 2.71   Show that a principle we can add to Principles 1–5 in order to get classical
theory is the requirement of perfect joint discriminability of all pure states for any
system.

2.10 Purification

In this section we will derive the purification principle for quantum theory, which in
the axiomatic derivation will become the distinctive quantum axiom. As we will see in
Chapter 7, some of the most "quantum traits" of the theory (e.g. entanglement, teleportation, the existence of steering states, reversibility, process tomography, no information
without disturbance) are direct consequences of the purification principle. Some of the

consequences of the purification principle that are explored here will be re-derived in Chapter 7 and in Part III without using Hilbert spaces, only as a consequence of principles.

2.10.1 Purification of States

We have seen in Exercise 2.21 that generally the marginal state $\rho \in \mathsf{St}_1(A)$ of a pure joint state $\Psi \in \mathsf{St}_1(AB)$ is not necessarily pure. What about the converse, namely: given a state $\rho \in \mathsf{St}(A)$, is there always a pure state $\Psi \in \mathsf{St}(AB)$ for some system B, having ρ as its marginal state? The situation is represented in the following diagrammatic equation:

$$\boxed{\rho}\!-\!\!A\quad = \quad \left(\Psi \begin{array}{c} -A \\ -B\boxed{e} \end{array}\right) .$$

We write $|\rho)_A = (e|_B|\Psi)_{AB}$ and call the state $|\Psi)_{AB}$ a *purification* of state $\rho \in \mathsf{St}_1(A)$. One can easily generalize the notion of purification to probabilistic states, since this simply corresponds to rescaling both states in (2.10.1) by a probability $0 < p \le 1$, and the state $|\Psi)_{AB}$ is deterministic iff $|\rho)_A$ is deterministic. The system B is called a *purifying system*, or *reference system*. One of the unique features of quantum theory is that a purification always exists for any state. This can be easily proved by considering a pure bipartite state corresponding to the vector $|\Psi\rangle\!\rangle \in \mathcal{H}_A \otimes \mathcal{H}_B$, with $\Psi \in \mathsf{Lin}(\mathcal{H}_B \to \mathcal{H}_A)$ and recalling that

$$\rho = (e|_B|\Psi)_{AB} = \mathrm{Tr}_B[|\Psi\rangle\!\rangle\langle\!\langle\Psi|] = \Psi\Psi^\dagger.$$

Using the polar decomposition for Ψ (see Appendix 2.1), we see then that all possible states purifying ρ are of the form $|\Psi\rangle\!\rangle\langle\!\langle\Psi|$, with

$$\Psi = \rho^{\frac{1}{2}}V^\dagger, \qquad V \in \mathsf{Lin}(\mathcal{H}_A \to \mathcal{H}_B), \qquad V^\dagger V = I_{\mathsf{Supp}\,\rho}, \tag{2.62}$$

namely V is a partial isometry with support equal to $\mathsf{Supp}\,\rho$. Since $|\rho^{\frac{1}{2}}V^\dagger\rangle\!\rangle = (I \otimes V^*)|\rho^{\frac{1}{2}}\rangle\!\rangle$, one can see that the isometry V simply embeds $\mathsf{Supp}\,\rho$ into the reference Hilbert space $\mathcal{H}_B$, namely it corresponds to associating an orthonormal set of vectors in $\mathcal{H}_B$ with an orthonormal basis in $\mathsf{Supp}\,\rho$. Indeed, for spectral decomposition $\rho = \sum_i |\lambda_i\rangle\langle\lambda_i|$, the purification can be written in the form $|\Psi\rangle\!\rangle\langle\!\langle\Psi|$ with

$$|\Psi\rangle\!\rangle = \sum_i |\lambda_i\rangle \otimes |u_i\rangle, \tag{2.63}$$

with $\{|u_i\rangle\}$ orthonormal set in $\mathcal{H}_B$, given by

$$|u_i\rangle = V^* \frac{|\lambda_i\rangle^*}{\|\lambda_i\|},$$

where we have written all vectors with respect to the canonical basis chosen for the definition of $|I\rangle\!\rangle_B$,[23] and we chose the specific purification in Eq. (2.62). One can also equivalently write the same purification as follows:

$$|\Psi\rangle\!\rangle = \sum_j |\psi_j\rangle \otimes |j\rangle, \tag{2.64}$$

---

[23] In the following, we will use the shorthand $|I\rangle\!\rangle_A$ to denote $|I\rangle\!\rangle_{\mathcal{H}_A \otimes \mathcal{H}_A}$.

where $\{|j\rangle\}$ is the canonical basis defining $|I\rangle\!\rangle_B$, and

$$|\psi_j\rangle = \sum_i V_{ji} \frac{|\lambda_i\rangle}{\|\lambda_i\|}, \quad V_{ji} = \langle j|V^*|\lambda_i\rangle^*. \tag{2.65}$$

Notice that the cardinality of the sum in Eq. (2.64) is generally larger than that in Eq. (2.63). The two different ways of writing the purification reflect the two isometrically equivalent decompositions $\rho = \sum_i |\lambda_i\rangle\langle\lambda_i| = \sum_j |\psi_j\rangle\langle\psi_j|$ in Eq. (2.18).

Exercise 2.72   Prove Eqs. (2.64) and (2.65).

From what said above, it follows that for quantum theory the following principle holds.

Principle 6 (Purification)   Every state has a purification. For fixed purifying system, every two purifications of the same state are connected by a reversible transformation on the purifying system.

In the principle we used the fact that for fixed purifying system B two different purifications are connected by a unitary $U \in \mathsf{Lin}(B)$ as follows:

$$|\Psi'\rangle\!\rangle = (I_A \otimes V'^*)|\rho^{\frac{1}{2}}\rangle\!\rangle =: (I_A \otimes (UV)^*)|\rho^{\frac{1}{2}}\rangle\!\rangle = (I_A \otimes U^*)|\Psi\rangle\!\rangle. \tag{2.66}$$

It is remarkable that the purification principle (Principle 6) does not hold in classical theory, whereas all the other Principles 1–5 hold (see Exercise 2.69).

Exercise 2.73   Show that the purification of a mixed state is necessarily an entangled state, hence Principle 6 does not hold for the classical theory.

A relevant property of purification is given by the following lemma.

Lemma 2.13   *A purification of a state also steers any of its refinements. Precisely, if $|\Psi\rangle_{AB}$ purifies $|\rho\rangle_A$, then it steers every state $\sigma \in \mathsf{RefSet}\ \rho$, meaning that there exists an effect $b_\sigma \in \mathsf{Eff}(B)$ providing the preparation of σ as follows:*

$$\tag{2.67}$$

Conversely, every effect $a \in \mathsf{Eff}(B)$ steers a state $\sigma_a \in \mathsf{RefSet}\ \rho$ (possibly including the zero) by the scheme of Eq. (2.67).

Proof   We recall that $\mathsf{RefSet}\ \rho$ is the set of density operators with support contained in $\mathsf{Supp}\ \rho$ (see Eq. (2.32)) scaled by a suitable positive constant. It is then sufficient to prove the statement just for pure states on $\mathsf{Supp}\ \rho$, since any mixed state is a sum of pure states, and the corresponding effect will be sum of effects corresponding to pure states. For any pure state $|\varphi\rangle = \sum_i v_i|\lambda_i\rangle \in \mathsf{Supp}\ \rho$ consider the rank-one effect $|v_\varphi\rangle\langle v_\varphi| \in \mathsf{Lin}_+(\mathcal{H}_B)$, with $|v_\varphi\rangle = \sum_i v_i^*|u_i\rangle$, and $|u_i\rangle$ given in Eq. (2.63). Indeed, we can immediately check that

$$|\varphi\rangle\langle\varphi| = \mathrm{Tr}_B[|\Psi\rangle\!\rangle\langle\!\langle\Psi|(I \otimes |v_\varphi\rangle\langle v_\varphi|)].$$

Finally, every effect $b \in \mathsf{Eff}(B)$ steers a state $\sigma_b \in \mathsf{St}(A)$ refining ρ through Eq. (2.67), since b refines the deterministic effect corresponding to ρ. Finally, it is easy to check that the sum of effects corresponding to a decomposition into pure states of a mixed σ is dominated by the identity. $\square$

For a generally mixed state ρ we can restrict to its support (see also Problem 2.24), and consider the operator ρ invertible. An explicit construction of the effect E_σ steering $\sigma \leq \rho$ is then given by the following effect

$$E_\sigma = (\rho^{-\frac{1}{2}}\sigma\rho^{-\frac{1}{2}})^*. \qquad (2.68)$$

It is immediate to check that $E_\sigma \leq I_B$, and one has

$$\mathrm{Tr}_B[(I \otimes E_\sigma)|\rho^{\frac{1}{2}}\rangle\!\rangle\langle\!\langle\rho^{\frac{1}{2}}|] = \sigma.$$

Remark   The purification of an internal state is a *steering state*, namely it allows us to prepare any state $\sigma \in \mathsf{St}(A)$ by application of a suitable effect $b_\sigma \in \mathsf{Eff}(B)$.

Lemma 2.13 in conjunction with the local discriminability principle (Principle 2) bear the following consequence.

Theorem 2.14 (Equality Upon Input of ρ vs Equality on Purifications)   *Let $\Psi \in \mathsf{St}_1(AC)$ be a purification of $\rho \in \mathsf{St}_1(A)$, and let $\mathcal{A}, \mathcal{A}' \in \mathsf{Transf}(A\to B)$ be two transformations. Then one has*

$$\mathcal{A}|\Psi)_{AC} = \mathcal{A}'|\Psi)_{AC} \qquad \Longleftrightarrow \qquad \mathcal{A} =_\rho \mathcal{A}', \qquad (2.69)$$

where $\mathcal{A} =_\rho \mathcal{A}'$ denotes that the two transformations $\mathcal{A}, \mathcal{A}'$ are equal upon input of ρ, namely they are equal when restricted to states in the refinement set of ρ.

Proof   Being Ψ a purification of ρ, due to Lemma 2.13 σ belongs to the refinement set of ρ iff there exists an effect $b \in \mathsf{Eff}(B)$ such that

Therefore, we have that $\mathcal{A} =_\rho \mathcal{A}'$ if and only if one has

$$\forall b \in \mathsf{Eff}(B):$$

This is equivalent to the requirement that the states $\mathcal{A}|\Psi)_{AC}$ and $\mathcal{A}'|\Psi)_{AC}$ cannot be discriminated by local tests, namely iff

$$\forall b \in \mathsf{Eff}(B), c \in \mathsf{Eff}(C):$$

$$(2.70)$$

By the local discriminability principle Eq. (2.70) is equivalent to $\mathcal{A}|\Psi)_{AC} = \mathcal{A}'|\Psi)_{AC}$. $\square$

Notice that in the above theorem the choice of the purification of ρ is arbitrary, namely identity (2.69) equally holds for any other purification $|\Phi)_{AC}$ of ρ.

2.10.2 Faithful States and Ancilla-assisted Tomography

According to Theorem 2.14 two transformations are equal upon input of ρ iff they act in the same way on a purification of ρ. If ρ is completely mixed, namely its refinement set spans the whole $\mathsf{St}_\mathbb{R}(A)$, then $\mathsf{Supp}\,\rho = \mathcal{H}_A$ is the full Hilbert space of system A, which means that the two transformations act in the same way on all possible input states, and thanks to the local discriminability principle this means that the two transformations are identical. This also means that two transformations acting in the same way over the purification $|\Psi_\rho\rangle\rangle$ of an internal state ρ are the same transformation. In other words, the action of a transformation $\mathcal{A}$ on the the purification $|\Psi_\rho\rangle\rangle$ of an internal state ρ completely characterizes the transformation $\mathcal{A}$. We will then call $|\Psi_\rho\rangle\rangle$ a *faithful state*, namely the output state $(\mathcal{A} \otimes \mathcal{I})|\Psi_\rho\rangle\rangle$ is in one-to-one correspondence with the transformation $\mathcal{A}$. Therefore, instead of running an informationally complete preparation test to perform a tomography of the transformation $\mathcal{A}$, we just use a single input state $|\Psi_\rho\rangle\rangle\langle\langle\Psi_\rho|$ with the following set-up:

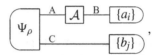

with two local informationally complete observation tests at the output. The above method is called *ancilla-assisted tomography*. An example of application is provided in the following exercise.

Exercise 2.74 For $d = \dim\mathcal{H}_A < \infty$, the state $\frac{1}{d}|I\rangle\rangle\langle\langle I|$ is faithful. Show how, apart from a phase factor, it is possible to recover all matrix elements of the operator $A \in \mathsf{Lin}(\mathcal{H})$ by performing an ancilla-assisted tomography. [*Hint:* use the identity $|A\rangle\rangle = (A \otimes I)|I\rangle\rangle$ and evaluate the matrix element of A_{ij} in terms of the expectations $\langle\langle\Phi|i_0,j_0\rangle\rangle\langle\langle i,j|\Phi\rangle\rangle$ with $|\Phi\rangle\rangle = \frac{1}{\sqrt{d}}(A \otimes I)|I\rangle\rangle$.]

Exercise 2.75 The Choi–Jamiołkowski isomorphism corresponds to the special case of faithful state $\Omega = \frac{1}{d}|I\rangle\rangle\langle\langle I| \in \mathsf{St}_1(AA)$. Does any other faithful state $\mathsf{St}_1(AA)$ induce a cone-isomorphism between $\mathsf{St}_+(BA)$ and $\mathsf{Transf}_+(A\to B)$?

In Problem 2.27 a result analogous to that of Exercise 2.74 is derived for a general input state Ψ, showing that the method works also for infinite dimensions.

We recall that the purification of an internal state is also a *steering state*, namely it allows us to prepare any state $\sigma \in \mathsf{St}(A)$ by the application of a suitable effect $b_\sigma \in \mathsf{Eff}(B)$.

2.10.3 No Information without Disturbance

We say that a test $\{\mathcal{A}_i\}_{i\in X}$ on system A is *non-disturbing upon input of* $\rho \in \mathsf{St}(A)$ if upon input of ρ the coarse-grained test is equivalent to the identity, i.e.

$$\widehat{\mathcal{A}} := \sum_{i\in X} \mathcal{A}_i =_\rho \mathcal{I}_A.$$

This means that the test does not disturb any state $\sigma \in \mathsf{RefSet}\,\rho$. If ρ is a completely mixed state, then the test is simply *non-disturbing*, which means that $\widehat{\mathcal{A}}$ is the identity on any state.

A test $\{\mathcal{A}_i\}_{i \in X}$ made of events $\mathcal{A}_i \in \mathsf{Transf}(A \to B)$ does not provide any information on the system when the occurrence probability of each outcome is independent of the state of the system, namely when the effect of each event is proportional to the deterministic effect:

$$\forall i \in X: \quad (e|_B \mathcal{A}_i = p_i (e|_A, \quad \text{(no-information test).}$$

Clearly the no-information feature of a test can be restricted upon input of ρ trivially as follows:

$$\forall i \in X: \quad (e|_B \mathcal{A}_i =_\rho p_i (e|_A, \quad \text{(no-information test upon input of } \rho \text{).}$$

It follows immediately that a necessary and sufficient condition for the test to provide no-information test upon input of ρ is

$$\forall i \in X: \quad \mathcal{A}_i =_\rho p_i \mathcal{D}, \quad \text{(no-information test upon input of } \rho \text{).}$$

with $\mathcal{D} \in \mathsf{Transf}(A \to B)$ deterministic.

We have now the following relevant theorem.

Theorem 2.15 (No Information without Disturbance)   *If a test $\{\mathcal{A}_i\}_{i \in X} \subseteq \mathsf{Transf}(A)$ is non-disturbing upon input of $\rho \in \mathsf{St}(A)$ then it does not provide information input of ρ.*

Proof   Let $\Psi \in \mathsf{St}(AB)$ be a purification of $\rho \in \mathsf{St}(A)$. By Theorem 2.14, the no-disturbance condition $\widehat{\mathcal{A}} =_\rho \mathcal{I}_A$ holds iff

$$\sum_{i \in X} \left(\Psi \begin{array}{c} \xrightarrow{A} \boxed{\mathcal{A}_i} \xrightarrow{A} \\ \xrightarrow{B} \end{array} \right. = \left(\Psi \begin{array}{c} \xrightarrow{A} \\ \xrightarrow{B} \end{array} \right. .$$

Since Ψ is pure, each term of the decomposition must be proportional to $|\Psi\rangle_{AB}$, namely $\mathcal{A}_i|\Psi\rangle_{AB} =_\rho p_i|\Psi\rangle_{AB} =_\rho (p_i \mathcal{I}_A)|\Psi\rangle_{AB}$, and again, by Theorem 2.14, this is equivalent to $\mathcal{A}_i =_\rho p_i \mathcal{I}_A$, and the test is a no-information one upon input of ρ. $\square$

2.10.4 Purification of Transformations

We now see how the notion of purification can be extended to transformations. We will say that a tranformation $\mathcal{A} \in \mathsf{Transf}(A \to B)$ is purified by the atomic transformation $\mathcal{P} \in \mathsf{Transf}(A \to BE)$ (the purifying system E is usually called *environment*) if

$$\xrightarrow{A} \boxed{\mathcal{A}} \xrightarrow{B} = \begin{array}{c} \xrightarrow{E} \boxed{e} \\ {}_A\boxed{\mathcal{P}}{}_B \end{array} . \tag{2.71}$$

In terms of a Kraus form of the map, one has explicitly

$$\mathcal{A}(\rho) = \sum_i A_i \rho A_i^\dagger = \mathrm{Tr}_E[A \rho A^\dagger], \quad A = \sum_i A_i \otimes |i\rangle \in \mathsf{Lin}(\mathcal{H}_A \to \mathcal{H}_B \otimes \mathcal{H}_E), \tag{2.72}$$

where $\{|i\rangle\} \subseteq \mathcal{H}_E$ is an orthonormal set. It is easy to see that the purifying transformation $\mathcal{P}(\rho) = A\rho A^\dagger$ is unique modulo an isometry over the environment E, corresponding to embed the minimal-dimension $\mathcal{H}_E$ (e.g. for the canonical Kraus) into a larger Hilbert space, and/or to change the orthonormal set $\{|i\rangle\} \in \mathcal{H}_E$.

Exercise 2.76   [Stinespring] Show that the Heisenberg-picture version of the purification of transformation in Eq. (2.71) is given by $\mathcal{A}^\dagger(X) = A^\dagger(X \otimes I_E)A$.

Similarly to what happens for quantum states, the purification of a quantum transformation also purifies its entire refinement set. This means that for each transformation $\mathcal{B} \in \mathsf{RefSet}\, \mathcal{A}$ there exists an effect $a_\mathcal{B}$ on the environment such that the following diagram holds for the purification $\mathcal{P}$ of $\mathcal{A}$:

Again, it is sufficient to prove the statement only for an atomic transformation $\mathcal{B} = B \cdot B^\dagger$ in the refinement set of $\mathcal{A} = \sum_i A_i \cdot A_i^\dagger$. As shown in Exercise 2.49 a quantum operation that belongs to the refinement set of $\mathcal{A}$ has Kraus operators that are linear combination of those of $\mathcal{A}$. Any atomic transformation $\mathcal{B}$ in the refinement set of $\mathcal{A}$ is then obtained using a rank-one effect $|v_\mathcal{B}\rangle\langle v_\mathcal{B}| \in \mathsf{Lin}_+(\mathcal{H}_E)$ with $|v_\mathcal{B}\rangle = \sum_i v_i^*|i\rangle$ as follows:

$$\mathcal{B}(\rho) = B\rho B^\dagger = \mathrm{Tr}_E[A\rho A^\dagger(I \otimes |v_\mathcal{B}\rangle\langle v_\mathcal{B}|)], \qquad B = \sum_i v_i A_i.$$

From Corollary 2.10 it follows that the purification of a deterministic transformation – i.e. a *quantum channel* – is isometric, namely

$$\mathcal{A} \in \mathsf{Transf}_1(A \to B) \longrightarrow \mathcal{A}(\rho) = \mathrm{Tr}_E[A\rho A^\dagger], \quad A^\dagger A = I_A.$$

Using the *Golden rule for quantum extensions* from Appendix 2.2

$$A : \mathcal{H}_A \hookrightarrow \mathcal{H}_B \otimes \mathcal{H}_E \simeq \mathcal{H}_A \otimes \mathcal{H}_F, \quad A = U(I_A \otimes |0\rangle_F), \tag{2.73}$$

with $|0\rangle_F \in \mathcal{H}_F$ any chosen unit vector and $U \in \mathsf{Lin}(\mathcal{H}_A \otimes \mathcal{H}_F)$ a suitable unitary, one proves that any deterministic quantum transformation can be achieved by a unitary interaction with an environment as follows:

$$\mathcal{A}(\rho) = \mathrm{Tr}_E[U(\rho \otimes |0\rangle\langle 0|)U^\dagger]. \tag{2.74}$$

Diagrammatically Eq. (2.74) would be represented as follows:

Exercise 2.77   Derive Eq. (2.74).

Exercise 2.78   [Extension of a quantum instrument] Consider a quantum instrument $\{\mathcal{T}_i\}_{i \in X}$ with $\mathcal{T}_i \in \mathsf{Transf}(A \to B)\; \forall i \in X$. Show that it is possible to achieve it as an indirect measurement as follows:

$$\mathcal{T}_i(\rho) = \mathrm{Tr}_E[U(\rho \otimes \sigma)U^\dagger(I_B \otimes Z_i)], \tag{2.76}$$

namely, by having the system A interacting with an environment F in the state $\sigma \in \mathsf{PurSt}_1(F)$ via the unitary operator $U \in \mathsf{Lin}(\mathcal{H}_A \otimes \mathcal{H}_F)$, and performing the PVM measurement $\{Z_i\}_{i \in X} \in \mathsf{Lin}(\mathcal{H}_E)$ over a generally different environment E afterwards. Find the relation between input and output Hilbert spaces $\mathcal{H}_A \otimes \mathcal{H}_F$ and $\mathcal{H}_B \otimes \mathcal{H}_E$, respectively, determine explicit forms of U and $\{Z_i\}_{i \in X}$, and show that the ancilla F can be taken in any desired pure state σ.

Exercise 2.79 [Stinespring form of quantum instrument] Show that a Quantum instrument $\{\mathcal{T}_i\}_{i \in X}$ with $\mathcal{T}_i \in \mathsf{Transf}(A \to B)$ $\forall i \in X$ in the Heinsenberg picture can be written as follows:

$$\mathcal{T}_i^\dagger(X) = V^\dagger(X \otimes Z_i)V,$$

with $V \in \mathsf{Lin}(\mathcal{H}_A \to \mathcal{H}_B \otimes \mathcal{H}_E)$ isometric on $\mathcal{H}_A$, hence the Heisenberg-picture form of a quantum channel is given by

$$\mathcal{T}^\dagger(X) = V^\dagger(X \otimes I_E)V.$$

Exercise 2.80 Let $\mathcal{A}$ be a quantum channel with Choi operator $R_{\mathcal{A}} \in \mathsf{Lin}_+(\mathcal{H}_{BA})$. Show that the minimal extension isometry for $\mathcal{A}$ is given by

$$A = (I_B \otimes (R_{\mathcal{A}}^{\mathsf{T}})^{\frac{1}{2}})(|I_B\rangle\!\rangle \otimes I_A).$$

Exercise 2.81 Consider a quantum instrument corresponding to an indirect von-Neumann measurement scheme with ancilla in a mixed state. Derive the Kraus operators in terms of the preparation of the ancilla, the unitary operator of the interaction, and the orthonormal basis of the von Neumann measurement.

2.10.5 Purifications and Dilations of Effects

The purification principle has been introduced for states, and has been later extended to transformations. However, one can immediately see that it cannot be extended also to effects. One would object that effects are just a special kind of transformations with trivial output system. However, the purification of a transformation is a transformation of which we need to discard a non-trivial environment system, as in Eq. (2.71), hence the purification of an effect would be a transformation, not an effect. What we seek here is a purification of an effect by another effect, namely a kind of "dual purification" where, instead of tracing out an output environment, we want to achieve the effect as a bipartite one, preparing an input environment into a suitable state. This would be the reverse of scheme (2.67).

By purifying an effect $(a|_A$ we then mean to have an atomic effect $(A|_{AB}$ and a suitable state σ_a for B such that the following diagram holds:

$$(2.77)$$

That such purification scheme is generally impossible is simply proved by the existence of a counter-example: the deterministic effect. The deterministic effect is indeed not atomic, since it is the coarse-graining of an observation test.

Lemma 2.16 *It is impossible to purify the deterministic effect, namely to find an atomic effect E and a state $\sigma \in \mathsf{St}_1(A)$ such that*

$$\begin{array}{c}\text{—}\,\text{A}\,\boxed{e}\end{array} = \left(\begin{array}{c}\boxed{\sigma}\!-\!\!\begin{array}{c}\text{A}\\ \text{B}\end{array}\!\boxed{E}\end{array}\right).$$

Proof In quantum theory an atomic effect is an expectation over a rank-one operator. Therefore the purifying effect $(E|_{AB}$ would correspond to a rank-one operator $|E\rangle\!\rangle\langle\!\langle E| \leq I_{AB}$, working as follows:

$$\forall \rho \in \mathsf{St}(A), \quad (e|\rho)_A = (E|_{AB}|\rho)_A|\sigma_a)_B = \mathrm{Tr}[(\rho \otimes \sigma)|E\rangle\!\rangle\langle\!\langle E|] = \mathrm{Tr}[\rho E\sigma^T E^\dagger]. \quad (2.78)$$

One has $\|E\|_2 \leq 1$, due to domination $|E\rangle\!\rangle\langle\!\langle E| \leq I_{AB}$, since $\langle\!\langle E|(I_{AB} - |E\rangle\!\rangle\langle\!\langle E|)|E\rangle\!\rangle = \|E\|_2^2(1 - \|E\|_2^2) \geq 0$. But Eq. (2.78) means that $E\sigma^T E^\dagger = I_A$, which requires both E and σ to be full-rank. Then $\sigma^T = (E^\dagger E)^{-1}$. But since $E^\dagger E$ is full rank with $\mathrm{Tr}[E^\dagger E] \leq 1$, all eigenvalues of $E^\dagger E$ are strictly smaller than 1, hence $\mathrm{Tr}\,\sigma = \mathrm{Tr}(E^\dagger E)^{-1} > 1$, a contradiction. $\square$

More generally, one is not guaranteed that there exists an atomic effect $(A|_{AB}$ achieving a purification as in Eq. (2.78) with $(a|_A$ in place of $(e|_A$ and for some state $|\sigma_a)_B$. This does not rule out existence of special cases where the purification is possible. Indeed, a rank-one resolution of the identity

$$\forall i \in X: \qquad \sum_{i \in X} |A_i\rangle\!\rangle\langle\!\langle A_i| = I_A \otimes I_B, \qquad \{A_i\}_{i \in X} \subseteq \mathsf{Lin}(\mathcal{H}_B \to \mathcal{H}_A),$$

would purify any POVM of the form

$$\forall i \in X: \qquad E_i = A_i \sigma^T A_i^\dagger, \qquad\qquad\qquad\qquad (2.79)$$

corresponding to the local state $\sigma \in \mathsf{St}_1(B)$ as follows:

$$\forall i \in X: \qquad E_i = \mathrm{Tr}_B[|A_i\rangle\!\rangle\langle\!\langle A_i|(I_A \otimes \sigma)]. \qquad\qquad (2.80)$$

Exercise 2.83 checks that E_i is indeed a POVM.

Exercise 2.82 Show that the following identities are equivalent, for both $\{A_i\}_{i \in X}$ and $\{B_i\}_{i \in X}$ sets of operators in $\mathsf{Lin}(\mathcal{H}_B \to \mathcal{H}_A)$ and $\forall X \in \mathsf{Lin}(\mathcal{H}_A)$ and $\forall Y \in \mathsf{Lin}(\mathcal{H}_B)$:

$$\sum_{i \in X} |A_i\rangle\!\rangle\langle\!\langle B_i| = I_A \otimes I_B, \qquad\qquad\qquad\qquad (2.81)$$

$$\sum_{i \in X} B_i^\dagger X A_i = \sum_{i \in X} A_i^\dagger X B_i = \mathrm{Tr}[X]I_B, \qquad\qquad (2.82)$$

$$\sum_{i \in X} B_i Y A_i^\dagger = \sum_{i \in X} A_i Y B_i^\dagger = \mathrm{Tr}[Y]I_A. \qquad\qquad (2.83)$$

Exercise 2.83 Using the results in Exercise 2.81, check that $\{E_i\}_{i \in X}$ in Eq. (2.79) is a POVM, and that Eq. (2.80) gives the POVM in Eq. (2.79).

POVM extensions and dilations In practice, rather than purifying effects, one is interested in achieving POVMs from PVMs by adding an ancillary system or, more generally, projecting from a higher-dimensional system (the former case is generally called *POVM extension*, the latter *POVM dilation*). In the following exercises we consider some of these cases, the most general one being the Naimark dilation.

Exercise 2.84 [Naimark dilation] Consider a POVM as a special case of instrument where the final system is the trivial system. Using the dilation of instruments in Exercise 2.77, show that every POVM $\{P_i\}_{i \in X}$ can be dilated to a PVM on a larger Hilbert space. More precisely, for a POVM $\{P_i\}_{i \in X}$ on $\mathcal{H}_A$, there exists a Hilbert space $\mathcal{H}_B \supseteq \mathcal{H}_A$, a PVM $\{Z_i\}_{i \in X}$ on $\mathcal{H}_B$, and an isometry $V \in \text{Lin}(\mathcal{H}_A \to \mathcal{H}_B)$ such that

$$P_i = V^\dagger Z_i V. \tag{2.84}$$

The PVM $\{Z_i\}$ is called the *Naimark dilation* of the POVM $\{P_i\}$.

Exercise 2.85 In some literature the Naimark dilation is written as

$$P_i = W Z_i W, \tag{2.85}$$

with $W \in \text{Lin}(\mathcal{H}_B)$ orthogonal projector on $\mathcal{H}_A$. Connect W with the isometry V in Eq. (2.84).

Exercise 2.86 Prove identity (2.84) by substituting the explicit forms of Z_i and V.

Exercise 2.87 Show that there are infinitely many inequivalent Naimark dilations with increasingly large dimension of the extended Hilbert space. The case of PVM made with eigenvectors of the POVM elements is called the *canonical Naimark dilation*. Prove that it has $\text{rank}(Z_i) = \text{rank}(P_i) \; \forall i \in X$. Every dilation with the same rank is called minimal.

Exercise 2.88 Notice that the dimension of the Hilbert space $\mathcal{H}_B \supseteq \mathcal{H}_A$ is generally not a multiple of $\dim \mathcal{H}_A$. When this is the case, then there exists an ancilla E such that $\mathcal{H}_B = \mathcal{H}_A \otimes \mathcal{H}_E$. In such case, show that the POVM can be achieved as an indirect measurement as follows:

$$P_i = (I \otimes \langle 0|_E) U^\dagger Z_i U (I \otimes |0\rangle_E) = \text{Tr}_E[(I \otimes |0\rangle\langle 0|) U^\dagger Z_i U]$$

namely the POVM is equivalent to a joint PVM $U^\dagger Z_i U$ on system+ancilla, with the ancilla prepared in the fixed pure state $|0\rangle$.

Exercise 2.89 [Minimal Naimark dilation] Determine the minimal dimension $\dim \mathcal{H}_B$ of the Naimark dilation.

Exercise 2.90 [Abelian POVM] Show that a POVM made of commuting positive operators is equivalent to an observable with error, namely that there is a nonvanishing probability of scrambling the outcome to a different one. Find a Naimark dilation.

2.11 Quantum No Cloning

An alternative route to proving the impossibility of discriminating two non-orthogonal states is the famous no-cloning theorem.

Theorem 2.17 (Quantum No Cloning) *No deterministic transformation can produce two identical copies of a state drawn from a set of two non-orthogonal pure states.*

Proof Denote by $\{|\varphi_i\rangle\}_{i=1,2}$ with $0 < |\langle\varphi_1|\varphi_2\rangle| < 1$ the two unit vectors corresponding to the states that we want to clone. A deterministic transformation $\mathcal{C} \in \mathsf{Transf}(A \to AA)$ achieving the cloning would work as follows:

$$\forall i = 1, 2: \quad \mathcal{C}(|\varphi_i\rangle\langle\varphi_i|) = |\varphi_i\rangle\langle\varphi_i| \otimes |\varphi_i\rangle\langle\varphi_i|.$$

If such a transformation exists, it would then be purifiable to an atomic one as $\mathcal{C}(\rho) = \mathrm{Tr}_E(A\rho A^\dagger)$ with a suitable environment E. In the purified form the transformation would act as

$$\forall i = 1, 2: \quad A|\varphi_i\rangle = |u_i\rangle_E \otimes |\varphi_i\rangle \otimes |\varphi_i\rangle.$$

However, being $\mathcal{C}$ deterministic, the operator A should be isometric, hence, by taking the scalar product of the two vectors in the last equation, we get

$$|\langle\varphi_1|\varphi_2\rangle| = |\langle\varphi_1|\varphi_2\rangle|^2 |\langle u_1|u_2\rangle|,$$

which implies either $|\langle\varphi_1|\varphi_2\rangle| = 0$ or $|\langle\varphi_1|\varphi_2\rangle| = |\langle u_1|u_2\rangle|^{-1} \geq 1$, contradicting the hypothesis.   □

Clearly the impossibility of preparing two copies of a state drawn from a set of two non-orthogonal states implies *a fortiori* the general impossibility of achieving any number $N \geq 2$ of copies of any state drawn from any non-orthogonal set.

The no-cloning theorem can be used to prove the impossibility of determining the state of a single quantum system.

Corollary 2.18 [24] *There exists no measurement scheme that allows one to determine the state of a single quantum system.*

Proof The statement can be proved by contradiction, assuming the possibility, in principle, of preparing any desired *known* quantum state. The argument is the following. If one could determine the quantum state of a single quantum system, then one could achieve perfect cloning via state preparation. Conversely, if perfect cloning were possible, then the quantum state of a single quantum system would be determined with arbitrary precision via quantum tomography.   □

After realizing that cloning and discrimination are equivalent tasks, it becomes now obvious that exact cloning of two non-orthogonal states is indeed possible, but only probabilistically, namely with non-zero failure probability. Indeed, one can use POVMs of the form given in Exercises 2.57 and 2.58 to build up a measure-and-prepare instrument preparing multiple copies of the *i*th state for outcome *i*, and failing to do so for the inconclusive outcome.

---

[24] D'Ariano and Yuen (1996).

2.12 The von Neumann Postulate: Do We Need It?

The von Neumann postulate in Quantum Mantra 5 has been designed as a solution to the problem of the apparent contradiction between the simple fact that we see a definite outcome in a quantum measurement, whereas we describe all quantum processes ultimately in terms of unitary interactions. The apparent contradiction has been well remarked by Schrödinger in his notorious *Schrödinger's cat paradox.*[25] The paradox relies on the fact that, after the unitary interaction, system and measuring apparatus are in an entangled state. This is due to the fact that the apparatus must report any possible outcome i by evolving in a state $|i\rangle$ when it measures the system in the state $|i\rangle$, hence if the system is e.g. in the superposition state $\frac{1}{\sqrt{2}}(|1\rangle + |2\rangle)$, due to linearity the system and the apparatus will be entangled in the state $\frac{1}{\sqrt{2}}(|1\rangle \otimes |1\rangle + |2\rangle \otimes |2\rangle)$. But we know that the outcome is a definite value i, which seems to contradict the assertion that the two systems are entangled. Schrödinger dramatized the issue describing the macroscopic apparatus as a cat with two states "dead" and "alive," corresponding to the states "decayed" or "not decayed" of a radioactive atom which plays the role of the measured system.

The von Neumann postulate enforces a solution to the Schroedinger's cat paradox by assuming that the entangled state "collapses" to either one of the two possibilities that are superimposed in the entangled state. This, however, raises the further problem of finding a criterion to assess when we should invoke the collapse in place of the unitary evolution, opening the Pandora's jar of interpretations of quantum theory, whose consideration is not in the purposes of the present book.

Do we need the von Neumann postulate in Quantum Mantra 5? After all, it seems that the postulate solves only an interpretational issue, since, as we will see in the following, unitary evolution and Born rule are sufficient to make all desired predictions correctly, including determining the state of the measured system after the measurement.

We describe the quantum measurement as an interaction between the quantum system A which undergoes the measurement, and the measuring apparatus, which we separate into a *pointer* P corresponding to the system on which we read the outcome, and an *environment* E, which represents every other constituent of the apparatus. Thanks to the purification theorem for channels, we can always assume that the interaction is unitary. We will show now that we can determine the state $\rho_i \in \mathsf{St}(A)$ of the measured system A for outcome i without invoking the von Neumann postulate.

By definition, the state $\rho_i \in \mathsf{St}(A)$ is the state that provides the correct joint probability of the outcome i along with the outcome of any forthcoming measurement on the same system A. We describe the forthcoming measurement in the most general way, associating a quantum effect E_j to the generic outcome j. Then, the joint probability is given by the Born rule

$$p(j, i) = \mathrm{Tr}[E_j \rho_i]. \tag{2.86}$$

---

[25] Schrödinger (1935a).

We have supposed that the reading of the outcome is an observable of the pointer P. Then, according to the Born rule the outcome i occurs with probability given by the expectation of $|i\rangle\langle i|$ over the state of the pointer. On the environment E, instead, we do not perform any measurement. Therefore, if we prepare the measured system A in the state ρ and the instrumental systems PE in the joint pure state $|0\rangle\langle 0|$ (we can always purify the state by extending E) the joint probability will be given by

$$p(j, i) = \mathrm{Tr}[(E_j \otimes |i\rangle\langle i|_{\mathrm{P}} \otimes I_{\mathrm{E}})U(\rho \otimes |0\rangle\langle 0|_{\mathrm{PE}})U^\dagger]$$
$$= \mathrm{Tr}\{E_j \mathrm{Tr}_{\mathrm{EP}}[(I_{\mathrm{A}} \otimes |i\rangle\langle i|_{\mathrm{P}} \otimes I_{\mathrm{E}})U(\rho \otimes |0\rangle\langle 0|_{\mathrm{PE}})U^\dagger]\} \tag{2.87}$$

Upon comparing Eq. (2.87) with Eq. (2.86), and using the fact that effects are separating for states and both equations must hold for any effect of A, we obtain

$$\rho_i = \mathrm{Tr}_{\mathrm{PE}}[(I_{\mathrm{A}} \otimes |i\rangle\langle i|_{\mathrm{P}} \otimes I_{\mathrm{E}})U(\rho \otimes |0\rangle\langle 0|_{\mathrm{PE}})U^\dagger].$$

Using our tensor notation we can write ρ_i as follows:

$$\rho_i = \sum_k [I_{\mathrm{A}} \otimes \langle i|_{\mathrm{P}} \otimes \langle k|_{\mathrm{E}}]\, U(\rho \otimes |0\rangle\langle 0|)U^\dagger \,[I_{\mathrm{A}} \otimes |i\rangle_{\mathrm{P}} \otimes |k\rangle_{\mathrm{E}}],$$

with $\{|k\rangle\}$ any orthonormal basis for $\mathcal{H}_{\mathrm{E}}$. We then have

$$\rho_i = \sum_k A_k^{(i)} \rho A_k^{(i)\dagger},$$

with $A_k^{(i)} \in \mathrm{Lin}(\mathrm{A})$ given by

$$A_k^{(i)} = [I_{\mathrm{A}} \otimes \langle i|_{\mathrm{P}} \otimes \langle k|_{\mathrm{E}}]\, U[I_{\mathrm{A}} \otimes |0\rangle_{\mathrm{PE}}]. \tag{2.88}$$

Therefore, the probabilistic state ρ_i after the measurement for outcome i is simply described by the quantum instrument $\mathcal{T}_i := \sum_k A_k^{(i)} \cdot A_k^{(i)\dagger}$, with $A_k^{(i)}$ in Eq. (2.88), which we remark has been derived using only the Born rule.

The above theoretical description of the measurement which does not invoke the von Neumann postulate cannot, however, provide information about the local state of the pointer P after the measurement. In order to recover also the state of P from the Born rule in the same fashion as we did for the system A, one needs to consider the pointer P itself as a measured system, and perform a measurement on it. But this will introduce another apparatus with another pointer P', opening in this way an infinite chain of apparatuses, one measuring the other. Such an infinite chain of measuring apparatuses is known as the *Wigner's friend chain*, after a thought experiment introduced by Wigner in relation to the Schrödinger's cat paradox. The thought experiment hypothesizes a friend of Wigner who performs the Schrödinger's cat experiment after Wigner leaves the laboratory. Only when Wigner comes back he is informed about the result of the experiment by his friend, who otherwise would have been entangled with the cat and the measured system. Here Wigner plays the role of pointer P', where his friend is P.

We conclude that unitary evolution and Born rule are sufficient to make all predictions about any measurement performed over a quantum system, including the evaluation of the state of the system after the measurement. What cannot be derived is only the state of the pointer. The latter ultimately represents our knowledge of the measurement outcome, and John von Neumann argued that the collapse ultimately occurs in our consciousness.[26] Von Neumann was well aware of the problem of Wigner's friend, and the possibility of shifting along the chain the *cut between what is quantum* (the measured system) *and what is classical* (the measured result) was considered by him as a "fundamental requirement of the scientific viewpoint,"[27] and his discussion of the measurement process was exactly aimed to show that quantum theory fulfills this requirement.

2.13 Quantum Teleportation

Technically speaking, *teleportation* means to prepare a remote system C in the same state as that of a local system A. How can we teleport a state that, due to the no-cloning theorem, cannot be known (see Corollary 2.18)? Moreover, even if we knew the quantum state, how can we transfer it exactly, if the amount of information to be transmitted is virtually infinite (in the easiest case we need to transmit a vector in $\mathbb{C}^2$)? The answer to the puzzle[28] is that teleportation can be achieved if an entangled state is available between system C and another system B identical to both A and C and located "close" to system A: the requirement for B to be close to A is dictated by the need of performing a joint measurement on AB.

The protocol consists of performing a Bell measurement jointly on systems A (the system to teleport) and B (one of the two entangled systems), and then performing a unitary transformation on C conditional on the outcome of the Bell measurement. Therefore, in order to achieve the task, we need two resources: (1) to have entanglement already established with the remote system C; and (2) to transmit classical information to the remote location. Since teleportation is equivalent to an identity transformation (apart from a swap of Hilbert spaces), according to the no information without disturbance (Theorem 2.15) the Bell measurement must provide no information on the state to be teleported: as we will see in Exercise 2.91, this is indeed the case.

Exercise 2.91 [Teleportation] Consider the following measurement scheme designed to transmit a state from system A to system C

In the above diagram, A, B, C are all d-dimensional systems ($d < \infty$). $\Omega = \frac{1}{d}|F\rangle\rangle\langle\langle F| \in \mathsf{St}_1(AB)$ is a maximally entangled state (i.e. F is unitary), $\{\mathcal{B}_i\}_{i\in X} \subseteq \mathsf{Eff}(AB)$, $X = \{0, 1, \ldots, d^2-1\}$, is a Bell measurement, with $\mathcal{B}_i = \mathrm{Tr}[B_i\cdot] \in \mathsf{Eff}(AB)$ and quantum effect $B_i := \frac{1}{d}|U_i\rangle\rangle\langle\langle U_i| \in \mathsf{Lin}(\mathcal{H}_{AB})$ with $\{U_i\}_{i\in X}$ a set of orthogonal

---

[26] von Neumann (1996).
[27] See p. 418 of von Neumann (1996).
[28] The teleportation protocol has been introduced by Bennett *et al.* (1993).

unitary operators (see Exercise 2.42). $\mathcal{V}_i = V_i \cdot V_i^\dagger$, $i \in X$, are unitary channels on C that are executed conditionally on the outcome i of the Bell measurement. Show that $\mathcal{T}_{AC} \in \mathsf{Transf}(A \rightarrow C)$ is the teleportation channel

$$\rho \in \mathsf{St}_1(A) \mapsto \mathcal{T}_{AC}\,\rho = \rho \in \mathsf{St}_1(C).$$

Solution

Without loss of generality we can consider the case in which we teleport a pure state $|\psi\rangle \in \mathcal{H}_A$. The conditional state at system C for outcome i at the input of transformation $\mathcal{V}_i \in \mathsf{Transf}(C)$ is given by

$$V_i \mathrm{Tr}_{AB}[(B_i \otimes I_C)(|\psi\rangle\langle\psi| \otimes \Omega)]V_i^\dagger = |v_i(\psi)\rangle\langle v_i(\psi)|_C, \qquad (2.90)$$

with

$$|v_i(\psi)\rangle = \tfrac{1}{d}V_i\langle\!\langle U_i|_{AB}|\psi\rangle_A|F\rangle\!\rangle_{BC} = \tfrac{1}{d}V_i F^{\mathsf{T}} U_i^\dagger \langle\!\langle I|_{AB}|I\rangle\!\rangle_{BC}|\psi\rangle_A$$
$$= \tfrac{1}{d}V_i F^{\mathsf{T}} U_i^\dagger |\psi\rangle_C$$

where we used identity (2.51), along with the *teleportation tensor*

$$T_{AC} := \langle\!\langle I|_{AB}|I\rangle\!\rangle_{BC}, \qquad T_{AC}|\psi\rangle_A = |\psi\rangle_C$$

which "teleports" the state $|\psi\rangle$ from system A to system C. From Eq. (2.90) it is readily seen that the state can be teleported if F is unitary – namely the state Ω is maximally entangled – and $V_i = U_i F^* \in \mathsf{Lin}(\mathcal{H}_C)$. The probability of outcome i is

$$p_i = \|v_i(\psi)\|^2 = \frac{1}{d^2},$$

independently of the state ψ at the input, namely the measurement gives no information on ψ.

In Exercise 2.93 it is shown that for mixed state Ω the scheme in Eq. (2.89) gives a non-atomic instrument, while in order to have $\mathcal{T}$ atomic the state Ω must be pure. It is also proved that when the state Ω is not maximally entangled it is impossible to achieve perfect teleportation deterministically. This implies that perfect teleportation is impossible in infinite dimensions, since a maximally entangled state Ω would not be normalizable.

Teleportation as Storage of Channels   The teleportation protocol can be used to store an unknown channel into a quantum state. However, the recovery of the channel can be achieved only probabilistically. The protocol is indeed a physical implementation of the Choi–Jamiołkowski isomorphism. It is as follows.

Consider a Bell measurement $P_i = \tfrac{1}{d}|U_i\rangle\!\rangle\langle\!\langle U_i|$ for $i = 0, \ldots, d^2 - 1$, with $\{U_i\}$ orthogonal unitary operators, and without loss of generality take $U_0 = I_d$. The channel $\mathcal{C} = \sum_i C_i \cdot C_i^\dagger$ is stored by applying it locally to two identical d-dimensional systems in the state $\Omega = \tfrac{1}{d}|I\rangle\!\rangle\langle\!\langle I|$

$$\Omega \mapsto \Omega_{\mathcal{C}} = \mathcal{C}^{\mathsf{T}} \otimes \mathcal{I}(\Omega) = \frac{1}{d}\sum_i |C_i^{\mathsf{T}}\rangle\!\rangle\langle\!\langle C_i^{\mathsf{T}}|.$$

Then, using Eq. (2.91) in Exercise 2.93 we see that the channel $\mathcal{C}$ is achieved using $V_0 = I$. Unfortunately the probability of restoring the channel is only $\frac{1}{d^2}$ (see also Exercise 2.94).

Exercise 2.92 Show that quantum teleportation is an instrument that provides no information about the teleported state, hence it satisfies the no information without disturbance.

Exercise 2.93 Show that for mixed state $\Omega = \sum_j |F_j\rangle\rangle \langle\langle F_j|$ the teleportation scheme in Eq. (2.89) gives the instrument

$$\mathcal{A}_i(\rho) = \sum_j A_j^{(i)} \rho A_j^{(i)\dagger}, \quad A_j^{(i)} = \tfrac{1}{d} V_i F_j^\mathsf{T} U_i^\dagger T_{\mathrm{AC}}. \tag{2.91}$$

Show that the state Ω must be pure in order to have exact teleportation. Prove that for a non-maximally entangled state Ω it is impossible to achieve perfect teleportation deterministically.

Exercise 2.94 Is it possible to improve the probability of restoring an unknown channel encoded on a state with a single use, by using a bipartite state different from Ω?

The above protocol raises the question whether it is possible to store a known channel into a quantum memory. This means preparing a suitable ancilla (the memory) in a state depending on the channel to be stored: the channel is then perfectly recovered acting on a system, by making the memory suitably interacting with the system. The following theorem excludes such a possibility even if restricting to unitary channels.

Theorem 2.19 (No-programming Theorem)[29] *There exists no channel that can program any unitary transformation on a system* A *by changing the state of an ancilla* B, *as in Eq. (2.92).*

$$\underset{\mathrm{A}}{\overset{\mathrm{B}}{\boxed{\begin{array}{c} v_i \\ \hline \mathcal{C} \end{array}}}}\ {}_{\mathrm{A}} \quad = \quad {}_{\mathrm{A}}\boxed{\mathcal{W}_i}{}_{\mathrm{A}} \ . \tag{2.92}$$

Proof Denote by W_1 and W_2 any couple of unitary transformations on $\mathcal{H}_\mathrm{A}$, and by $|v_1\rangle$ and $|v_2\rangle$ the corresponding program states in $\mathcal{H}_\mathrm{B}$. Then, by isometric purification of the channel, we write the transformation achieving the programming of W_1 and W_2 on any state $\psi \in \mathcal{H}_\mathrm{A}$ as follows:

$$\forall \psi \in \mathcal{H}_\mathrm{A}, \ i = 1, 2 \quad V(|\psi\rangle_\mathrm{A} \otimes |v_i\rangle_\mathrm{B}) = W_i |\psi\rangle_\mathrm{A} \otimes |\omega_i(\psi)\rangle_\mathrm{BC}, \tag{2.93}$$

with V isometry. Linearity of the transformation requires the states $|\omega_i(\psi)\rangle$ to be independent on ψ. Therefore, by taking the scalar product between the left and the right sides of the two identities, we have

$$\langle v_1 | v_2 \rangle = \langle \psi | W_1^\dagger W_2 | \psi \rangle \langle \omega_1 | \omega_2 \rangle.$$

Suppose now that $\langle \omega_1 | \omega_2 \rangle \neq 0$. Then we have

$$\langle \psi | W_1^\dagger W_2 | \psi \rangle = \frac{\langle v_1 | v_2 \rangle}{\langle \omega_1 | \omega_2 \rangle},$$

and since the right hand side of the last identity is independent of ψ we have $W_1^\dagger W_2 \propto I$. This implies that the two operators are the same unitary operator apart from a phase factor.

---

[29] Nielsen and Chuang (1997).

Therefore, we need to have $\langle \omega_1 | \omega_2 \rangle = 0$. Then, in order to program a set of unitaries with cardinality N, we need a set of mutually orthogonal program states with the same cardinality N. We conclude that we cannot program the continuum of all possible unitary transformations on an ancilla with a separable Hilbert space. $\square$

2.14 Inverting Transformations

In Corollary 2.12 we have seen that the only transformations that are invertible (and also reversible) are the unitary ones. However, one may ask when a transformation can be inverted on a restricted set of states, that is a Hilbert subspace of the full Hilbert space $\mathcal{H}_A$ of system A. One can conveniently regard such a subspace as the support $\mathsf{Supp}\,\rho$ of a quantum state $\rho \in \mathsf{St}(A)$, corresponding to restrict states to a face of the convex set of states of the quantum system. As we will see, with such a restriction generally it becomes possible to physically invert a non-unitary quantum channel $\mathcal{C}$, corresponding to mathematically invert the channel with another channel, or, more generally, with a quantum operation $\mathcal{R}$ that works deterministically on $\mathcal{C}(\rho)$. It happens that such a physical problem is exactly the same as the problem of *error correction* in quantum information, where the Hilbert space $\mathsf{Supp}\,\rho$ becomes the *encoding space*, namely the Hilbert space where we can encode the quantum information (e.g. storing one side of a bipartite entangled state), in such a way that we can then correct the errors due to the lack of isolation from the environment – the main practical problem in building a quantum computer. For this reason, in the following we will often say "correcting errors" in place of "inverting the transformation," and use the adjective *correctable* and *invertible* interchangeably.

Consider a state $\rho \in \mathsf{St}(A)$ and a channel $\mathcal{C} \in \mathsf{Transf}_1(A \to B)$. The channel $\mathcal{C}$ is:

1. *correctable* upon input of ρ if there exists a channel $\mathcal{R} \in \mathsf{Transf}_1(B \to A)$ such that

$$\mathcal{R}\mathcal{C} =_\rho \mathcal{I}_A;$$

2. a *deletion channel* upon input of ρ if there is a fixed state $\sigma \in \mathsf{St}_1(B)$ such that $\mathcal{C} =_\rho |\sigma)_B(e|_A$;

3. *purification-preserving* for ρ if there is a recovery channel $\mathcal{R} \in \mathsf{Transf}_1(B \to A)$ such that $\mathcal{R}\mathcal{C}|\Psi_\rho)_{AR} = |\Psi_\rho)_{AR}$, with Ψ_ρ arbitrary purification of ρ;

4. *correlation-erasing* for $\rho \in \mathsf{St}(A)$ if there is a state $\sigma \in \mathsf{St}(B)$ such that $\mathcal{C}|\Psi_\rho)_{AR} = |\sigma)_B|\rho^\#)_R$, with Ψ_ρ arbitrary purification of ρ, and $\rho^\#$ the complementary state $|\rho^\#)_R :=(e|_A|\Psi_\rho)_{AR}$.

Notice that Definitions 3 and 4 can be restated for a single purification of ρ, since if they hold for one purification they automatically hold for any one.

In quantum theory the interplay between the above four definitions provides the basic underlying structure of error correction. We recall that definition 1 means that $\mathcal{C}$ can be inverted on all states having support contained in $\mathsf{Supp}\,\rho$, and according to Theorem 2.14 this is equivalent to saying that $\mathcal{R}\mathcal{C}$ leaves any purification of ρ invariant when applied locally on the system. This gives the following corollaries.

Corollary 2.20 *A channel is correctable upon input of ρ if and only if it is purification-preserving for ρ.*

Corollary 2.21 *A channel is correlation-erasing for ρ iff it is a deletion channel upon input of ρ.*

The above Corollaries 2.20 and 2.21 are immediate consequence of the definitions, along with Theorem 2.14.

In the following, for an isometric purification V of the channel C, we call *complementary channel* $C^{\#}$ the channel obtained upon tracing-out the system instead of the environment, namely

$$\begin{array}{c} \underset{A}{\longrightarrow}\boxed{C^{\#}}\underset{E}{\longrightarrow} \;=\; \underset{A}{\longrightarrow}\boxed{V}\begin{array}{l}\overset{E}{\longrightarrow}\\ \underset{B}{\longrightarrow}\!\!\!\;\overline{e)}\end{array} \end{array} ,$$

where $V = V \cdot V^{\dagger}$, $V \in \mathrm{Lin}(\mathcal{H}_A \to \mathcal{H}_B \otimes \mathcal{H}_E)$ isometry. In equations:

$$C(\rho) := \mathrm{Tr}_E[V\rho V^{\dagger}] \iff C^{\#}(\rho) := \mathrm{Tr}_B[V\rho V^{\dagger}].$$

The complementary channel is a unique modulo isometric mapping of a minimal environment E into a new environment E$'$ with $\mathcal{H}'_E \supseteq \mathcal{H}_E$.

Another simple fact about error correction, which follows from purification, is the following.

Lemma 2.22 *If a quantum channel $C \in \mathsf{Transf}_1(\mathrm{A} \to \mathrm{B})$ is correctable upon input of $\rho \in \mathsf{St}(\mathrm{A})$ with recovery channel $\mathcal{R}$, and $\mathcal{D} \in \mathsf{RefSet}\, C$ is a quantum operation in the refinement set of C, then $\mathcal{D}$ is correctable upon input of ρ, with recovery channel $\mathcal{R}$, i.e. $\mathcal{R}\mathcal{D} =_\rho p\mathcal{I}_A$ for some probability $p > 0$. Also, any quantum operation $\mathcal{F}$ in the refinement set of $\mathcal{R}$ corrects the channel.*

Proof By definition, since $\mathcal{D}$ is in the refinement set of C, there is a quantum instrument $\{C_i\}_{i \in \mathrm{X}}$ such that $\mathcal{D} \equiv C_{i_0}$ and $C = \sum_{i \in \mathrm{X}} C_i$. Since C is correctable with recovery channel $\mathcal{R}$, one has $\mathcal{I}_A =_\rho \mathcal{R}C = \sum_{i \in \mathrm{X}} \mathcal{R}C_i$. This means that the test $\{\mathcal{R}C_i\}_{i \in \mathrm{X}}$ is non-disturbing upon input of ρ. By the "no information without disturbance" property (Theorem 2.15) one then has $\mathcal{R}C_i =_\rho p_i\mathcal{I}_A$ for every $i \in \mathrm{X}$. A similar reasoning can be applied to refinements of $\mathcal{R}$ instead of C. $\square$

2.14.1 Complementarity between Correctable and Deletion Channels

We now discuss some necessary and sufficient conditions for correctability of channels. The simplest case is that of a channel from a system to itself: in such case Corollary 2.10 asserts that a quantum channel is correctable if and only if it is unitary. We will now provide necessary and sufficient conditions for error correction in the general case of channels in $\mathsf{Transf}_1(\mathrm{A} \to \mathrm{B})$. This will be reconsidered again in Chapter 8.

We first need a simple lemma.

Lemma 2.23   *Two purifications* $\Psi_\rho \in \mathsf{St}(AB)$ *and* $\Psi'_\rho \in \mathsf{St}(AB')$ *of the same mixed state* $\rho \in \mathsf{St}(A)$ *are always connected by a quantum channel* $\mathcal{C} \in \mathsf{Transf}(B \to B')$ *as follows*

$$(2.94)$$

Proof   One can make the two purifying systems equal by picking two pure states $\alpha \in \mathsf{St}_1(B)$ and $\alpha' \in \mathsf{St}_1(B')$ to compose in parallel with the purifications obtaining

$$|\Phi_\rho\rangle_{ABB'} = |\Psi_\rho\rangle_{AB} \otimes |\alpha'\rangle_{B'}, \qquad |\Phi'_\rho\rangle_{ABB'} = |\Psi'_\rho\rangle_{AB'} \otimes |\alpha\rangle_B,$$

and since they are purifications of the same state they are connected by a reversible transformation according to the purification principle, namely

$$|\Phi'_\rho\rangle_{ABB'} = |\Psi'_\rho\rangle_{AB'} \otimes |\alpha\rangle_B = (I_A \otimes U_{BB'})|\Phi_\rho\rangle_{ABB'} = |\Psi_\rho\rangle_{AB} \otimes |\alpha'\rangle_{B'},$$

with $U \in \mathsf{Lin}(BB')$ unitary. Diagrammatically we have

$$(2.95)$$

with $\mathcal{U} = U_{BB'} \cdot U_{BB'}^\dagger$. By tracing out B on both sides, using Eq. (2.75) we then obtain $|\Psi'_\rho\rangle = (\mathcal{I}_A \otimes \mathcal{C})|\Psi_\rho\rangle$, where $\mathcal{C}$ is the quantum channel defined by

$$(2.96)$$

$\square$

Exercise 2.95   Write the explicit operator form of Eq. (2.95), including the trace over the environment B.

We are now in position to prove a relevant theorem for quantum error correction.

Theorem 2.24 (Factorization of Reference and Environment)   *A quantum channel* $\mathcal{C} \in$ $\mathsf{Transf}_1(A \to B)$ *is correctable upon input of* ρ *if and only if there are purifications* $\mathcal{V} \in \mathsf{Transf}(A \to BE)$ *of* $\mathcal{C}$ *and* $|\Psi_\rho\rangle\!\rangle_{AR}$ *of* ρ *such that environment* E *and reference* R *remain uncorrelated after application of* $\mathcal{V}$. *Diagrammatically,*

$$(2.97)$$

where σ *is some state of* E *and* $\rho^\#$ *is the complementary state of* ρ *on system* R.

Proof Suppose that $\mathcal{C}$ is correctable upon input of ρ with some recovery channel $\mathcal{R}$. Then, by Corollary 2.20 we have

and, inserting two arbitrary purification schemes for $\mathcal{C}$,

This means that $\mathcal{R}V|\Psi_\rho\rangle\!\rangle_{AR}$ has $|\Psi_\rho\rangle\!\rangle_{AR}$ as marginal state. Therefore, since $|\Psi_\rho\rangle\!\rangle$ is pure, the only way to obtain it as marginal is by tensor product with a state of E, namely one has

where σ is some state on E. Applying the deterministic effect on A and using the fact that $\mathcal{R}$ is a channel, we then obtain Eq. (2.97).

Conversely, suppose that Eq. (2.97) holds for some purification $\mathcal{V} = V \cdot V^\dagger$ and some purification $|\Psi\rangle\!\rangle_{AR}$. Then take a purification of σ, say $\Psi_\sigma \in \mathsf{St}_1(EF)$. Since $V|\Psi_\rho\rangle\!\rangle_{AR}$ and $|\Psi_\rho\rangle\!\rangle_{AR}|\Psi_\sigma\rangle\!\rangle_{EF}$ are both purifications of $\rho_R^\# \otimes \sigma_E$, according to Lemma 2.23 they must be connected by a channel connecting the two environments, namely

for some channel $\mathcal{D} \in \mathsf{Transf}_1(B \to FA)$. Upon discarding the environments EF, the last diagram is equivalent to the correctability condition for the channel over any purification of ρ, namely upon input of ρ. □

In Section 8.6 we will see that since the statement that "a purification of a pure state is the tensor product with another pure state" holds for any operational probabilistic theory that is causal and obeys local discriminability, the above theorem holds more generally than in quantum theory, and is a direct consequence only of these two postulates along with purification.

Corollary 2.25 (Necessary Condition for Invertibility of Channels) *A necessary condition for correctability of channel $\mathcal{C} \in \mathsf{Transf}_1(A \to B)$ upon input of ρ is*

$$\dim \mathcal{H}_B \geq \operatorname{rank} \rho \, \operatorname{rank} \mathcal{C}^\#(\rho). \tag{2.98}$$

Proof Invertibility of the channel is equivalent to have reference and environment uncorrelated at the output of a purification of the channel, which means that their joint state ρ_{ER} is factorized, hence one has $\operatorname{rank} \rho_{ER} = \operatorname{rank} \rho_E \operatorname{rank} \rho_R$. On the other hand, the joint state of system BRE at the channel output remains pure, hence system B provides a purification refence system for RE. This means that one should have

$$\dim \mathcal{H}_B \geq \operatorname{rank} \rho_{ER} = \operatorname{rank} \rho_E \operatorname{rank} \rho_R$$

But $\operatorname{rank} \rho_R = \operatorname{rank} \rho$, and, according to Eq. (2.72) $\operatorname{rank} \rho_E$ is the rank of the output of the complementary channel, namely $\operatorname{rank} \mathcal{C}^\#(\rho)$. $\square$

Corollary 2.26 (Complementarity of Purification-Preserving and Correlation-Erasing Channels)   *A channel $\mathcal{C} \in \mathsf{Transf}_1(A \to B)$ is purification-preserving for $\rho \in \mathsf{St}(A)$ if and only if its complementary channel $\mathcal{C}^\# \in \mathsf{Transf}_1(A \to E)$ is correlation-erasing for ρ.*

Proof   By Corollary 2.20, $\mathcal{C}$ is purification-preserving for ρ iff it is correctable upon input of ρ and, by Theorem 2.24, iff Eq. (2.97) holds. But Eq. (2.97) by Theorem 2.14 implies that $\mathcal{C}^\#$ is a correlation-erasing channel. $\square$

Corollary 2.27 (Complementarity of Correctable and Deletion Channels)   *A quantum channel $\mathcal{C} \in \mathsf{Transf}_1(A \to B)$ is correctable upon input of $\rho \in \mathsf{St}(A)$ if and only if its complementary channel $\mathcal{C}^\# \in \mathsf{Transf}_1(A \to E)$ is a deletion channel upon input of ρ.*

Proof   Direct consequence of Theorem 2.24 and Corollaries 2.26 and 2.21. $\square$

Theorem 2.28   *The following are equivalent conditions for the channel $\mathcal{C}$ to be correctable upon input of ρ:*

1. *$\mathcal{C}$ is correctable by the same channel $\mathcal{R}$ upon any input state σ with $\mathsf{Supp}\, \sigma \subseteq \mathsf{Supp}\, \rho$, and when acting locally on any purification of σ.*
2. *The same channel $\mathcal{R}$ inverts probabilistically all quantum operations with Kraus operators that are linear combinations of the Kraus operators of $\mathcal{C}$. Vice versa, all quantum operations with Kraus operators that are linear combinations of the Kraus operators of $\mathcal{R}$ invert probabilistically the same channel $\mathcal{C}$.*
3. *The complementary channel of $\mathcal{C}$ applied to a state σ with $\mathsf{Supp}\, \sigma \subseteq \mathsf{Supp}\, \rho$ deterministically prepares a fixed state.*
4. *[Knill–Laflamme] The following conditions hold*

$$PE_i^\dagger E_j P = \alpha_{ij} P, \tag{2.99}$$

with $\{E_i\}$ Kraus operators for $\mathcal{C}$, $P = \mathsf{Proj}\,\mathsf{Supp}\, \rho$, and α deterministic quantum state.

Proof

1. This is just the definition of channel correctable upon input of ρ.
2. For any Kraus decomposition $\{\mathcal{C}_i\}_{i \in X}$ of $\mathcal{C}$ and $\{\mathcal{R}_j\}_{j \in Y}$ of $\mathcal{R}$ the test $\{\mathcal{R}_j \mathcal{C}_i\}_{i \in X, j \in Y}$ is non-disturbing upon input ρ, and Theorem 2.15 (no information without disturbance) guarantees that one has $\mathcal{R}_j \mathcal{C}_i =_\rho p_{ij} \mathcal{I}_A$, p_{ij} a probability, namely any quantum operation

in the refinement set of the channel C is corrected probabilistically upon input of ρ by any quantum operation in the refinement set of R. According to Exercise 2.49 any quantum operation in the refinement set of a given quantum operation has Kraus operators that are linear combination of its Kraus.

3. This is just Corollary 2.27.

4. Suppose that identity (2.99) holds. Corresponding to a Kraus decomposition $\{E_j\}$ for the channel C, the complementary channel is given by

$$C^{\#}(\sigma) := \mathrm{Tr}_B[V\sigma V^{\dagger}],$$

where $V \in \mathrm{Lin}(\mathcal{H}_A \to \mathcal{H}_B \otimes \mathcal{H}_E)$ is the isometry $V = \sum_j E_j \otimes |j\rangle$. For a state σ on $\mathsf{Supp}\,\rho$ one has

$$C^{\#}(\sigma) = \sum_{ij} \mathrm{Tr}[E_i\sigma E_j^{\dagger}]|i\rangle\langle j| = \sum_{ij} \mathrm{Tr}[\sigma P E_j^{\dagger} E_i P]|i\rangle\langle j| = \mathrm{Tr}[\sigma]\sum_{ij} \alpha_{ij}|i\rangle\langle j|,$$

where P projects over $\mathsf{Supp}\,\rho$, and we have used the Knill–Laflamme conditions (2.99). Therefore, the map $C^{\#}$ is the deletion channel preparing the state $\alpha = \sum_{ij} \alpha_{ij}|i\rangle\langle j|$ when applied to any state on $\mathsf{Supp}\,\rho$, namely

$$C^{\#}(\sigma) = \mathrm{Tr}[\sigma]\alpha, \quad \forall \sigma : \mathsf{Supp}\,\sigma \subseteq \mathsf{Supp}\,\rho. \tag{2.100}$$

Vice versa, if the ancillary map of the channel C satisfies Eq. (2.100), then, upon applying $C^{\#}$ to $|\psi\rangle\langle\varphi|$ with $|\psi\rangle, |\varphi\rangle \in \mathsf{Supp}\,\rho$ one obtains

$$\langle\varphi|E_j^{\dagger}E_i|\psi\rangle = \langle\varphi|\psi\rangle\alpha_{ij},$$

which is the Knill–Laflamme condition.

$\square$

Notice that the error correction or channel inversion can be always performed without performing measurements, via a unitary interaction with an ancilla, as in Eq. (2.74).

Exercise 2.96 Using the Knill–Laflamme condition, prove that the only channels that are invertible over $\mathcal{H}$ are the unitary ones, and in that case the state α is pure.

Exercise 2.97 Show by direct calculation that the Knill–Laflamme conditions (2.99) imply that the reference and the environment remain uncorrelated.

Exercise 2.98 Consider the unitary matrix $U = \{u_{ij}\}$ diagonalizing the matrix α in the Knill–Laflamme condition, namely $U^{\dagger}\alpha U = \mathsf{Diag}\,[d_j]$, and take the new Kraus operators $F_n = \sum_j u_{jn}E_j$. Show that the operators $V_j \in \mathrm{Lin}(\mathsf{Supp}\,\rho \to \mathcal{H})$ defined as

$$V_j := \frac{F_j P}{\sqrt{d_j}},$$

are partial isometries over $\mathsf{Supp}\,\rho$ having orthogonal ranges, namely

$$V_i^{\dagger}V_j = \delta_{ij}P.$$

Therefore, a channel is correctable upon input of ρ if it admits a Kraus form with Kraus operators that (apart from a multiplicative factor that is the square root of a probability) act on $\mathsf{Supp}\,\rho$ as isometries with orthogonal ranges.

Exercise 2.99 Show that for any ρ on Hilbert space $\mathcal{H}$ there always exists a channel $\mathcal{C}$ with rank k that is correctable upon input of ρ, and satisfying condition rank ρ rank $\mathcal{C} \leq \dim(\mathcal{H})$.

In Exercise 2.99 the error-correcting code is called *non-degenerate* if α is full-rank, otherwise it is called *degenerate*.

2.14.2 Correction via Classical Communication with Environment

In this section we briefly discuss a more general kind of correction, in which the environment is not completely unaccessible, but rather some operations on it are allowed. In particular we consider error-correction in a scenario where one is allowed to measure the environment and then to perform transformations on the system conditioned on the outcomes. For simplicity we will focus here on the case of a single round of forward classical communication from the environment to the output system.

Environment-Correctable Channels A channel $\mathcal{C} \in \mathsf{Transf}_1(A \to B)$ is *environment-correctable* upon input of ρ if for every purification $\mathcal{V} \in \mathsf{Transf}(A \to BE)$ there is an observation test $\{a_i\}_{i \in X}$ on E and a collection $\{\mathcal{R}_i\}_{i \in X} \subset \mathsf{Transf}(B \to A)$ of recovery channels such that

$$\sum_{i \in X} \quad \underset{A}{\overset{}{\boxed{\mathcal{V}}}} \begin{matrix} \overset{E}{\underline{\quad}}\boxed{a_i} \\ \overset{B}{\underline{\quad}}\boxed{\mathcal{R}_i}\overset{A}{\underline{\quad}} \end{matrix} \quad =_\rho \quad \overset{A}{\underline{\quad}}\boxed{\mathcal{I}}\overset{A}{\underline{\quad}} \ .$$

If ρ is a completely mixed state, we simply say that $\mathcal{C}$ is *environment correctable*.

The following theorem states that environment-correctable channels are nothing but randomizations of correctable channels.

Theorem 2.29 (Characterization of Environment-Correctable Channels) *A channel $\mathcal{C} \in \mathsf{Transf}_1$ $(A \to B)$ is environment-correctable upon input of $\rho \in \mathsf{St}_1(A)$ if and only if $\mathcal{C}$ is the coarse-graining of a test $\{\mathcal{C}_i\}_{i \in X}$ where each transformation $\mathcal{C}_i$ is correctable upon input of ρ. In particular, if ρ is completely mixed, then $\mathcal{C}$ is a randomization of invertible channels.*

Proof Suppose that $\mathcal{C}$ is environment-correctable upon input of ρ. Then, for any purification $|\Psi_\rho\rangle_{AR}$ of ρ we have

$$\sum_{i \in X} \quad \boxed{\Psi_\rho} \begin{matrix} \overset{A}{\underline{\quad}}\boxed{\mathcal{V}}\begin{matrix}\overset{E}{\underline{\quad}}\boxed{a_i}\\ \overset{B}{\underline{\quad}}\boxed{\mathcal{R}_i}\overset{A}{\underline{\quad}}\end{matrix} \\ \underline{\qquad\qquad\qquad}_R \end{matrix} \quad = \quad \boxed{\Psi_\rho}\begin{matrix}\overset{A}{\underline{\quad}}\\ \underline{\quad}_R\end{matrix} \ .$$

Since Ψ_ρ is pure, each term in the sum must be proportional to it. Defining the test $\{\mathcal{C}_i\}_{i \in X}$ by $\mathcal{C}_i := (a_i|_E \mathcal{V}$, and using Theorem 2.14, we then obtain $\mathcal{R}_i \mathcal{C}_i =_\rho p_i \mathcal{I}_A$. Therefore, $\mathcal{C}$ is the coarse-graining of a test where each transformation is correctable upon input of ρ. Moreover, if ρ is completely mixed, using the fact that each $\mathcal{R}_i$ is a channel, we obtain

$$(e|_A \mathcal{R}_i \mathcal{C}_i = (e|_B \mathcal{C}_i = p_i(e|_A,$$

namely each C_i must be proportional to a channel, say $C_i = p_i D_i$, with channel D_i correctable upon input of ρ. Conversely, suppose that $C = \sum_{i \in X} C_i$ for some test $\{C_i\}$ where each transformation C_i is correctable upon input of ρ. Dilating such a test, we then obtain a channel $V \in \mathsf{Transf}(A \to BE)$ and an observation test $\{a_i\}_{i \in X}$ on E such that

$$\begin{array}{c} \underset{A}{\quad} \boxed{V} \genfrac{}{}{0pt}{}{\overset{E}{\longrightarrow} \boxed{a_i}}{\underset{B}{\longrightarrow}} \end{array} = \quad \overset{A}{\longrightarrow} \boxed{C_i} \overset{B}{\longrightarrow} \;,$$

for every outcome $i \in X$. Since each C_i is correctable upon input of ρ, knowing the outcome $i \in X$, we can perform the recovery channel for C_i, thus correcting channel C. In the special case when ρ is completely mixed, all channels C_i must be invertible, hence proportional to a unitary channel (see Corollary 2.12). □

It is worth noticing that all results of the present subsection are just consequences of the preceding results about error correction, which, as already noticed, follow only from causality, local discriminability, and purification. Therefore the results of the present subsection more generally hold for any operational probabilistic theory satisfying these principles (see Section 8.6).

2.15 Summary

In this chapter we have introduced the mathematical structure of quantum theory for a finite number of finite-dimensional systems, starting from three simple Hilbert-space postulates within a causal context. We have derived in form of theorems the six principles which will be taken as axioms for the derivation of the theory in Part IV of the book. The two-way derivation will establish the complete equivalence between quantum theory and the six Principles, within the informational framework given in Chapter 3. In the present chapter we have derived and illustrated the general mathematical structure of the theory, along with its most relevant results and theorems, including some from the theory of open quantum systems and quantum information theory.

Notes

Probabilistic and Deterministic States   It is customary in courses of quantum mechanics to take states as "normalized," with the density operator having unit trace, or with unit vectors representing pure states. Here we more conveniently take the state as subnormalized, since in this way the density operator ρ contains also the information about the probability of preparation of the state as $p(\rho) = \mathrm{Tr}[\rho]$.

Infinite Dimensions, Infinitely Many Systems   In this book we consider only Hilbert spaces with finite dimensions. In quantum mechanics a system describing a particle in space has an infinite-dimensional Hilbert space. In such a case we require the Hilbert space to be *separable*, namely admitting a countable orthonormal basis. This is a very important request, since it guarantees the existence of an expansion over a countable orthonormal basis. By definition, all infinite-dimensional separable Hilbert spaces are isometrically isomorphic to $\ell^2(\mathbb{Z})$. It is sometimes argued that non-separable Hilbert spaces also are of interest, e.g. in quantum field theory, when infinite tensor products of Hilbert spaces need to be considered. However, such a situation is more conveniently addressed within the framework of von Neumann algebras.

One of the main differences between finite-dimensional and infinite-dimensional Hilbert spaces is the fact that an infinite-dimensional Hilbert space $\mathcal{H}$ can be isomorphic to a proper subspace $\mathcal{K} \subsetneq \mathcal{H}$. This also leads to the fact that an operator that is isometric on $\mathcal{H}$ is not necessarily unitary, in stark contrast with the finite dimensional case. The canonical example of a non-unitary isometry is the *shift operator* $S = \sum_{i=1}^{\infty} |n+1\rangle\langle n|$, for $\{|n\rangle\}_{n=1,\dots,\infty}$ orthonormal basis for $\mathcal{H}$. The operator S is isometric, since $S^\dagger S = I$, but not unitary, since $SS^\dagger = I - |1\rangle\langle 1|$.

Banach Spaces of Operators For dim $\mathcal{H} = \infty$ we have additional convergence requirements for the spaces of states $\mathsf{St}_{\mathbb{R}}(A)$ and effects $\mathsf{Eff}_{\mathbb{R}}(A)$, the former required to be trace-class, and the latter required to be bounded. While in finite-dimensions both spaces are just linear spaces of operators $\mathsf{St}_{\mathbb{R}}(A) \simeq \mathsf{Eff}_{\mathbb{R}}(A) \simeq \mathsf{Lin}(\mathcal{H}_A)$, for dim $\mathcal{H}_A = \infty$ the two spaces become non-isomorphic Banach spaces, since $\mathsf{St}_{\mathbb{R}}(A) \simeq \mathsf{T}_1(\mathcal{H}_A)$ are trace-class operators, and $\mathsf{St}_{\mathbb{R}}(A) \simeq \mathsf{Bnd}(\mathcal{H}_A)$ are bounded operators. In the infinite-dimensional setting, the space of effects is still dual to the space of states, namely $\mathsf{Bnd}(\mathcal{H}_A) = \mathsf{T}_1(\mathcal{H}_A)^\vee$, but the converse is not true anymore: one has $\mathsf{T}_1(\mathcal{H}_A) \supsetneq \mathsf{Bnd}(\mathcal{H}_A)^\vee$. We say that $\mathsf{T}_1(\mathcal{H}_A)$ is the *predual* of $\mathsf{Bnd}(\mathcal{H}_A)$. Notice also that $\mathsf{Bnd}(\mathcal{H}_A)$ is a Banach algebra, a structure that $\mathsf{T}_1(\mathcal{H}_A)$ lacks. The phenomenon of the predual is due to the fact that generally on an infinite-dimensional Banach space $\mathcal{B}$ one has linear functionals $f : \mathcal{B} \to \mathbb{C}$ that cannot be written as scalar product with a vector in another Banach space. It is customary to denote by X^* the dual of X, and by X_* the predual of X. In general, if $\mathsf{X} = \mathsf{Y}_*$, then $\mathsf{Y} = \mathsf{X}^*$, and one has $\mathsf{X} = \mathsf{Y}_* \subseteq \mathsf{Y}^* = \mathsf{X}^{**}$. A typical example is that of the Banach space of the bounded sequences ℓ_∞ and the functionals made with the Banach space of summable sequences ℓ_1. One has that $\ell_\infty = \ell_1^*$, but $\ell_1 = \ell_{\infty*} \subsetneq \ell_\infty^*$. However, it should be noticed that without the axiom of choice there are models in which $\ell_1 = \ell_\infty^*$.[30]

The Original Notions of POVM and PVM for Continuum Probability Space The name *positive-operator valued measure* (POVM) historically originated from considering observation tests with outcomes in a general probability space Ω. In such case the events are associated with elements of the σ-algebra $\sigma(\Omega)$ associated to the probability space Ω. Then, the POVM is defined as a map $P : \Delta \in \sigma(\Omega) \mapsto P_\Delta \in \mathsf{Lin}(\mathcal{H}_A)$ satisfying the requirements:

1. $P_\Delta \geq 0$ for all $\Delta \in \sigma(\Omega)$;
2. $P_\emptyset = 0$ and $P_\Omega = I$;

---

[30] This is the case of the Solovay model (Solovay, 1970).

3. $P_{\cup_n B_n} = \sum_n P_{B_n}$, $\{B_n\}$ disjoint sequence in $\sigma(\Omega)$, and the series converges in the strong operator topology.

Born's rule becomes $p_\Delta = \text{Tr}[\rho P_\Delta]$. A special case of POVM is the PVM (projector-valued measure), in which P_Δ is an orthogonal projector, $P_\Delta^2 = P_\Delta$. This is what is also generally called a *spectral measure*, and corresponds to the spectral resolution of an observable X, which is a self-adjoint operator over $\mathcal{H}$, i.e. $X = \int_{\text{Sp}(X)} x P(\mathrm{d}x)$ and $P_\Delta = \int_\Delta P(\mathrm{d}x)$. The notions of POVM and PVM were introduced for dealing with the general situation of continuum probability space, and the notion of *observable* was historically associated to the measurement of a classical mechanical variable, such as the position or the momentum. The connection of the self-adjoint operator with the classical variable was provided by the so-called *quantization rules*. Later, the notion of POVM became popular in the context of quantum estimation theory (Helstrom, 1976; Holevo, 1982), for measuring parameters that do not have a corresponding observable, such as a phase-shift.

Similarly to the case of the POVM and PVM, also for the quantum test, or quantum instrument, we can consider continuum-outcome space Ω, *quantum instrument* associating events with elements of a σ-algebra $\sigma(\Omega)$ of Ω. Then, the quantum instrument is defined as a map $\mathcal{T}$ from $\Delta \in \sigma(\Omega)$ to the completely positive maps transforming trace class operators on $\mathcal{H}_A$ and satisfying the following requirements:

1. $\mathcal{T}_\Delta$ is completely positive and trace-non-increasing for all $\Delta \in \sigma(\Omega)$;
2. $\mathcal{T}_\emptyset = 0$ and $\mathcal{T}_\Omega$ is trace-preserving (i.e. a quantum channel);
3. $\mathcal{T}_{\cup_n B_n} = \sum_n \mathcal{T}_{B_n}$, $\{B_n\}$ disjoint sequence in $\sigma(\Omega)$, and the series converges ultra-weakly.

The realization of such a general quantum instrument in terms of an indirect von Neumann–Lüders measurement has been proven in Ozawa (1984). This theorem is especially interesting for infinite dimensions. Indeed, for finite dimension one can prove that any quantum instrument with continuum-outcome space can be achieved as the continuum convex combination of discrete-outcome atomic instruments (Chiribella *et al.*, 2007). The proof is relatively easy for compact outcome space, whereas it becomes rather technical for non-compact ones (Chiribella *et al.*, 2010b). The convex structure of POVMs can be found in Parthasarathy (1999) and D'Ariano *et al.* (2005).

A final comment about historical nomenclature related to purification. In the literature the words *extension* and *dilation* are used for what we now call *purification*, for instruments and POVM. We need, however, to keep the nomenclature extension/dilation whenever it refers to a quantum operation/effect that is not atomic. Moreover, the world *dilation* is more appropriately used for Hilbert-space extensions that are not tensor products, but rather direct sum enlargements.

Informationally Complete POVM and Quantum Tomography   The notion of *informationally complete POVM*, after the first appearence in specialized books (Busch *et al.*, 1991), became popular with the birth of quantum tomography, which was boosted by the practical feasibility in quantum optical labs using homodyne detection. The homodyne method first came out just as an application of the usual back-projection in X-ray tomography (Vogel and Risken, 1989; Smithey *et al.*, 1993), before the first analytical method appeared in D'Ariano *et al.* (1994). Here the analytics are made much more complicated due to the inifinite dimension. There is an extensive literature on the subject, and here we just mention the book by Paris and Rehacek (2004). The ancilla assisted tomography method first came out in D'Ariano and Lo Presti (2001); Leung (2001), and later gave rise to the notion of faithful state (D'Ariano and Lo Presti, 2003) that has been considered in preliminary studies of informational axiomatization of quantum theory (D'Ariano, 2006a,b, 2007a,b, 2010). For the reader interested in quantum estimation theory and optimization of apparatus design, an example is provided by a thorough method to optimize tomography of an apparatus (Bisio *et al.*, 2009a,b), based on the use of quantum combs (Chiribella *et al.*, 2008a) that are not considered in this book (for a thorough review the reader is addressed to Bisio *et al.*, 2012).

Quantum tomography can be regarded as a universal kind of measurement (D'Ariano, 1997, 2005), and as such it can be used for estimating the ensemble average of any complex operator. The optimal data-processing for such a measurement can be in found D'Ariano and Perinotti (2007).[31]

The Double-ket Notation in Infinite Dimensions   The isomorphism **Vec** introduced in Section 2.8.2 first appeared in D'Ariano *et al.* (2000). In infinite dimensions it is just the isomorphism $\mathsf{HS}(\mathcal{K} \to \mathcal{H}) \simeq \mathcal{H} \otimes \mathcal{K}$ between the Hilbert space of Hilbert–Schmidt operators from $\mathcal{K}$ to $\mathcal{H}$ and the Hilbert space $\mathcal{H} \otimes \mathcal{K}$. In infinite dimensions the vector $|I\rangle\!\rangle_{\mathcal{K} \otimes \mathcal{K}}$ remains a notational tool to express isomorphism **Vec** (the vector is a linear functional on $\mathsf{HS}(\mathcal{K} \to \mathcal{H})$: see also Holevo (2011)). For infinitely many systems, in the general context of von Neumann and C*-algebras the correspondence (2.52) is a special case of the GNS construction of representations of C*-algebras (Haag and Haag, 1996).

About the No-cloning Theorem   The no-cloning theorem is universally considered to be the start of the new field of quantum information. Indeed, the impossibility of cloning non-orthogonal states is the main idea underlying quantum cryptography, and moreover, the no-cloning theorem prevents the use of redundancy error correction techniques on quantum states. Thus the theorem is conceptually involved in the two main tasks of quantum information: quantum cryptography and quantum error correction. The theorem is then used in a number of different results. The version of the no-cloning theorem in Theorem 2.17 is a variant of the proof of Yuen (1986). The first proof was originally given in Dieks (1982) and Wootters and Zurek (1982), where it has been shown that cloning violates linearity of quantum mechanics.[32] The problem of cloning was raised by Herbert (1982), where a protocol for superluminal (actually instantaneous!) communication was

---

[31] When estimating the ensemble average of an observable on an ensemble of equally prepared identical quantum systems, it is not obvious that among all kinds of measurements performed jointly on the copies, the optimal unbiased estimation is achieved by the usual procedure, namely: performing independent measurements of the observable on each system and averaging the measurement outcomes. The proof is non-trivial and can be found in D'Ariano *et al.* (2006).

[32] The proof by linearity of Wootters and Zurek (1982) applies to a minimum total number of *three* states, where the orthogonality proof of Yuen (1986) shows that any *two* non-orthogonal states cannot be cloned.

proposed, based on local measurements on two entangled systems, and using stimulated laser emission as a cloning device at the receiver (for an historical review on the many attempts of performing superluminal communication with similar setups the reader is addressed to Kaiser (2011)). The no-cloning theorem, however, does not prove the impossibility of communicating without interaction and using only locally performed operations, since the error due to non-ideal cloning can be in principle corrected by error-correcting techniques. Apart from some arguments against the possibility of a communication by local operations on an entangled state which appeared in Eberhard (1978), the first thorough impossibility proof appeared in Ghirardi *et al.* (1980), while in relation to non-ideal quantum cloning as in the original Herbert proposal we recall Bruss *et al.* (2000). As will be shown in Chapter 5, the impossibility of communication without interaction using a shared entangled state is just an immediate consequence of the causality principle.

About the Impossibility of Determining the State of a Single Quantum System
The equivalence of the no-cloning theorem with the impossibility of determining the state of a single quantum system proposed in Corollary 2.18 has been established in D'Ariano and Yuen (1996), after a sequel of attempts by several authors exploring concrete measurement schemes based on vanishingly weak quantum non-demolition measurements (Alter and Yamamoto, 1995), weak measurements on "protected" states (Aharonov and Vaidman, 1993; Aharonov *et al.*, 1993), measurements that are "logically reversible" (Ueda and Kitagawa, 1992), and "physically reversible" (Imamoglu, 1993; Royer, 1994, 1995). In each of these schemes the conclusion is that it is practically impossible to measure the wave function of a single system, either because the weakness of the measuring interaction prevents one from gaining information on the wave function (Alter and Yamamoto, 1995) (which is related to the theorem of no information without disturbance 2.15), or because the method of protecting the state (Aharonov and Vaidman, 1993; Aharonov *et al.*, 1993) actually requires some a priori knowledge on the state (as suggested in Royer, 1994, 1995, and Alter and Yamamoto, 1995), or because quantum measurements can be physically inverted only with vanishingly small probability of success (Royer, 1994, 1995).

Choi–Jamiołkowski Isomorphism in Infinite Dimensions   The Choi–Jamiołkowski cone isomorphism is based on the application of the CP map of the quantum operation to the rank-one operator $|I\rangle\!\rangle\langle\!\langle I|$, which, apart from a factor $\frac{1}{d}$, is a maximally entangled state. The original papers (Choi, 1972, 1975; Jamiolkowski, 1972) were in finite dimensions and the two isomorphisms defined by Choi and Jamiołkowski differ by a partial transposition, which made the Jamiołkowski operator of a CP map non-positive. The Choi–Jamiołkowski operator in Eq. (2.55) in infinite dimensions can be rigorously interpreted as a positive semidefinite form over an appropriate dense subspace (Holevo, 2011). An alternative approach to the Choi–Jamiołkowski isomorphism in infinite dimension was proposed by Belavkin and Staszewski (Belavkin and Staszewski, 1986).

About the von Neumann Postulate   Postulating the von Neuman projection in a quantum measurement is still a popular mantra, which seems to require the projection as a necessary ingredient of the measurement of an observable. This is misleading, since as seen in Section 2.12 the projection postulate is needed only for the state of the instrument pointer (which von Neumann ultimately identified with the observer consciousness). Ozawa (1997) advocated this point, along with the possibility of using the Born rule to determine the state of the measured system at the output of the measurement, corresponding to the customary notion of conditional state.

Other Historical Notes and Literature The original reference of the quantum error-correction theory that includes Eq. (2.99) is Knill and Laflamme (1997). The iff condition of Theorem 2.24 has been presented in the quantum case in Schumacher (1996), Schumacher and Nielsen (1996), and Barnum *et al.* (1998). The possibility of correcting a channel using information from the environment was first presented in Gregoratti and Werner (2003). Such a kind of error correction is always possible only for qubits and qutrits, as shown in Buscemi *et al.* (2005), whereas for dimension $d > 3$ there are situations where recovery is impossible, even with complete access to the environment. All general theorems in this chapter that do not use Hilbert spaces but just the operational framework are derived from Chiribella *et al.* (2010a, 2011).

Historical Notes and Literature About Some Exercises and Problems Regarding Problem 2.32, Gorini, Kossakowski, and Sudarshan (1976) have shown that Eq. (2.114) defines the most general generator of a quantum dynamical semigroup for the case of a finite-dimensional Hilbert space $\mathcal{H}$. Independently Lindblad (1976) proved that Eq. (2.114) is the most general bounded operator for any separable Hilbert space if one admits countable set of indices and $\sum_k V_k^\dagger V_k \in \mathsf{Bnd}(\mathcal{H})$. There exists no such characterization for the case of unbounded generator $\mathcal{L}$. However, there is a class of formal expressions (2.114) which rigorously define quantum dynamical semigroups (Davies, 1977). Moreover, all known examples of semigroup generators are of the standard form (2.114) with (possibly unbounded) operators V_k, or even more singular objects, such as field operators, and with the sums that might be replaced by integrals (see also Alicki and Lendi, 1987).

Appendix 2.1 Polar Decomposition

Let us start from the polar decomposition of an operator.

Theorem 2.30 *Let T be an operator from the Hilbert space $\mathcal{H}$ to the Hilbert space $\mathcal{H}'$. Then, T can be written in the polar decomposition as $T = V|T|$, where V is an isometry with $\mathsf{Supp}(V) = \mathsf{Rng}(|T|)$. Such a decomposition is unique.*

Proof We start from the proof of uniqueness, which also provides an explicit construction of the operators $|T|$ and V. Suppose that T can be written as $T = V|T|$ with $|T|$ positive semidefinite and V partial isometry with $\mathsf{Supp}(V) = \mathsf{Rng}(|T|)$. Then we must have $T^\dagger T = |T| V^\dagger V |T| = |T|^2$, having used the fact that $V^\dagger V$ is the projector on $\mathsf{Rng}(|T|)$.

The equation $T^\dagger T = |T|^2$ forces $|T|$ to be the square root of $T^\dagger T$. Explicitly, we can diagonalize $T^\dagger T$ as

$$T^\dagger T = \sum_{i=1}^{r} t_i |t_i\rangle\langle t_i|, \qquad t_i > 0 \,\forall i \in \{1,\dots,r\}. \tag{2.101}$$

Note that the eigenvectors $\{|t_i\rangle \mid i = 1,\dots,r\}$ are a basis for $\mathsf{Rng}\,(|T|)$, since $|T| = \sqrt{T^\dagger T} = \sum_{i=1}^{r} \sqrt{t_i}\,|t_i\rangle\langle t_i|$. Now, the condition $T = V|T|$ implies $T|t_i\rangle = V|T||t_i\rangle = \sqrt{t_i}\,V|t_i\rangle =: \sqrt{t_i}\,|v_i\rangle$, or, equivalently

$$V|t_i\rangle = \frac{T|t_i\rangle}{\sqrt{t_i}} \qquad \forall i \in \{1,\dots,r\}. \tag{2.102}$$

The above condition determines uniquely the action of V on $\mathsf{Rng}\,(|T|)$. Combined with the condition $\mathsf{Supp}\,(V) = \mathsf{Rng}\,(|T|)$, it forces the operator V to have the form $V := \sum_{i=1}^{r} |v_i\rangle\langle t_i|$. Since the vectors $\{|v_i\rangle \mid i = 1,\dots,r\}$ are orthonormal, one has $V^\dagger V = \sum_{i=1}^{r} |t_i\rangle\langle t_i|$ meaning that $V^\dagger V$ is the projector on $\mathsf{Rng}\,(|T|)$. $\qquad\qquad\square$

The polar decomposition yields the *singular value decomposition* of the operator T: using Eqs. (2.101) and (2.102) one obtains

$$T = \sum_{i=1}^{r} \sqrt{t_i}\,|v_i\rangle\langle t_i|, \qquad t_i > 0,\,\forall i \in \{1,\dots,r\}.$$

The quantities $\sqrt{t_i}$ are usually referred to as *singular values* of the operator T and are conventionally ordered in decreasing order, and denoted as $\sigma_i(T)$. The generalization of the polar decomposition to operators on infinite-dimensional Hilbert spaces can be found in Kato (1980) for closed operators. When the operator $T^\dagger T$ has discrete spectrum, the polar decomposition leads to a singular value decomposition for T.

Appendix 2.2 The Golden Rule for Quantum Extensions

In mathematics, an *embedding* is one instance of some mathematical structure contained within another instance of the same structure, such as a group that is a subgroup, or an Hilbert space that is an Hilbert subspace. When some object X is said to be embedded in another object Y, the embedding is given by some injective and structure-preserving map $f : X \to Y$. The precise meaning of "structure-preserving" depends on the kind of mathematical structure of which X and Y are instances. In the terminology of category theory, a structure-preserving map is called a *morphism*. The fact that a map is an embedding is often indicated by the use of a "hooked arrow," as: $f : X \hookrightarrow Y$ (which is generally used for injective maps).

For Hilbert spaces the morphisms are *isometries*, namely injective linear maps that preserve the scalar product. Therefore, an isometric embedding $V : \mathcal{H} \hookrightarrow \mathcal{E}$ of the Hilbert space $\mathcal{H}$ in the Hilbert space $\mathcal{E}$ such that $\mathcal{H} \subseteq \mathcal{E}$ is represented by a linear operator $V \in \mathsf{Lin}(\mathcal{H} \to \mathcal{E})$ such that

$$\langle \psi | V^\dagger V | \varphi \rangle = \langle \psi | \varphi \rangle, \quad \forall | \psi \rangle, | \varphi \rangle \in \mathcal{H},$$

namely V is an isometry from $\mathcal{H}$ to $\mathcal{E}$. Using the singular value decomposition one can see that it is always possible to write an isometry $V \in \mathsf{Lin}(\mathcal{H} \to \mathcal{E})$ as follows:

$$V = \sum_n |\upsilon_n\rangle\langle n|,$$

where $\{|n\rangle\}$ is any orthonormal basis for $\mathcal{H}$ (e.g. the canonical one), and $\{|\upsilon_n\rangle\}$ is an orthonormal set in $\mathcal{E}$. If we now fix the set $\{|\upsilon_n\rangle\}$ in $\mathcal{E}$, we can rewrite all possible isometric embeddings $V : \mathcal{H} \hookrightarrow \mathcal{E}$ as follows:

$$V = U \sum_n |\upsilon_n\rangle\langle n|,$$

where $U \in \mathsf{Lin}(\mathcal{E})$ is a unitary operator. Notice that the unitary operators U are in one-to-one correspondence with the isometries V modulo local unitary transformations over $\mathsf{Rng}\,(V)^\perp$. A special case of isometric embedding is that of a Hilbert space $\mathcal{H}$ into a tensor product $\mathcal{H} \otimes \mathcal{A}$, or more generally into $\mathcal{K} \otimes \mathcal{B}$, with $\dim(\mathcal{H} \otimes \mathcal{A}) = \dim(\mathcal{K} \otimes \mathcal{B})$. In such case we can conveniently choose the orthonormal set

$$|\upsilon_n\rangle = |n\rangle \otimes |0\rangle \in \mathcal{H} \otimes \mathcal{A}$$

$|0\rangle \in \mathcal{A}$ denoting any given reference pure state. In this way any isometric embedding in a tensor product has the form $V = U(I_\mathcal{H} \otimes |0\rangle)$, since

$$\sum_n |\upsilon_n\rangle\langle n| = \sum_n (|n\rangle \otimes |0\rangle)\langle n| = \sum_n |n\rangle\langle n| \otimes |0\rangle = I_\mathcal{H} \otimes |0\rangle.$$

The above result is the key for deriving all extension (and dilation) theorems in quantum theory, and we will refer to it as the *Golden rule for quantum extensions*:

Golden rule: any isometric embedding of an Hilbert space $\mathcal{H}$ in the tensor product $\mathcal{K} \otimes \mathcal{B}$, with $\dim(\mathcal{H} \otimes \mathcal{A}) = \dim(\mathcal{K} \otimes \mathcal{B})$ can be written as follows:

$$V : \mathcal{H} \hookrightarrow \mathcal{K} \otimes \mathcal{B}, \quad V = U(I_\mathcal{H} \otimes |0\rangle), \tag{2.103}$$

with $|0\rangle \in \mathcal{A}$ any pure state, and $U \in \mathsf{Lin}(\mathcal{H} \otimes \mathcal{A})$ unitary.

Problems

2.1 Show that a self-adjoint trace-class operator X is a pure quantum state iff one has[33]

$$\mathrm{Tr}X^2 = \mathrm{Tr}X^3 = 1.$$

2.2 Check the operator form of the *polarization identity*

$$X^\dagger Y = \tfrac{1}{4} \sum_{k=0,3} i^k (X^\dagger + i^k Y^\dagger)(X + (-i)^k Y), \tag{2.104}$$

---

[33] This derivation has been named *Remarkable Theorem* by C. Fuchs, according to whom it is due to Flammia (unpublished 2004).

and using the tensor notation in Section 2.8.2 show that the identity in Exercise 2.1 is just a special case of Eq. (2.104).

2.3 Denote by $E \in \text{Lin}(\mathcal{H} \otimes \mathcal{K})$ the unitary operator swapping $\mathcal{H}$ with $\mathcal{K}$. Prove the following identities, which hold for any operators $A \in \text{Lin}(\mathcal{K} \to \mathcal{H})$ and $B \in \text{Lin}(\mathcal{H} \to \mathcal{K})$

$$\text{Tr}_{\mathcal{H}}[(A \otimes B)E] = BA, \ \ \text{Tr}_{\mathcal{K}}[(A \otimes B)E] = AB. \tag{2.105}$$

2.4 Show that for any $A \in \text{Lin}(\mathcal{H}_1 \to \mathcal{H}_2)$ and $B \in \text{Lin}(\mathcal{K}_1 \to \mathcal{K}_2)$ one has the identity

$$E_{\mathcal{H}_2,\mathcal{K}_2}(A \otimes B)E_{\mathcal{H}_1,\mathcal{K}_1} = B \otimes A, \tag{2.106}$$

where $E_{\mathcal{H},\mathcal{K}}$ is the operator swapping $\mathcal{H} \leftrightarrow \mathcal{K}$.

2.5 A Euclidean space is a real vector space with a positive definite inner product. A *self-dual theory* is theory where the cones $\text{St}_+(A)$ and $\text{Eff}_+(A)$ can be embedded in a Euclidean space and the embedded cones coincide. Show that quantum theory is self-dual.

2.6 Prove the following overlap formula for deterministic qubit states

$$\text{Tr}[\rho_\mathbf{n} \, \rho_{\mathbf{n}'}] = \tfrac{1}{2}(1 + \mathbf{n} \cdot \mathbf{n}'). \tag{2.107}$$

Show that for pure states this is equivalent to the square modulus of the scalar product $|\langle \psi | \psi' \rangle|^2$ of the corresponding vectors in the Hilbert space. How does it generalize to probabilistic states?

2.7 Write a simple expression for the *purity* $\text{Tr}[\rho^2]$ of a qubit deterministic state in terms of the length of the corresponding Bloch vector $\mathbf{n}$ (see also Problem 2.1).

2.8 [Projective unitary representation of a group] A unitary representation $\{U_g\}_{g \in G}$ of a group G is a group homomorphism, namely it satisfies $U_g U_f = U_{gf}$. If we now consider a slightly modified definition for the composition of the unitary operators by introducing a phase factor as follows:

$$U_g U_h = \omega(g, h) U_{gh},$$

we obtain what is known as a *projective unitary representation* of the group G. Determine the constraints that the phase factor $\omega(g, h)$ must obey in order to reflect associativity of the group composition (assume that $U_e = I$).

2.9 [Shift-and-multiply group] Show that the operators in Exercise 2.43 form a projective unitary representation of the group $\mathbf{Z}_d \times \mathbf{Z}_d$, with composition law

$$U_j U_l = U_{j \oplus l} e^{i\varphi(j,l)}, \tag{2.108}$$

where $(p, q) \oplus (r, s) = (p \oplus r, q \oplus s)$, and $\varphi(j, l)$ determines the cocycle.

2.10 [The group of Pauli matrices and the Klein group] Show that the Pauli matrices in Exercise 2.4 form a group, which can be regarded as the projective representation of the Abelian group of the rotations of a π angle around three orthogonal axes. Show that this group is the special case of the group in Problem 2.9 for $d = 2$.

2.11 [Pauli matrices and Minkowski metric] From the general four-vector $n = (n_0, \mathbf{n})$ with $\mathbf{n} = (n_x, n_y, n_z)$ we can build the Hermitian matrix

$$\mathbf{N} = \sum_{i=0}^{3} n^i \sigma_i = \begin{pmatrix} n^0 + n^3 & n^1 - in^2 \\ n^1 + in^2 & n^0 - n^3 \end{pmatrix}.$$

Show that $\det(\mathbf{N}) = g_{ij} n^i n^j$, where $g_{ij} = \mathsf{Diag}\,[1, -1, -1, -1]$.

2.12 [Pauli matrices and Lorentz transformations] Show that the matrix

$$\Lambda_i^j(\mathbf{A}) = \frac{1}{2} \mathrm{Tr}[\mathbf{A}\sigma_i \mathbf{A}^\dagger \sigma_j]$$

for $\mathbf{A} \in \mathrm{SL}(2, \mathbb{C})$ is a matrix from the defining representation of the Lorentz group, i.e. the group of 4-by-4 matrices Λ such that $\Lambda^T G \Lambda = G$, where $G = \mathsf{Diag}\,(1, -1, -1, -1)$.

2.13 [Pauli matrices as Lie algebra generators] Show that the Pauli matrices span a *Lie algebra*, namely an algebra closed under commutation $[\cdot, \cdot]$ (such composition is also called *Lie product*). Take the Pauli matrices as a basis for the algebra, and write their commutation relations, which provide the Lie product for all elements of the algebra.

2.14 [Adjoint representation of SU(2)] Using the identity

$$e^A B e^{-A} = B + [A, B] + \frac{1}{2!}[A, [A, B]] + \cdots = e^{\mathrm{Ad}A} B,$$

where $(\mathrm{Ad}\,A)B := [A, B]$, check the rotation formula

$$\exp(\tfrac{1}{2} i\boldsymbol{\sigma} \cdot \mathbf{n}\theta) \begin{bmatrix} \sigma_x \\ \sigma_y \\ \sigma_z \end{bmatrix} \exp(-\tfrac{1}{2} i\boldsymbol{\sigma} \cdot \mathbf{n}\,\theta) = \mathbf{R_n}(\theta) \begin{bmatrix} \sigma_x \\ \sigma_y \\ \sigma_z \end{bmatrix}, \qquad (2.109)$$

with $\mathbf{R_n}(\theta)$ denoting the orthogonal matrix representing a rotation of a three-dimensional vector by an angle θ around the axis $\mathbf{n}$ counterclockwise. This is also called the *adjoint representation* of SU(2), generated by the action of the group over its Lie algebra viewed as a representation space. Show that it can also be equivalently written as

$$\mathbf{n}(U^\dagger X U) = \mathbf{R_n}(-\theta)\mathbf{n}(X),$$

where $\mathbf{n}(X) := \mathrm{Tr}[X\boldsymbol{\sigma}]$.

2.15 [SU(2) as double-covering of SO(3)] Prove the group isomorphism $\mathrm{SO}(3) = \mathrm{SU}(2)/\mathbb{Z}_2$.

2.16 Show that the Pauli matrices including the identity span the Lie algebra of the whole group matrix group $\mathrm{GL}(2, \mathbb{C})$, of which SU(2) is a real compact subgroup.

2.17 Consider the following observables:

$$S_1 = \sigma_{1x}\sigma_{2y}\sigma_{3y},$$
$$S_2 = \sigma_{1y}\sigma_{2x}\sigma_{3y},$$
$$S_3 = \sigma_{1y}\sigma_{2y}\sigma_{3x},$$
$$S_4 = \sigma_{1x}\sigma_{2x}\sigma_{3x}.$$

Show that any joint eigenvector of any three of them is eigenvector of all of them. Find all the joint eigenvectors of the four observables. Then write the unitary operator mapping the usual factorized canonical basis (of eigenvectors of local σ_z) to that of joint eigenvectors of the above operators.

2.18 Check that the following is a rank-one POVM

$$P_j = |\psi_j\rangle\langle\psi_j|, \ j = 1, 2, 3,$$

with

$$|\psi_1\rangle = \frac{1}{\sqrt{6}}(|0\rangle - \sqrt{3}|1\rangle), \ |\psi_2\rangle = \frac{1}{\sqrt{6}}(|0\rangle + \sqrt{3}|1\rangle), \ |\psi_3\rangle = \sqrt{\frac{2}{3}}|0\rangle,$$

and construct a minimal Naimark extension for it.

2.19 [Bhatia extension of effect][34] An example of Naimark dilation is given by the projective extension Z of the effect $0 \leq P \leq I$

$$Z = \begin{pmatrix} P & \sqrt{P(I-P)} \\ \sqrt{P(I-P)} & I-P \end{pmatrix}. \tag{2.110}$$

Check that Z is an orthogonal projector. Upon considering the Hilbert space dilation with a qubit ancilla, write the Naimark extension of P to Z.

2.20 [Quantum roulette wheel] An example of (generally non-commutative) POVM is the following:

$$P_m = \sum_{i=1}^{M} \zeta_i Z_m^{(i)}, \qquad m = 1, \ldots, N,$$

where

$$\zeta_i \geq 0, \quad \sum_{i=1}^{M} \zeta_i = 1, \quad Z_m^{(i)} Z_n^{(i)} = \delta_{mn} Z_m^{(i)}, \quad \sum_{m=1}^{N} Z_m^{(i)} = I, \ \forall i.$$

The POVM (2.20) describes a measuring apparatus where one out of M different observables is selected at random at every measurement step, with ζ_i as the probability of the i-th observable. Find a Naimark dilation.

2.21 [Existence of minimal informationally complete observation test] Show that it is always possible to construct a minimal informationally complete observation test for system A out of a set of its effects.[35]

2.22 Consider a convex cone C_+ and its real span C_R. Clearly, each element $x \in C_R$ can be decomposed as $x = x_- - x_-$, with $x_\pm \in C_R$, but such a decomposition is generally not unique. If the cone is self-dual (namely $C_+ = C_+^\vee$ coincide as embedded in C_R Euclidean space), then such decomposition is the unique *Jordan decomposition*[36]

$$x = x_+ - x_-, \quad x_\pm \in C_+, \quad \langle x_+, x_- \rangle = 0,$$

[34] Bhatia (1997).
[35] D'Ariano (2007b).
[36] Béllissard and Iochum (1978).

where $\langle \cdot, \cdot \rangle$ is the scalar product on $\mathbf{C}_{\mathbb{R}}$ which coincides with the cone-duality pairing. The decomposition allows to define the unique absolute value of x as $|x| = x_+ + x_-$. Show that in quantum theory the Jordan decomposition holds and is unique, and the absolute value is the operator absolute value for Hermitian operators.

2.23 [Iff condition for CP] Prove that a linear map $\mathcal{E} : \mathsf{Lin}(\mathcal{K}) \longrightarrow \mathsf{Lin}(\mathcal{H})$ is CP iff for all sequences of vectors $\{|h_i\rangle\} \in \mathcal{H}$ and $\{|k_i\rangle\} \in \mathcal{K}$, for $i = 1, ..., n$ for every n one has

$$\sum_{ij=1}^{n} \langle h_i | \mathcal{E}(|k_i\rangle\langle k_j|) | h_j \rangle \geq 0.$$

2.24 Write explicitly the effect E_σ of Eq. (2.68) for a general purification of a non-invertible state ρ.

2.25 Show that if a state $R = \sum_i |A_i\rangle\!\rangle \langle\!\langle A_i|$ is faithful, then the CP map $\mathcal{R} = \sum_i A_i \cdot A_i^\dagger$ is invertible.

2.26 Show that for $\rho \in \mathsf{St}_1(A)$, one has $\mathsf{RefSet}_1(\rho) = \mathsf{Span}_+ \mathsf{RefSet}\, \rho \cap \mathsf{St}_1(A)$, where $\mathsf{Span}_+ \mathsf{X}$ denotes the conic span of X.

2.27 Derive an analogous result as that of Exercise 2.74 for a general input state Ψ, and show that it holds also in infinite dimensions.

2.28 Let $X \in \mathsf{Lin}_+(\mathcal{H})$ be a positive operator and $\mathcal{E}$ a completely positive identity-preserving map, namely $\mathcal{E} \in \mathsf{Lin}_+(\mathsf{Lin}(\mathcal{H}) \to \mathsf{Lin}(\mathcal{H}))$, with $\mathcal{E}(I) = I$. Show that the conditions $\mathcal{E}(X) = X$ and $\mathcal{E}(X^{\frac{1}{2}}) = X^{\frac{1}{2}}$ hold iff X commutes with any Kraus operator of $\mathcal{E}$.[37]

2.29 [Non-PVM repeatable measurement] Show that the binary instrument $\{\mathcal{A}_i\}_{i=0,1}$ with

$$\mathcal{A}_i = A_i \cdot A_i^\dagger,$$

$$A_0 = \sum_{n=1}^{\infty} |4n\rangle\langle 2n| + \frac{1}{\sqrt{2}}|2\rangle\langle 0|, \quad A_1 = \sum_{n=0}^{\infty} |4n+1\rangle\langle 2n+1| + \frac{1}{\sqrt{2}}|3\rangle\langle 0|, \tag{2.111}$$

with $\{|n\rangle\}_{n=0}^{\infty}$ orthonormal basis for $\mathcal{H}_A$, is repeatable, but has non-orthogonal POVM.

2.30 Show that two purifications of the same state with different environments are connected by a partial isometry between the two respective Hilbert spaces. Write the quantum channel connecting the two environments as in Eq. (2.96).

2.31 [Closed systems evolve with the Schrödinger equation] Show that an isolated quantum system A must undergo a unitary time-homogeneous evolution of the form $U(t) = \exp(-iHt)$, with $H \in \mathsf{Herm}(\mathcal{H}_A)$, and t denotes time. Correspondingly the evolution of the time dependent density matrix $\rho_t \in \mathsf{St}(A)$ must be of the form

$$i\frac{d\rho_t}{dt} = [H, \rho_t] \quad \text{(quantum Liouville equation)} \tag{2.112}$$

and for the system prepared in a pure state one has $\rho_t = |\psi_t\rangle\langle\psi_t|$ with $|\psi_t\rangle \in \mathcal{H}_A$ satisfying the differential equation

---

[37] Lindblad (1999).

$$i\hbar\frac{d|\psi_t\rangle}{dt} = H|\psi_t\rangle \quad \text{(Schrödinger equation).} \tag{2.113}$$

Therefore, the Schrödinger equation for generic Hamiltonian H is just the most general evolution of an isolated system.

2.32 [Open systems evolve with the Lindblad equation] Show that a finite-dimensional quantum system undergoing a time-homogeneous deterministic transformation evolves according to the *master equation* in the Lindblad form (also called the *Lindblad equation*)[38]

$$\frac{d\rho_t}{dt} = \mathcal{L}\rho := -i[H,\rho] + \frac{1}{2}\sum_k (2V_k\rho V_k^\dagger - V_k^\dagger V_k\rho - \rho V_k^\dagger V_k), \tag{2.114}$$

where $V_k \in \mathsf{Lin}(\mathcal{H}_A)\ \forall k$, and $H \in \mathsf{Herm}(\mathcal{H}_A)$.

Solutions to Selected Problems and Exercises

Exercise 2.1

By Eq. (2.6) we can express any matrix-element $\langle y|A|x\rangle$ of $A \in \mathsf{Lin}(\mathcal{H})$ in terms of its diagonal elements. Then we use the fact that $\langle\mu|A|\mu\rangle = |c|^2\langle\lambda|A|\lambda\rangle$ for any $|\mu\rangle = c|\lambda\rangle \in \mathcal{H}$.

Exercise 2.5

A set S is convex if it is closed under convex combinations, namely for each two points of the set $x_1, x_2 \in \mathsf{S}$ the full segment $\{px_1 + (1-p)x_2|p \in [0,1]\}$ belongs to the set. The set $\mathsf{St}(A)$ is convex, since convex combinations preserve both operator positivity and trace-domination.

Exercise 2.7

Convex combinations preserve both operator positivity and identity-domination. Another way of proving the statement is to regard effects as probability functionals over states, and show that these are closed under convex combination.

Exercise 2.9

The density operator can be expanded in a unique way on the orthonormal basis of Pauli matrices, which are self-adjoint. Since $\mathrm{Tr}\rho = 1$ and I_2 is the only basis element with non-zero trace, its coefficient must be $\frac{1}{2}$, as in Eq. (2.12). Moreover, since ρ is self-adjoint, $\mathbf{n} \in \mathbb{R}^3$. Positivity of ρ means that $1 + \langle v|\mathbf{n}\cdot\boldsymbol{\sigma}|v\rangle \geq 0, \forall|v\rangle \in \mathbb{C}^2$, in particular for the eigenvectors $|\pm\rangle$ of $\mathbf{n}\cdot\boldsymbol{\sigma}$ corresponding to the eigenvalues $\pm\|\mathbf{n}\|$ (the reader must check), hence $\|\mathbf{n}\| \leq 1$. Pure states correspond to the surface of the Bloch ball, namely to Bloch vectors with $\|\mathbf{n}\| = 1$, whereas mixed states have $\|\mathbf{n}\| < 1$. The totally mixed state $\rho = \frac{1}{2}I_2$ corresponds to the center of the Bloch ball.

---

[38] See the historical note in this chapter.

Exercise 2.10

The set of generally probabilistic qubit states $\mathsf{St}(A)$ is given by the following truncated cone in $\mathbb{R}^4$, with base of the cone given by the Bloch ball

$$\rho_{\mathbf{n}} = \tfrac{t}{2}(I_2 + \mathbf{n} \cdot \boldsymbol{\sigma}), \quad \mathbf{n} \in B^3, \, t \in [0,1].$$

Exercise 2.11

This is just the condition $0 \leq E \leq I_2$, which is equivalent to having both eigenvalues of E bounded between 0 and 1.

Exercise 2.12

The operator $\sigma_a = \mathrm{Tr}_A[\sigma(E_a \otimes I_B)]$ belongs to $\mathsf{Lin}(\mathcal{H}_B)$ by definition. It is positive, since

$$\forall |\lambda\rangle \in \mathcal{H}_B: \quad \langle\lambda|\mathrm{Tr}_A[\sigma(E_a \otimes I_B)]|\lambda\rangle = \sum_i \langle e_i|\langle\lambda|\sigma|e_i\rangle|\lambda\rangle \geq 0,$$

where we used the decomposition of $E_a = \sum_i |e_i\rangle\langle e_i| \geq 0$. In the same way we prove that

$$\mathrm{Tr}\,\sigma_a = \mathrm{Tr}[\sigma(E_a \otimes I_B)] \leq \mathrm{Tr}[\sigma(I_A \otimes I_B)] = \mathrm{Tr}\sigma = 1.$$

Exercise 2.17

The statement of Lemma 2.2 can be restated as follows. Let X a subspace of the linear space B, and $\gamma \in \mathsf{B} \setminus \mathsf{X}$. Then there exists $y \in \mathsf{X}^\perp$ such that $(y|\gamma) \neq 0$. This can be simply proved as follows.

Let $\{\xi_i\}_{i=1}^n$ be a complete linearly independent set in X. Then since X is linearly closed, also $\{\xi_i\}_{i=1}^n \cup \{\gamma\}$ is linearly independent. Let us extend the linearly independent set to a complete one in B as $\{\xi_i\}_{i=1}^n \cup \{\gamma\} \cup \{\zeta_j\}_{j=1}^{\dim \mathsf{B}-n-1}$. One can then define $y \in \mathsf{B}^\vee$ by posing

$$(y|\xi_i) = 0, \quad \forall 1 \leq i \leq n,$$
$$(y|\zeta_j) = c_j, \quad \forall 1 \leq j \leq \dim \mathsf{B} - n - 1,$$
$$(y|\gamma) = k \neq 0.$$

The functional y can then be extended by linearity to B. One then has $y \in \mathsf{X}^\perp$, and $(y|\gamma) \neq 0$.

We now prove Lemma 2.3. One can easily prove that independently of the set A the set $\mathsf{A}^\perp$ is a vector space. Indeed, let $a, a' \in \mathsf{A}^\perp$, and $\lambda, \mu \in \mathbb{R}$. Then one has

$$(\lambda a + \mu a'|\alpha) = \lambda(a|\alpha) + \mu(a'|\alpha) = 0, \quad \forall a \in \mathsf{A},$$

namely $\lambda a + \mu a' \in \mathsf{A}^\perp$. Analogously, one can prove that $\mathsf{Span}(\mathsf{A})^\perp = \mathsf{A}^\perp$. Indeed, for $\alpha, \alpha' \in \mathsf{A}$ and $\lambda, \mu \in \mathbb{R}$ one has

$$(a|\lambda\alpha + \mu\alpha') = \lambda(a|\alpha) + \mu(a|\alpha') = 0, \quad \forall a \in \mathsf{A}^\perp.$$

The former argument implies that $(\mathsf{A}^\perp)^\perp \subseteq \mathsf{B}$ is a subspace. Moreover, since for every $\alpha \in \mathsf{A}$ one has $(a|\alpha) = 0$ for every $a \in \mathsf{A}^\perp$, it is $\mathsf{A} \subseteq (\mathsf{A}^\perp)^\perp$. Thus,

$$\mathsf{Span}(\mathsf{A}) \subseteq (\mathsf{A}^\perp)^\perp. \tag{2.115}$$

On the other hand, according to Lemma 2.17 for any $\beta \in (\mathsf{A}^\perp)^\perp$ one has $\beta \in \mathsf{Span}(\mathsf{A})$.

Exercise 2.18

States are separating for effects, and the set of states belongs to the conic span of pure states. It follows that pure states are separating for effects. Similarly atomic effects separate states. An alternative solution is the following. $A_1 = A_2 \in \mathsf{Lin}_+(A)$ iff $\langle \lambda |A_1|\lambda \rangle = \langle \lambda |A_2|\lambda \rangle$ $\forall |\lambda\rangle \in \mathcal{H}_A$. Therefore $A_1 \neq A_2$ iff there exists a vector $|\mu\rangle \in \mathcal{H}$ such that $\langle \mu |A_1|\mu \rangle \neq \langle \mu |A_2|\mu \rangle$. This implies that pure states separate effects. In a similar way one proves that atomic effects separate states.

Exercise 2.19

Using the expansion

$$X = \frac{1}{2} \sum_i \mathrm{Tr}[\sigma_i X]\sigma_i = \frac{1}{2}\left[I\,\mathrm{Tr}X + \sum_{\alpha=x,y,z} \mathrm{Tr}(\sigma_\alpha X)\sigma_\alpha \right],$$

we can obtain the expectation of any operator by taking the expected value of both sides, namely

$$\langle X \rangle = \frac{1}{2}\left[\mathrm{Tr}X + \sum_{\alpha=x,y,z} \mathrm{Tr}(\sigma_\alpha X)\langle \sigma_\alpha \rangle \right].$$

In particular, the matrix element $\langle u|\rho|v \rangle$ of the density operator ρ is the expectation of $X = |v\rangle\langle u|$, hence

$$\langle u|\rho|v \rangle = \frac{1}{2}\left(\langle u|v \rangle + \sum_{\alpha=x,y,z} \langle u|\sigma_\alpha|v \rangle \langle \sigma_\alpha \rangle \right).$$

Exercise 2.20

The rank of the tensor product of positive operators is the product of the respective ranks. A rank-one positive operator that is the tensor product of positive operators must necessarily be the product of rank-one positive operators, hence it is necessarily of the form $|\lambda\rangle\langle\lambda|$ with $|\lambda\rangle$ tensor product of vectors. A non-trivial convex combination of rank-one factorized operators has rank strictly larger than one. The operator R in Eq. (2.23) has rank one, but is not of the form $|\lambda\rangle\langle\lambda|$ with $|\lambda\rangle$ tensor product. We then conclude that R is not separable.

Exercise 2.30

Proof of Eq. (2.33). Let $\rho - \sigma \geq 0$. Then taking the expectation of both sides for $\psi \in \mathsf{Ker}\,\rho$ one has $-\langle \psi |\sigma|\psi \rangle \geq 0$. Since $\sigma \geq 0$, it must be $\psi \in \mathsf{Ker}\,\sigma$, and then $\mathsf{Ker}\,\rho \subseteq \mathsf{Ker}\,\sigma$. Finally, since $\mathsf{Supp}\,X = (\mathsf{Ker}\,X)^\perp$, we have $\mathsf{Supp}\,\sigma \subseteq \mathsf{Supp}\,\rho$. Vice versa, if $\mathsf{Supp}\,\sigma \subseteq \mathsf{Supp}\,\rho$ one can write

$$\rho^{-\frac{1}{2}}\sigma\rho^{-\frac{1}{2}} \leq \lambda P_{\mathsf{Supp}\,\rho},$$

where ρ^{-1} inverts ρ on $\mathsf{Supp}\,\rho$, $P_{\mathsf{Supp}\,\rho}$ is the projection on $\mathsf{Supp}\,\rho$, and $\lambda > 0$ is the maximum eigenvalue of $\rho^{-\frac{1}{2}}\sigma\rho^{-\frac{1}{2}}$. Thus, we have

$$\lambda^{-1}\sigma \leq \rho.$$

Exercise 2.31

The proof is an immediate application of the definition (2.29) of $\mathsf{RefSet}_1\,\rho$ and of Exercise 2.30.

Exercise 2.33

Explicitly we can write the action of the dual map $\mathcal{A}^\dagger$ in terms of the action of the map $\mathcal{A}$ by extending identity (2.40) linearly to $\mathsf{Lin}(\mathcal{H}_A)$ and $\mathsf{Lin}(\mathcal{H}_B)$, respectively, and expanding over e.g. the orthonormal basis $\{|i\rangle\langle j|\}$ for $\mathsf{Lin}(\mathcal{H}_A)$ ($\{|i\rangle\}$ is an orthonormal basis for $\mathcal{H}_A$), as follows

$$\mathrm{Tr}[E\,\mathcal{A}(|i\rangle\langle j|)] = \mathrm{Tr}[|i\rangle\langle j|\,\mathcal{A}^\dagger(E)], \quad E \in \mathsf{Lin}(\mathcal{H}_B).$$

We then have

$$\mathcal{A}^\dagger(E) = \sum_{ij} \mathrm{Tr}[E\,\mathcal{A}(|i\rangle\langle j|)]|j\rangle\langle i|. \tag{2.116}$$

Clearly both $\mathcal{A}$ and $\mathcal{A}^\dagger$ are linear over operator spaces. Precisely one has

$$\mathcal{A} \in \mathsf{Lin}[\mathsf{Lin}(\mathcal{H}_A) \to \mathsf{Lin}(\mathcal{H}_B)] \leftrightarrow \mathcal{A}^\dagger \in \mathsf{Lin}[\mathsf{Lin}(\mathcal{H}_B) \to \mathsf{Lin}(\mathcal{H}_A)].$$

$\mathcal{A}^\dagger$ is the dual map of $\mathcal{A}$ and $\mathsf{Lin}[\mathsf{Lin}(\mathcal{H}_A) \to \mathsf{Lin}(\mathcal{H}_B)]^\vee = \mathsf{Lin}[\mathsf{Lin}(\mathcal{H}_B) \to \mathsf{Lin}(\mathcal{H}_A)]$ (Hilbert spaces are supposedly finite dimensional).

Exercise 2.34

The map is obviously positive, since by Eq. (2.116) one can check that $\mathcal{T} = \mathcal{T}^\dagger$, and by the solution of Exercise 2.33 one has

$$\langle v|\mathcal{T}(\rho)|v\rangle = \langle v^*|\rho|v^*\rangle \geq 0, \quad \forall |v\rangle \in \mathcal{H},$$

where $|v^*\rangle$ denotes the vector $|v\rangle$ where all coefficients with respect to the fixed basis have been complex-conjugated. However, the map is not CP, as a trivial consequence of Exercises 2.23 and 2.24. Another example of the fact that the map doesn't preserve positivity when applied locally to a joint state is provided by the case of the singlet state. Apply the map locally to a pair of systems in the "singlet" state of two qubits

$$|\Psi\rangle = \frac{1}{\sqrt{2}}(|0\rangle \otimes |1\rangle - |1\rangle \otimes |0\rangle).$$

Transposing with respect to the orthonormal basis $\{|0\rangle, |1\rangle\}$ and using identity Eq. (2.37) gives

$$\mathcal{T} \otimes \mathcal{I}(|\Psi\rangle\langle\Psi|) =$$
$$\tfrac{1}{2}(|0\rangle\langle 0| \otimes |1\rangle\langle 1| + |1\rangle\langle 1| \otimes |0\rangle\langle 0| - |1\rangle\langle 0| \otimes |1\rangle\langle 0| - |0\rangle\langle 1| \otimes |0\rangle\langle 1|).$$

The expectation over the state $|\Phi\rangle = \frac{1}{\sqrt{2}}(|0\rangle \otimes |0\rangle + |1\rangle \otimes |1\rangle)$ is negative

$$\langle\Phi|[(\mathcal{T} \otimes \mathcal{I})|\Psi\rangle\langle\Psi|]|\Phi\rangle = -\tfrac{1}{2},$$

hence the map $\mathcal{T} \otimes \mathcal{I}$ is not positive, and $\mathcal{T}$ is not CP.[39]

Exercise 2.35

For any extension, and positive operator R on the extended Hilbert space one has

$$\mathcal{A} \otimes \mathcal{I}(R) = \sum_i (A_i \otimes I) R (A_i \otimes I)^\dagger = \sum_i \left| (A_i \otimes I) R^{\frac{1}{2}} \right|^2 \geq 0.$$

Exercise 2.44

It is easy to see that the second of identities (2.55) gives back the map $\mathcal{A}$. One has

$$\mathrm{Tr}_2[(I_B \otimes \rho^\mathsf{T}) R_{\mathcal{A}}] = \mathrm{Tr}_2[(I_B \otimes \rho^\mathsf{T}) \mathcal{A} \otimes \mathcal{I}_A(|I\rangle\!\rangle\langle\!\langle I|)] =$$
$$\mathrm{Tr}_2[\mathcal{A} \otimes \mathcal{I}_A(|\rho\rangle\!\rangle\langle\!\langle I|)] = \mathcal{A}[\mathrm{Tr}_2(|\rho\rangle\!\rangle\langle\!\langle I|)] = \mathcal{A}(\rho),$$

where we used identities (2.50), and the trivial commutations

$$(I_B \otimes X)(\mathcal{A} \otimes \mathcal{I}_A) = (\mathcal{A} \otimes \mathcal{I}_A)(I_B \otimes X),$$
$$\mathrm{Tr}_2[\mathcal{A} \otimes \mathcal{I}_A X_{AB}] = \mathcal{A}\,\mathrm{Tr}_2 X_{AB}.$$

Exercise 2.45

Clearly, if $(U^\dagger U)_{jk} = \delta_{jk}$, then

$$\sum_i A'_i \rho A_i'^\dagger = \sum_{ijk} U_{ij}^* U_{ik} A_k \rho A_j^\dagger = \sum_{jk} (U^\dagger U)_{jk} A_k \rho A_j^\dagger = \sum_j A_j \rho A_j^\dagger.$$

Conversely, the requirement of maintaining the same range of the map for all input states forces the two sets of operators to be linearly related, whereas uniqueness of the expansion follows from linear independence of $\{A_i\}$.

Exercise 2.47

It is sufficient to show the two assertions for the atomic cases. Let's regard a state $\rho = |\lambda\rangle\langle\lambda| \in \mathsf{St}(A)$, for $|\lambda\rangle \in \mathcal{H}_A$ and $\|\lambda\| \leq 1$, as a transformation $\rho \in \mathsf{Transf}(I \to A)$ from the trivial system I with Hilbert space $\mathcal{H}_I = \mathbb{C}$ to the system A with Hilbert space $\mathcal{H}_A$. The state transforms a nonnull probability $0 < p \leq 1$ (considered as a state of I) into the state $A p A^\dagger = p|\lambda\rangle\langle\lambda|$, and the Kraus operator (apart from a phase) is $A = |\lambda\rangle \in \mathsf{Lin}(\mathbb{C} \to \mathcal{H}_A)$. One has $|I\rangle\!\rangle \in \mathbb{C} \otimes \mathbb{C} \equiv \mathbb{C}$, corresponding to $|I\rangle\!\rangle\langle\!\langle I| = 1 \in \mathbb{C}$, and the Choi–Jamiołkowski operator is thus the state itself $R_\lambda = |\lambda\rangle\langle\lambda|$.

As regards an atomic $a = \mathrm{Tr}[\cdot|\lambda\rangle\langle\lambda|] \in \mathsf{Eff}(A) \equiv \mathsf{Transf}(\mathcal{H}_A \to \mathbb{C})$ the Kraus operator is given by $A = \langle\lambda| \in \mathsf{Lin}(\mathcal{H}_A \to \mathbb{C})$, leading to $\mathsf{St}(A) \ni \rho \mapsto A\rho A^\dagger = \langle\lambda|\rho|\lambda\rangle$. Now one has $|I\rangle\!\rangle \in \mathcal{H}_A^{\otimes 2}$, and the Choi–Jamiołkowski operator is given by

$$(\langle\lambda| \otimes I)|I\rangle\!\rangle\langle\!\langle I|(|\lambda\rangle \otimes I) = |\lambda\rangle^{**}\langle\lambda|.$$

---

[39] Notice that for self-adjoint operators ρ transposition is equivalent to complex conjugation $\rho^\mathsf{T} = \rho^*$ with respect to the same basis. In QT the complex conjugation of the time-dependent density matrix represents its time-reversal. Time-reversal is therefore unphysical.

Exercise 2.48

The convex set of states is the cone $\mathsf{Lin}_+(\mathcal{H}_{\mathrm{BA}})$ truncated by the $d_{\mathrm{A}}^2 d_{\mathrm{B}}^2 - 1$ hyperplane orthogonal to the cone axis. On the other hand, the convex set of transformations is the intersection of the cone $\mathsf{Lin}_+(\mathcal{H}_{\mathrm{BA}})$ with the convex set

$$\mathrm{Tr}_{\mathrm{B}}[R] \leq I_{\mathrm{A}},$$

which has dimensionality d_{A}^2. Therefore, the two convex sets are different convex truncations of the same convex cone.

Exercise 2.51

The completeness of the quantum instrument is conveniently written as follows:

$$\sum_i \mathcal{T}_i^{\dagger}(I) = I. \tag{2.117}$$

By definition, the instrument is repeatable if the joint probability of two measurements in cascade is given by $p(i_2, i_1) = \delta_{i_1 i_2} p(i_1)$. By the polarization identity, this is equivalent to

$$\mathcal{T}_{i_1}^{\dagger} \mathcal{T}_{i_2}^{\dagger}(I) = \delta_{i_1 i_2} \mathcal{T}_{i_1}^{\dagger}(I). \tag{2.118}$$

For Kraus decomposition $\mathcal{T}_i = \sum_n A_n^{(i)} \cdot A_n^{(i)\dagger}$, one has

$$\sum_{nm} A_m^{(i_2)\dagger} A_n^{(i_1)\dagger} A_n^{(i_1)} A_m^{(i_2)} = \delta_{i_1 i_2} \sum_n A_n^{(i_1)\dagger} A_n^{(i_1)}. \tag{2.119}$$

Since all terms in the double sum in Eq. (2.119) are positive, we must have

$$\forall i_1, i_2, \ i_1 \neq i_2: \qquad \forall n, m \quad A_m^{(i_2)\dagger} A_n^{(i_1)\dagger} A_n^{(i_1)} A_m^{(i_2)} = 0,$$

and since $A^{\dagger} A = 0 \implies A = 0$ for any operator A, one has

$$\forall i_1, i_2, \ i_1 \neq i_2: \qquad \forall n, m \quad A_n^{(i_1)} A_m^{(i_2)} = 0. \tag{2.120}$$

On the other hand, if Eq. (2.120) is satisfied along with the normalization condition (2.117), one has

$$\mathcal{T}_{i_1}^{\dagger} \mathcal{T}_{i_2}^{\dagger}(I) = \sum_m A_m^{(i_1)\dagger} \left(\sum_n A_n^{(i_2)\dagger} A_n^{(i_2)} \right) A_m^{(i_1)} = \delta_{i_1 i_2} \sum_{mn} A_m^{(i_1)\dagger} A_n^{(i_1)\dagger} A_n^{(i_1)} A_m^{(i_1)}$$

$$= \delta_{i_1 i_2} \mathcal{T}_{i_1}^{\dagger} \mathcal{T}_{i_1}^{\dagger}(I) = \delta_{i_1 i_2} \sum_i \mathcal{T}_{i_1}^{\dagger} \mathcal{T}_i^{\dagger}(I) = \delta_{i_1 i_2} \mathcal{T}_{i_1}^{\dagger}(I),$$

and condition (2.120) implies the repeatability condition (2.118), hence it is a necessary and sufficient condition for repeatability.

Exercise 2.55

The set of states $\{\rho_i\}_{i \in \mathsf{X}}$ is perfectly discriminable iff there exists a POVM $\{P_i\}_{i \in \mathsf{X}}$ such that $\mathrm{Tr}(P_j \rho_i) = \delta_{ij}$. For every couple of outcomes i, j the identity $\mathrm{Tr}(P_j \rho_i) = 0$ implies that $\mathsf{Supp}\, \rho_i \cap \mathsf{Supp}\, P_j = \{0\}$, whereas $\mathrm{Tr}(P_i \rho_i) = 1$ implies that $P_i = Z_{\mathsf{Supp}\, \rho_i} + W$, where $Z_{\mathsf{Supp}\, \rho_i}$ denotes the orthogonal projector over support of ρ_i and $W \geq 0$ with support orthogonal to $\mathsf{Supp}\, \rho_i$. It follows that $\mathsf{Supp}\, \rho_i \subseteq \mathsf{Supp}\, P_i$. Moreover one should have

$\operatorname{Supp}\rho_i \cap \operatorname{Supp}\rho_j = \{0\}$, otherwise we cannot have $Z_{\operatorname{Supp}\rho_i} + Z_{\operatorname{Supp}\rho_j} \le I_A$, hence we cannot have $P_i + P_j \le I_A$, and one cannot satisfy the normalization condition for the POVM. Finally, when $\oplus_{i\in X}\operatorname{Supp}\rho_i = \mathcal{H}$ one has $\operatorname{Tr}[W\rho_j] = 0$ for $j \ne i$, hence P_i are orthogonal projectors.

Exercise 2.56

In terms of the joint probability p_{ij} of obtaining outcomes i followed by j, repeatability is defined as follows:

$$p_{ij} = \delta_{ij}p_i.$$

Writing the atomic instrument in the form $\mathcal{A}_i(\rho) = A_i\rho A_i^\dagger$, we have

$$A_iA_j = 0, \quad \text{for } i \ne j,$$
$$(A_i^\dagger)^2 A_i^2 = A_i^\dagger A_i.$$

First we notice that Eq. (2.121) implies that

$$\operatorname{Rng}A_i \subseteq \operatorname{Supp}A_i. \tag{2.121}$$

In fact, suppose by contradiction that a vector $|\psi\rangle \in \mathcal{H}_A$ exists such that

$$A_i|\psi\rangle = |v\rangle + |\psi'\rangle, \quad |v\rangle \in \operatorname{Ker}A_i, \ |\psi'\rangle \in \operatorname{Supp}A_i.$$

Then, since $\|A_i\| \le 1$, using Eq. (2.121), i.e. $\|A_i^2|\psi\rangle\| = \|A_i|\psi\rangle\|$ for all $|\psi\rangle \in \mathcal{H}_A$, we have

$$\|\psi'\|^2 \ge \|A_i\psi'\|^2 = \|A_i^2\psi\|^2 = \|A_i\psi\|^2 = \|v\|^2 + \|\psi'\|^2,$$

and this is possible if and only if $|v\rangle = 0$. Therefore, we have $\operatorname{Rng}A_i \subseteq \operatorname{Supp}A_i$. Moreover, Eq. (2.121) implies that $\|A_i^2\psi\| = \|A_i\psi\|$, namely for $\varphi = A_i\psi \in \operatorname{Rng}A_i$ one has $\|A_i\varphi\| = \|\varphi\|$, which means that A_i is isometric on its range. On finite-dimensional Hilbert spaces $\mathcal{H}_A$, for any operator A one has that $\dim\operatorname{Rng}A = \dim\operatorname{Supp}A = \operatorname{rank}A$, hence Eq. (2.121) implies that

$$\operatorname{Rng}A_i \equiv \operatorname{Supp}A_i, \tag{2.122}$$

and A_i is isometric on its support, namely it is a partial isometry, or else

$$A_i^\dagger A_i = Z_i,$$

with Z_i orthogonal projector on $\operatorname{Supp}A_i$. On the other hand, Eq. (2.121) implies that

$$\operatorname{Rng}A_j \subseteq \operatorname{Ker}A_i, \quad i \ne j. \tag{2.123}$$

namely $\operatorname{Supp}A_i \subseteq (\operatorname{Rng}A_j)^\perp$ for $i \ne j$, and, using Eq. (2.122), we obtain

$$\operatorname{Supp}A_i \cap \operatorname{Supp}A_j = \{0\}, \quad \text{for } i \ne j.$$

Therefore, one has

$$I = \sum_{i\in X} A_i^\dagger A_i \equiv \bigoplus_{i\in X} A_i^\dagger A_i \equiv \bigoplus_{i\in X} Z_i,$$

namely $\{A_i^\dagger A_i\}_{i\in X}$ is a PVM.

We remark that the above condition for repeatability holds only for atomic instruments in infinite dimensions. An example of repeatable non-atomic instrument that has non-orthogonal POVM is a measure-and-prepare instrument of the form

$$\mathcal{T}_i = \mathrm{Tr}[P_i \cdot]\rho, \quad \mathrm{rank}\,\rho > 1, \, \mathsf{Supp}\,\rho \subseteq \mathsf{Supp}\,P_{i_0}, \, \mathsf{Supp}\,P_{i_0} \cap \mathsf{Supp}\,P_i = 0 \text{ for } i \neq i_0.$$

In infinite dimensions, an atomic instrument that is repeatable and has non-orthogonal POVM is given by[40]

$$A_1 = \sqrt{p_1}|2\rangle\langle 0| + \sqrt{p_2}|4\rangle\langle 1| + \sum_{n=1}^{\infty} |2(n+2)\rangle\langle 2n|$$

$$A_2 = \sqrt{1-p_1}|3\rangle\langle 0| + \sqrt{1-p_2}|5\rangle\langle 1| + \sum_{n=1}^{\infty} |2(n+2)+1\rangle\langle 2n+1|,$$

with corresponding non-orthogonal POVM

$$P_1 = p_1|0\rangle\langle 0| + p_2|1\rangle\langle 1| + \sum_{n=1}^{\infty} |2n\rangle\langle 2n|,$$

$$P_2 = (1-p_1)|0\rangle\langle 0| + (1-p_2)|1\rangle\langle 1| + \sum_{n=1}^{\infty} |2n+1\rangle\langle 2n+1|.$$

It is easy to check completeness, using the identities

$$W_1^\dagger W_1 = p_1|0\rangle\langle 0| + p_2|1\rangle\langle 1|, \quad W_2^\dagger W_2 = (1-p_1)|0\rangle\langle 0| + (1-p_2)|1\rangle\langle 1|.$$

Exercise 2.57

Denote the two vectors as $|\psi_1\rangle$ and $|\psi_2\rangle$, with $\langle\psi_1|\psi_2\rangle \neq 0$. To discriminate the two pure states one needs a POVM $\{P_i\}_{i=1,2}$ with the two outcomes in one-to-one correspondence with the two states. In a perfect discrimination, the outcomes $k = 1, 2$ would correspond to the state $|\psi_1\rangle, |\psi_2\rangle$, respectively. This corresponds to set the probabilities

$$\langle\psi_1|P_1|\psi_1\rangle = \langle\psi_2|P_2|\psi_2\rangle = 1. \tag{2.124}$$

However, since $P_1 + P_2 = I$, one has $\langle\psi_1|P_1 + P_2|\psi_1\rangle = 1$, which, along with Eq. (2.124) implies that $\langle\psi_1|P_2|\psi_1\rangle = 0$, namely $P_2^{\frac{1}{2}}|\psi_1\rangle = 0$. Writing

$$|\psi_2\rangle = \alpha|\psi_1\rangle + \beta|\varphi\rangle,$$

with $|\varphi\rangle$ orthogonal to $|\psi_1\rangle$ and $|\beta| < 1$, one obtains

$$\langle\psi_2|P_2|\psi_2\rangle = \langle\psi_2|P_2^{\frac{1}{2}}P_2^{\frac{1}{2}}|\psi_2\rangle = |\beta|^2\langle\varphi|P_2|\varphi\rangle \leq |\beta|^2 < 1,$$

which contradicts Eq. (2.124).

Exercise 2.58

The error probability given state $|\psi_k\rangle$ is given by

$$p(i \neq k|k) = \alpha \sum_{i \neq k} |\langle\varphi_i|\psi_k\rangle|^2 = 0,$$

[40] Buscemi *et al.* (2004).

thanks to the duality condition, hence the overall error probability is zero, namely the discrimination is perfect. The allowed values of α are those satisfying the positivity condition for $P_?$, namely

$$\alpha \sum_{i \in X} |\varphi_i\rangle\langle\varphi_i| \leq I. \tag{2.125}$$

This condition is satisfied provided that

$$\alpha \leq \left[\max \mathrm{Eig} \left(\sum_{i \in X} |\varphi_i\rangle\langle\varphi_i| \right) \right]^{-1}.$$

The maximum probability of inconclusive result is

$$p_?^{\max} = 1 - \alpha \max_n |\langle\phi_n|\psi_n\rangle|^2,$$

and is minimized by

$$\alpha_{\mathrm{opt}} := \left[\max \mathrm{Eig} \left(\sum_{i \in X} |\varphi_i\rangle\langle\varphi_i| \right) \right]^{-1}.$$

The same value of α also minimizes the average probability of inconclusive result, given by

$$1 - \alpha \sum_n q_n |\langle\phi_n|\psi_n\rangle|^2,$$

where q_n denotes the prior probability of the state $|\psi_n\rangle$.

Exercise 2.59

The matrix $\sum_{i=1,2} |\varphi_i\rangle\langle\varphi_i| = \begin{bmatrix} \frac{3}{2} & \frac{1}{2} \\ \frac{1}{2} & \frac{1}{2} \end{bmatrix}$ has eigenvalues $\lambda = \frac{\sqrt{2}\pm 1}{\sqrt{2}}$, hence $\alpha_{\mathrm{opt}} = \sqrt{2}/(1 + \sqrt{2})$.

Exercise 2.64

Every sub-Markov matrix $\mathbf{M}$ can be seen as a conic combination of $m \times n$ matrices with only one non-null element equal to 1. Thus, atomic transformations can be labeled as $\mathbf{M}^{(ij)}$, and act as $\mathbf{M}^{(ij)} \mathbf{e}^{(k)} = \delta_{jk} \mathbf{e}'^{(i)}$, where the column vector $\mathbf{e}^{(i)} \in S_n$ has elements $e_j^{(i)} = \delta_{ij}$, and similarly for $\mathbf{e}'^{(i)} \in S_m$.

Exercise 2.65

The atomic transformations $\mathbf{M}^{(ij)}$ must annihilate every state $\mathbf{x} \in S_n$ that has $x_j = 0$, and prepare the pure state $\mathbf{e}'^{(i)}$. Thus, in the Kraus representation it must have a single Kraus operator proportional to $|i\rangle\langle j|$.

Exercise 2.72

A change from the canonical basis $\{|j\rangle\}$ to the orthonormal basis $\{|v_i\rangle\}$ corresponds to writing the vector $|I\rangle\rangle$ as follows:

$$|I\rangle\rangle = \sum_i |v_i\rangle \otimes |v_i\rangle^* = \sum_i |v_i\rangle^* \otimes |v_i\rangle.$$

By using the eigenvectors $\{|\lambda_i\rangle/\|\lambda_i\|\}$ of ρ one obtains

$$|\Psi\rangle\rangle = (\rho^{\frac{1}{2}} \otimes V^*)|I\rangle\rangle = \sum_i \rho^{\frac{1}{2}} \frac{|\lambda_i\rangle}{\|\lambda_i\|} \otimes V^* \frac{|\lambda_i\rangle^*}{\|\lambda_i\|} = \sum_i |\lambda_i\rangle \otimes V^* \frac{|\lambda_i\rangle^*}{\|\lambda_i\|}$$

$$= \sum_{ij} |\lambda_i\rangle \otimes |j\rangle \frac{\langle j|V^*|\lambda_i\rangle^*}{\|\lambda_i\|} = \sum_j \left(\sum_i \langle j|V^*|\lambda_i\rangle^* \frac{|\lambda_i\rangle}{\|\lambda_i\|} \right) \otimes |j\rangle.$$

Exercise 2.74[41]

From identities (2.52) we see that the quantum operation matrix A is imprinted over the state, and can be recovered as follows:

$$\frac{1}{\sqrt{d}}|I\rangle\rangle \to |\Phi\rangle\rangle = \frac{A \otimes I|I\rangle\rangle}{\|A\|_2},$$

where $\|A\|_2 = \text{Tr}[A^\dagger A]^{\frac{1}{2}}$ is the Hilbert–Schmidt norm of A. Then, one has

$$A = \Phi\|A\|_2, \quad \|A\|_2^2 = dp_A, \tag{2.126}$$

where p_A denotes the occurrence probability of the quantum operation on the state $\frac{1}{\sqrt{d}}|I\rangle\rangle$. The matrix Φ of the output state can be written in terms of measurable ensemble averages as follows:

$$\Phi_{ij} \equiv \langle\langle i,j|\Phi\rangle\rangle = e^{i\theta} \frac{\langle\langle\Phi|i_0,j_0\rangle\rangle\langle\langle i,j|\Phi\rangle\rangle}{\sqrt{\langle\langle\Phi|i_0,j_0\rangle\rangle\langle\langle i_0,j_0|\Phi\rangle\rangle}},$$

where i_0, j_0 are suitable fixed integers, and $e^{i\theta}$ is an irrelevant (unmeasurable) overall phase factor corresponding to $\theta = \arg(\langle\langle i_0,j_0|\Phi\rangle\rangle)$. Using Eq. (2.126) we can write the matrix A_{ij} in terms of output ensemble averages as follows:

$$A_{ij} = \kappa \langle E_{ij}\rangle, \tag{2.127}$$

where the operator $E_{ij}(\psi)$ is given by

$$E_{ij} = |i_0\rangle\langle i| \otimes |j_0\rangle\langle j|, \tag{2.128}$$

and the proportionality constant is given by

$$\kappa = e^{i\theta} \sqrt{\frac{dp_A}{\langle\langle\Phi|i_0,j_0\rangle\rangle\langle\langle i_0,j_0|\Phi\rangle\rangle}}. \tag{2.129}$$

Since A_{ij} is written only in terms of output ensemble averages, it can be estimated through quantum tomography, apart from the overall arbitrary phase θ.

Exercise 2.77

The derivation is just the substitution of the Golden rule for quantum extension in Eq. (2.73):

$$\mathcal{A}(\rho) = \text{Tr}_E[A\rho A^\dagger] = \text{Tr}_E[U(I_A \otimes |0\rangle)\rho(I_A \otimes \langle 0|)U^\dagger] = \text{Tr}_E[U(\rho \otimes |0\rangle\langle 0|)U^\dagger].$$

---

[41] D'Ariano and Lo Presti (2001).

Exercise 2.78

From the definitions of the various ingredients, it follows that the Hilbert spaces are constrained by the following identity

$$\dim(\mathcal{H}_A \otimes \mathcal{H}_F) = \dim(\mathcal{H}_B \otimes \mathcal{H}_E).$$

Upon writing the instrument in the Kraus form

$$\forall i \in X: \qquad \mathcal{T}_i(\rho) = \sum_{l \in X_i} T_l^{(i)} \rho (T_l^{(i)})^\dagger, \qquad T_l^{(i)} \in \mathsf{Lin}(\mathcal{H}_A \rightarrow \mathcal{H}_B),$$

it is easy to check that the instrument can be rewritten as follows:

$$\mathcal{T}_i(\rho) = \mathrm{Tr}_E[T \rho T^\dagger (I_B \otimes Z_i)], \tag{2.130}$$

$$T = \sum_{i \in X} \sum_{l \in X_i} T_l^{(i)} \otimes |i, l\rangle, \quad Z_i = \sum_{l \in X_i} |i, l\rangle \langle i, l|, \tag{2.131}$$

and the completeness of the instrument implies that $T^\dagger T = I_A$. Then, we just use the golden rule for quantum extensions in Appendix 2.1:

$$T : \mathcal{H}_A \hookrightarrow \mathcal{H}_B \otimes \mathcal{H}_F, \quad T = U(I_A \otimes |0\rangle_F), \tag{2.132}$$

with $|0\rangle_F \in \mathcal{H}_F$ any pure state and $U \in \mathsf{Lin}(\mathcal{H}_A \otimes \mathcal{H}_F)$ depending on T and $|0\rangle_F$.

Exercise 2.80

$$(R_A^\mathsf{T})^{\frac{1}{2}} = \sum_i \frac{1}{\|A_i^*\|} |A_i^*\rangle\rangle \langle\langle A_i^*|$$

$$(I_B \otimes (R_A^\mathsf{T})^{\frac{1}{2}})(|I_B\rangle\rangle \otimes I_A) = I_B \otimes \sum_i \frac{|A_i^*\rangle\rangle}{\|A_i^*\|} \langle\langle A_i^*|_{BA}|I\rangle\rangle_{BB} = \sum_i A_i \otimes |i\rangle,$$

where $|n\rangle := |A_n^*\rangle\rangle/\|A_n^*\|$ are orthonormal.

Exercise 2.81

For an ancilla prepared in a mixed state, say $\sigma = \sum_n |s_n\rangle\langle s_n|$ $(\sum_n \|s_n\|^2 = 1)$ we have

$$\rho_i = \mathrm{Tr}_C[U(\rho \otimes \sigma)U^\dagger (I \otimes |i\rangle\langle i|)] = \sum_n A_i^{(n)} \rho A_i^{(n)\dagger}, \tag{2.133}$$

with

$$A_i^{(n)} = (I \otimes \langle i|)U(I \otimes |s_n\rangle),$$

which is a general quantum operation.

Exercise 2.82

Using the canonical basis for $|I\rangle\rangle$ we take the matrix elements of both sides of identity (2.81)

$$\sum_{i\in X} (\langle n|\otimes\langle m|)|A_i\rangle\rangle \langle\langle B_i|(|l\rangle\otimes|k\rangle) = \delta_{nl}\delta_{mk}$$

and obtain

$$\sum_{i\in X} \langle n|A_i|m\rangle\langle k|B_i^{\dagger}|l\rangle = \delta_{nl}\delta_{mk}.$$

The same identity is obtained by taking the matrix element $\langle k|\dots|m\rangle$ of both sides of the first of identities (2.82) for $X = |l\rangle\langle n|$. The second of identities (2.82) is obtained by taking the adjoint of the first identity. A similar derivation can be easily found for the identities in Eq. (2.83).

Exercise 2.83

The operators E_i are positive by construction. Using the identities (2.83) one can check that they are also a resolution of the identity. The last check is a trivial application of the formalism in Section 2.8.2.

Exercise 2.84

The POVM can be regarded as a special case of instrument where the final system is the trivial system, namely with Hilbert space $\mathcal{H}_B = \mathbb{C}$. Explicitly the operation is given by

$$\mathcal{P}_i(\rho) = \text{Tr}[P_i\rho].$$

Notice that the Heisenberg picture will have as input the trivial space $\mathcal{H}_B = \mathbb{C}$, and the Heisenberg-picture version of the map is given by

$$\mathcal{P}_i^{\dagger} \equiv P_i,$$

where we omitted the c-number operator over the trivial Hilbert space $\mathcal{H}_B = \mathbb{C}$. Using any rank-one refinement of P_i, e.g.

$$P_i = \sum_{n\in Y_i} |f_n^{(i)}\rangle\langle f_n^{(i)}|, \tag{2.134}$$

one has

$$\mathcal{P}_i(\rho) = \text{Tr}[P_i\rho] = \sum_{n\in Y_i} \langle f_n^{(i)}|\rho|f_n^{(i)}\rangle.$$

One can see how here the vectors $|f_n^{(i)}\rangle$ play the role of the Kraus operators of the quantum operation. The dilation isometry $V \in \text{Lin}(\mathcal{H}_A \to \mathcal{H}_B)$ is then given by

$$V = \sum_{i\in X} V_i = \sum_{i\in X}\sum_{n\in Y_i} \langle f_n^{(i)}|\otimes|n,i\rangle \equiv \sum_{i\in X}\sum_{n\in Y_i} |n,i\rangle\langle f_n^{(i)}|, \quad V^{\dagger}V = I_A, \tag{2.135}$$

with $\{|n, i\rangle\}$ orthonormal basis for $\mathcal{H}_C$, and everything is defined as in Eq. (2.130). Then, it follows that

$$Z_i = \sum_{n \in Y_i} |n, i\rangle\langle n, i|. \tag{2.136}$$

Using the Heisenberg picture of the instrument and omitting the identity over the trivial system, one has

$$P_i = V^\dagger Z_i V,$$

namely the statement.

Exercise 2.89

From Eqs. (2.134) and (2.136) we can see that $\operatorname{rank} P_i \leq \operatorname{rank} Z_i$, and the dilation is minimal for $\operatorname{rank} Z_i = \operatorname{rank} P_i$, e.g. when we take $\{|f_n^{(i)}\rangle\}$ as the eigenvectors of P_i. The dimension of the dilated Hilbert space $\mathcal{E} \supseteq \mathcal{H}$ will then be bounded as follows (see Exercise 2.84):

$$\dim \mathcal{E} \geq \sum_{e \in X} \operatorname{rank} P_e.$$

Exercise 2.90

If a POVM $\{P_k\}$ is commuting, then we can jointly diagonalize all elements P_k (for $k = 1, \ldots, M$) as follows:

$$P_k = \sum_m p(k|m)|m\rangle\langle m|, \qquad k = 1, \ldots, M, \tag{2.137}$$

where $\{|m\rangle\}$ denote the common orthonormal eigenvectors, and the eigenvalues $p(k|m) \geq 0$, with the completeness condition $\sum_{k=1}^M p(k|m) = 1$, are written in this way, since they can be interpreted as the conditional probability of measuring k given the "true" value was m. From this diagonalization we can see that commuting POVMs are just trivial generalizations of the observable concept, with an additional randomization of the measurement outcome.

A Naimark dilation of the commuting POVM can be obtained on the extended space $\mathcal{H} \otimes \mathbb{C}^M$, in terms of the orthogonal projections

$$Z_k = \sum_m |m\rangle\langle m| \otimes U_m^\dagger|k\rangle\langle k|U_m,$$

with the ancilla prepared in any fixed state $|\omega\rangle \in \mathbb{C}^M$, and $\{|k\rangle\}$ any orthonormal basis for $\mathbb{C}^M$, $\{U_m\}$ being any family of unitary operators achieving the isometries

$$U_m|\omega\rangle = \sum_{k=1}^M \sqrt{p(k|m)}|k\rangle.$$

The Naimark dilation can be easily checked through the following steps:

$$(I \otimes \langle\omega|)Z_k(I \otimes |\omega\rangle) = \sum_m |m\rangle\langle m||\langle\omega|U_m^\dagger|k\rangle|^2 = \sum_m |m\rangle\langle m| \, p(k|m) \equiv P_k.$$

Exercise 2.93

The instrument in Eq. (2.91) is easily derived by following the same derivation of Exercise 2.91, in particular using Eq. (2.90) with a single Kraus element of the channel. The instrument is not atomic for mixed state Ω. From Eq. (2.89) we see that in order to have $\mathcal{T}$ atomic the state Ω must be pure, since otherwise $\Omega = \sum_j \Omega_j$ one must have all Ω_j mutually proportional, which in the present case is true iff the operators F_j are mutually proportional – in contradiction with the mixedness hypothesis of Ω. Using Corollary 2.29 one can also prove that for non-maximally entangled state Ω it is impossible to achieve perfect teleportation deterministically.

Exercise 2.96

In the considered case one has $P = I$, hence the Knill–Laflamme condition rewrites

$$E_j^\dagger E_i = \alpha_{ij} I.$$

For $i = j$ one has

$$\forall i \; E_i = \alpha_i V \qquad \alpha_i := \sqrt{\alpha_{ii}} e^{i\phi_i},$$

ϕ_i arbitrary phase, with V isometric, hence unitary, since we are considering operators on a finite-dimensional Hilbert space $\mathcal{H}$. For $i \neq j$ one has $\alpha_{ij} = \alpha_i^* \alpha_j$, hence the state α is pure.

Exercise 2.97

Consider the state $\rho \in \mathsf{St}(A)$ upon which the channel is invertible, and denote by R the purifying system and by E the environment for purification of the channel. Consider the isometry $V = \sum_i E_i \otimes |i\rangle_\mathrm{E}$, corresponding to the Kraus operators $\{E_i\}$. By the Golden rule (2.103) V describes the unitary interaction with a fixed state of the environment E, i.e. $V|\psi\rangle_\mathrm{A} = U(|\psi\rangle_\mathrm{A} \otimes |0\rangle_\mathrm{E})$. The action of the isometry on a purification $|F_\rho\rangle\!\rangle_\mathrm{AR}$ of ρ gives

$$(V \otimes I_\mathrm{R})|F_\rho\rangle\!\rangle\langle\!\langle F_\rho|(V^\dagger \otimes I_\mathrm{R}) = \sum_{ij} |E_i F_\rho\rangle\!\rangle\langle\!\langle E_j F_\rho| \otimes |i\rangle\langle j| =: |\Psi\rangle\langle\Psi|_\mathrm{ARE}.$$

The marginal state ρ_RE of the output state $|\Psi\rangle\langle\Psi|_\mathrm{ARE}$ is given by

$$\rho_\mathrm{RE} = \sum_{ij} F_\rho^{\mathsf{T}} E_i^* E_j^* F_\rho^* \otimes |i\rangle\langle j| = \sum_{ij} F_\rho^{\mathsf{T}} (P E_i^\dagger E_j P)^* F_\rho^* \otimes |i\rangle\langle j|,$$

and using the Knill–Laflamme conditions (2.99) one has

$$\rho_\mathrm{RE} = (F_\rho^\dagger F_\rho)^* \otimes \sum_{ij} \alpha_{ij}^* |i\rangle\langle j| = \rho^* \otimes \alpha^*,$$

which is factorized. The state of the environment α^* after the interaction is thus the complex conjugate state of the Knill–Laflamme state.

Exercise 2.98

The Knill–Laflamme conditions (2.99) for the environment-state in the diagonalized form become

$$P F_n^\dagger F_m P = \delta_{nm} d_n P,$$

which means that the following operators

$$V_m := \frac{F_m P}{\sqrt{d_m}}$$

are isometric over $\mathsf{Supp}\, P$ and have orthogonal ranges. Notice that the special Kraus decomposition $\{F_m\}$ is not necessarily the canonical one, but it can be connected to it via the isometric transformation that diagonalizes the environment state α

$$PF_n^\dagger F_m P = \sum_{ij} u_{in}^* u_{jm} PE_i^\dagger E_j P = \sum_{ij} u_{in}^* u_{jm} \alpha_{ij} P = \delta_{nm} d_n P.$$

Notice that the isometric matrix $\{u_{ij}\}$ is unitary only if the Kraus decomposition $\{F_m\}$ is minimal. Since the Knill–Laflamme condition is equivalent to correctability of the channel upon input of ρ, the existence of a Kraus decomposition providing a set of orthogonal isometries over $\mathsf{Supp}\, \rho = \mathsf{Supp}\, P$ is an equivalent condition for correctability upon input of ρ.

Exercise 2.99

We will explicitly construct such a channel. Take any probability distribution $\{p_i\}_{i=1,\dots,k}$, and any set of k isometries $V_1, V_2, \dots, V_k \in \mathsf{Lin}(\mathsf{Supp}\, \rho \to \mathcal{H})$ with orthogonal ranges: the ranges can be taken orthogonal since $\dim(\mathcal{H}) \geq k \operatorname{rank} \rho$. The channel satisfying Knill–Laflamme is given by the random isometry

$$\mathcal{C} = \sum_j p_j V_j \cdot V_j^\dagger.$$

Indeed, one has

$$\sqrt{p_i p_j} V_i^\dagger V_j \equiv \sqrt{p_i p_j} P V_i^\dagger V_j P = \delta_{ij} p_i P,$$

which is a special case of the Knill–Laflamme condition for $\alpha = \mathsf{Diag}\,[p_i]$.

Another way of proving that the map is correctable upon input ρ is to show that the following map is a channel correcting $\mathcal{C}$

$$\mathcal{R} = \sum_j V_j^\dagger \cdot V_j.$$

Indeed, $\mathcal{R}$ is evidently CP, but it is also trace-preserving, since for all $X \in \mathsf{Lin}(\mathcal{H})$ one has

$$\mathrm{Tr}[\mathcal{R}(X)] = \sum_j \mathrm{Tr}[V_j^\dagger X V_j] = \mathrm{Tr}\left[X \sum_j V_j V_j^\dagger\right] = \mathrm{Tr}\left[X \sum_j P_{\mathsf{Rng}\,(V_j)}\right] = \mathrm{Tr}[X].$$

We emphasize that once we found a correctable channel, then all other channels with Kraus operators that are linear combinations of the Kraus operators E_i of our channel are also correctable upon the same input. We will then just say that the inverting map $\mathcal{R}$ corrects all errors corresponding to a quantum operation $\mathcal{E}(\rho) = E\rho E^\dagger$, with E any linear combination of E_i.

Problem 2.1

Since the operator is self-adjoint, then it is diagonalizable with discrete spectrum, say $\{x_i\}$ and one has

$$\mathrm{Tr}X^2 = \sum_i x_i^2 = 1 \implies |x_i| \le 1 \implies 1 - x_i \ge 0.$$

Now, we have

$$0 = \mathrm{Tr}X^2 - \mathrm{Tr}X^3 = \sum_i x_i^2(1 - x_i) \implies x_i = 0 \text{ or } x_i = 1,$$

and the condition that $\mathrm{Tr}X^2 = 1$ implies that $x_i = \delta_{ii_0}$.

The proof holds in the general infinite-dimensional case, by considering that the operator X is compact (it is Hilbert–Schmidt since the trace of its square is bounded), then it is diagonalizable with discrete spectrum.

Problem 2.6

One has

$$\mathrm{Tr}[\rho\rho'] = \tfrac{1}{4}\mathrm{Tr}[I + n_\alpha n'^\alpha I] + \text{traceless terms} = \tfrac{1}{2}(1 + \mathbf{n} \cdot \mathbf{n}').$$

Upon substituting $\rho = |\psi\rangle\langle\psi|$ and $\rho' = |\psi'\rangle\langle\psi'|$ in the trace we find $|\langle\psi|\psi'\rangle|^2$. For probabilistic states the overlap (2.107) simply becomes $\frac{tt'}{2}(1 + \mathbf{n} \cdot \mathbf{n}')$, with $t = \mathrm{Tr}\,\rho_{\mathbf{n}}$ and $t = \mathrm{Tr}\,\rho_{\mathbf{n}'}$.

Problem 2.7

The purity is a special case of identity (2.107), corresponding to

$$\mathrm{Tr}[\rho^2] = \tfrac{1}{2}(1 + \|\mathbf{n}\|^2).$$

Therefore, pure states correspond to $\|\mathbf{n}\| = 1$, i.e. points on the sphere surface. This is in agreement with the convex structure of the sphere, with the surface being the set of extremal points.

Problem 2.8

The phase factor $\omega(g, h)$ (called *cocycle*) must satisfy the following constraints for associativity (called *Jacobi identities*):

$$\omega(gh, l)\,\omega(g, h) = \omega(g, hl)\omega(h, l), \qquad g, h, l \in G. \tag{2.138}$$

The projective representation of a group $\mathbf{G}$ can be lifted to a linear representation of a different group $\mathbf{G}'$, which is a *central extension* of $\mathbf{G}$.

Problem 2.10

One has

$$\sigma_\alpha^2 = I, \ \alpha = x, y, z.t, \quad \sigma_x\sigma_y = i\sigma_z \text{ and cyclic permutations}, \tag{2.139}$$

and the cocycle $\omega(\alpha, \beta)$ is given by

$$\omega(\alpha, \beta) = \begin{bmatrix} 1 & i & -i & 1 \\ -i & 1 & i & 1 \\ i & -i & 1 & 1 \\ 1 & 1 & 1 & 1 \end{bmatrix}.$$

One can check that $\omega(\alpha, \beta)$ satisfies the cocycle conditions (2.138). The group without cocycle is the Abelian group of the rotations of a π angle around three orthogonal axes, which is the *Klein group* $\mathbf{Z}_2 \times \mathbf{Z}_2$, the direct product of two copies of the cyclic group of order 2 (which is the smallest non-cyclic group).

Problem 2.12

A transformation of the form

$$\mathbf{N} \rightarrow \mathbf{N}' = \mathbf{A}\mathbf{N}\mathbf{A}^\dagger$$

preserves Hermiticity, hence it defines a transformation of the vector components

$$n^i \rightarrow m^i = \Lambda^i_j n^j,$$

with real matrix Λ. Moreover, the transformation preserves the determinant, since $\mathbf{A} \in SL(2, \mathbb{C})$, thus it preserves the Minkowski norm of n. Therefore, Λ is a matrix in the Lorentz group, and has the form

$$\Lambda^j_i(\mathbf{A}) = \frac{1}{2}\mathrm{Tr}[\mathbf{A}\sigma_i\mathbf{A}^\dagger\sigma_j].$$

Problem 2.13

Using Eq. (2.139) we find the commutation relations

$$[\sigma_\alpha, \sigma_\beta] = 2if^\gamma_{\alpha\beta}\sigma_\gamma, \tag{2.140}$$

$f^\gamma_{\alpha\beta}$ denoting the structure constants of the Lie algebra. From Eq. (2.139) we obtain the completely antisymmetric tensor with $f^z_{xy} = f^x_{yz} = f^y_{zx} = 1$. By taking real combinations of the basis in Eq. (2.140) we build up a real Lie algebra. A relevant result of the theory of Lie groups is that from a Lie algebra via the exponential map we generate the corresponding Lie group. In the present case this is the group SU(2) of unitary matrices for dimension $d = 2$ with unit determinant, given by the exponentials

$$U_\mathbf{n}(\theta) = \exp\left(-\frac{1}{2}i\boldsymbol{\sigma} \cdot \mathbf{n}\,\theta\right) = \cos(\theta/2) - i\boldsymbol{\sigma} \cdot \mathbf{n}\sin(\theta/2),$$

with $\mathbf{n}$ the unit vector of the rotation axis, and θ the rotation angle. The group coincides with its own defining representation as unitary matrices in dimension 2. The corresponding Lie algebra is denoted with the lowercase letters $su(2)$.

Problem 2.15

The adjoint representation in Eq. (2.109) establishes a group homomorphism SU(2) $\rightarrow$ SO(3). The mapping is not one-to-one since the kernel is not trivial, and this is related to the different periodicities of the two groups, namely 2π periodicity for SO(3) and 4π

periodicity for SU(2). Indeed, from Eq. (2.109) we see that $U_n(2\pi) = -I$, however, the corresponding SO(3) matrix is $\mathbf{R_n}(2\pi) = \mathbf{R_n}(0) = I$. Thus, two distinct elements map onto the identity of SO(3): this is the kernel of the homomorphism, namely $Z_2 = \{I_2, -I_2\}$, and one has the isomorphism SO(3) $=$ SU(2)$/Z_2$. As a consequence of this the group SU(2) is said to be the *double-covering* of SO(3). The two groups SO(3) and SU(2) differ in their global properties; however, they share the same Lie algebra, since one can write the rotation matrix as

$$\mathbf{R_n}(\theta) = \exp(-\tfrac{i}{2}\theta\,\mathbf{J}\cdot\mathbf{n}),$$

where the angular momentum three-dimensional matrices J_x, J_y, J_z satisfy the same commutation relations as the Pauli matrices in Eq. (2.140).

Problem 2.17

We notice that the product of any three operators is equal (apart from a sign) to the fourth one, hence any joint eigenvector of the three is also eigenvector of them all. The joint eigenvectors are simply given by

$$|\Psi^s_{ijk}\rangle = \frac{1}{\sqrt{2}}(|i\rangle \otimes |j\rangle \otimes |k\rangle + \sigma_z^{\otimes 3}|-i\rangle \otimes |-j\rangle \otimes |-k\rangle) = U|i\rangle \otimes |j\rangle \otimes |k\rangle, \quad i,j,k = \pm,$$

where $|\pm\rangle$ here denote the eigenvectors of σ_z, and $\sigma_z^{\otimes 3}$ is the parity operator, whereas U is the unitary operator

$$U = \frac{1}{\sqrt{2}}(P_+ - P_- + \sigma_x^{\otimes 3}),$$

$P_\pm$ denoting the orthogonal projectors over parity eigen-spaces, namely

$$P_\pm = \frac{1}{2}(I \pm \sigma_z^{\otimes 3}). \tag{2.141}$$

Using identity (2.141) we rewrite U in the manifestly unitary form

$$U = \frac{1}{\sqrt{2}}(\sigma_z^{\otimes 3} + \sigma_x^{\otimes 3}).$$

Problem 2.18

Construct the compression operator F

$$F = \sum_{i=1}^{3} |\psi_i\rangle\langle i|,$$

where $\{|i\rangle\}$ is any orthonormal basis. The matrix form of F is given by

$$F = \frac{1}{\sqrt{6}}\begin{bmatrix} 1 & 1 & 2 \\ -\sqrt{3} & \sqrt{3} & 0 \end{bmatrix}.$$

The Naimark extension is given by

$$P_i = F|i\rangle\langle i|F^\dagger,$$

and one can easily verify it using the matrix form.

Problem 2.19

Upon considering the Hilbert-space extension $\mathcal{E} = \mathcal{H} \otimes \mathbb{C}^2$ with $\{|0\rangle, |1\rangle\}$ canonical orthonormal basis for $\mathbb{C}^2$, one has

$$P = (I_{\mathcal{H}} \otimes \langle 0|)Z(I_{\mathcal{H}} \otimes |0\rangle), \qquad I_{\mathcal{H}} - P = (I_{\mathcal{H}} \otimes \langle 0|)(I_{\mathcal{E}} - Z)(I_{\mathcal{H}} \otimes |0\rangle).$$

One can also easily check that Z in Eq. (2.110) is an orthogonal projector.

Problem 2.20

A Naimark's dilation of the POVM (2.20) can be obtained as follows. Extend the Hilbert space $\mathcal{H}$ as $\mathcal{H} \otimes \mathcal{A}$, where $\mathcal{A} \simeq \mathbb{C}^N$. A set of orthogonal projectors on the extended Hilbert space is given by

$$Z_m = \sum_{l=1}^{M} Z_m^{(l)} \otimes |l\rangle\langle l|,$$

where $\{|l\rangle\}$ denotes any orthonormal basis for $\mathcal{A}$. Then, prepare the ancilla in the state

$$|\omega\rangle = \sum_{l=1}^{M} \varsigma_i^{1/2} |l\rangle .$$

One can immediately check that

$$(I \otimes \langle \omega|)Z_m(I \otimes |\omega\rangle) = \sum_{i=1}^{M} \varsigma_i Z_m^{(i)} \equiv P_m,$$

or, equivalently, one has the indirect measurement model

$$P_m = \mathrm{Tr}_{\mathcal{A}}[(I_{\mathcal{H}} \otimes |\omega\rangle\langle\omega|)Z_m].$$

This example also shows that a random observable can be simulated by a quantum ancilla in a pure state, the ancilla playing the role of a device randomizing the observable, a sort of "quantum roulette" wheel. The same construction as in this problem can be generalized to a random choice between POVMs.

Problem 2.21

We need to show that it is possible to have the set summing to e_{A}. The proof is by induction. The cardinality of the set is $\dim \mathsf{Eff}_{\mathbb{R}}(\mathrm{A}) < \infty$. If $\dim \mathsf{Eff}_{\mathbb{R}}(\mathrm{A}) = 1$ one has $\mathsf{Eff}(\mathrm{A}) = \{e_{\mathrm{A}}\}$ and we are done (the theory provides no information from observations, since all effects are proportional to the deterministic effect e_{A}). For $\dim \mathsf{Eff}_{\mathbb{R}}(\mathrm{A}) > 1$ there exists at least a test $\{l_i\}_{i\in\mathrm{X}}$ with $|\mathrm{X}| \geq 2$ linearly independent effects (otherwise all effects are proportional to e_{A}). If this is the only available test, again we are done. Otherwise, pick out a new binary test $\{x, y\}$ from the set of available tests (using coarse-graining we can make any test binary). If $x \in \mathsf{Span}_{\mathbb{R}}\{l_i\}_{i\in\mathrm{X}}$ discard the test. Otherwise for $x \notin \mathsf{Span}_{\mathbb{R}}\{l_i\}_{i\in\mathrm{X}}$, then necessarily also $y \notin \mathsf{Span}_{\mathbb{R}}\{l_i\}_{i\in\mathrm{X}}$ [since if there exists coefficient λ_i such that $y = \sum_{i\in\mathrm{X}} \lambda_i l_i$, then $x = \sum_{i\in\mathrm{X}}(1 - \lambda_i)l_i$]. Now, consider the observation test

$$\left\{ \tfrac{1}{2}y, \tfrac{1}{2}(l_1 + x), \tfrac{1}{2}l_2, \ldots, l_n \right\}$$

which is a coarse-graining of the randomization of the two tests $\{l_i\}$ and $\{x, y\}$ with probability $\frac{1}{2}$. Such new observation test has now $|X| + 1$ linearly independent effects (since y is linearly independent of the l_i and one has $y = \sum_{i \in X} l_i - x = \sum_{i=2}^{|X|} l_i + l_1 - x$). By iterating the above procedure we reach cardinality $|X| = \dim \mathsf{Eff}_\mathbb{R}(A)$, and we have so realized an informationally complete test.

Problem 2.23

Since $\mathcal{E}$ is CP iff $R_\mathcal{E} \geq 0$, this means that $\mathcal{E}$ is CP iff

$$\langle \Psi | R_\mathcal{E} | \Psi \rangle \geq 0, \qquad \forall | \Psi \rangle \in \mathcal{K} \otimes \mathcal{H}.$$

Upon writing $|\Psi\rangle = \sum_{i=1}^{n} |k_i\rangle \otimes |h_i\rangle^*$, for any sequence of vectors $|k_i\rangle \in \mathcal{K}$ and $|h_i\rangle \in \mathcal{H}$, one has

$$0 \leq \langle \Psi | R_\mathcal{E} | \Psi \rangle = \sum_{rs} \sum_{ij=1}^{n} \langle k_i | \mathcal{E}(|r\rangle \langle s|) | k_j \rangle \, \langle h_i | r \rangle^* \, \langle s | h_j \rangle^* = \sum_{ij=1}^{n} \langle k_i | \mathcal{E}(|h_i\rangle \langle h_j|) | k_j \rangle,$$

namely the statement. Notice the interesting fact that one checks the complete positivity of the map without using its extension!

Problem 2.24

Consider the effect on $\mathcal{H}_\mathrm{B}$ $P_\sigma = [\Psi^\ddagger \sigma (\Psi^\ddagger)^\dagger]^*$, where $\Psi^\ddagger$ is given by

$$\Psi^\ddagger = V \sum_{j, \lambda_j \neq 0} |j\rangle \lambda_j^{-\frac{1}{2}} \langle j|,$$

$\{\lambda_j\}$ and $\{|j\rangle\}$ denote the eigenvalues and eigenvectors of ρ, respectively ($\Psi^\ddagger$ is called the *Moore–Penrose pseudoinverse* of Ψ). It is immediate to check that $0 \leq P_\sigma \leq I_\mathrm{B}$, since

$$0 \leq \Psi^\ddagger \sigma (\Psi^\ddagger)^\dagger \leq \Psi^\ddagger \rho (\Psi^\ddagger)^\dagger = VV^\dagger \leq I_\mathrm{B}.$$

One can also check that

$$\Psi \Psi^\ddagger = P_{\mathsf{Supp}\,\rho} \in \mathsf{Lin}(\mathcal{H}_\mathrm{A}), \qquad \Psi^\ddagger \Psi = VV^\dagger \in \mathsf{Lin}(\mathcal{H}_\mathrm{B}),$$

$VV^\dagger$ being an orthogonal projector over a space isomorphic to $\mathsf{Supp}\,\rho$. Finally, one has

$$= (b_\sigma |_\mathrm{B} | \Psi) = \mathrm{Tr}_2[(I \otimes P_\sigma) |\Psi\rangle\rangle \langle\langle \Psi|] = \Psi P_\sigma^* \Psi^\dagger$$

$$= \Psi \Psi^\ddagger \sigma (\Psi^\ddagger)^\dagger \Psi^\dagger = \Psi \Psi^\ddagger \sigma (\Psi \Psi^\ddagger)^\dagger = \sigma$$

since $\mathsf{Supp}\,(\sigma) \subseteq \mathsf{Supp}\,\rho$, being $\sigma \in \mathsf{RefSet}\,\rho$. Notice that if we take a full preparation test $\{\sigma_j\}_{j \in X}$ refining ρ, we have

$$\sum_j P_j = [\Psi^\ddagger \rho (\Psi^\ddagger)^\dagger]^* = (VV^\dagger)^* \leq I_\mathrm{B},$$

and the POVM is generally not complete: its coarse-graining gives the orthogonal projector on a space isomorphic to $\mathsf{Supp}\,\rho$. However we can always complete it by adding a term $P_* = I_\mathrm{B} - (VV^\dagger)^*$ that occurs with zero probability over the purifying state.

Problem 2.25

If we apply a channel to the state R we have

$$\mathcal{C} \otimes \mathcal{I}(R) = \mathcal{C} \otimes \mathcal{R}^{\mathsf{T}}(|I\rangle\!\rangle\langle\!\langle I|) = \mathcal{I} \otimes \mathcal{R}^{\mathsf{T}}(R_{\mathcal{C}}),$$

where $\mathcal{R}^{\mathsf{T}} = \sum_i A_i^{\mathsf{T}} \cdot A_i^*$. The state R is faithful iff the Choi–Jamiołkowski operator $R_{\mathcal{C}}$ of the channel can be obtained from $\mathcal{C} \otimes \mathcal{I}(R)$, namely by inversion of the CP map $\mathcal{R}^{\mathsf{T}}$, which is invertible iff $\mathcal{R}$ is invertible.

Problem 2.27[42]

The solution proceeds in the same way as in that of Exercise 2.74, with the difference that Eq. (2.126) now becomes

$$A = \Phi \Psi^{-1} \sqrt{p_A(\Psi)}, \quad p_A = \|A\Psi\|_2,$$

hence Eq. (2.127) becomes

$$A_{ij} = e^{i\theta} \sqrt{\frac{p_A(\Psi)}{\langle\!\langle \Phi|i_0, j_0\rangle\!\rangle \langle\!\langle i_0, j_0|\Phi\rangle\!\rangle}} |i_0\rangle\langle i| \otimes |j_0\rangle\langle j|. \tag{2.142}$$

Everything is well defined in infinite dimensions.

Problem 2.28

Let $\mathcal{E}(\cdot) = \sum_i E_i \cdot E_i^\dagger$ be a Kraus representation for $\mathcal{E}$. Then, one has

$$\sum_i [X^{\frac{1}{2}}, E_i] [X^{\frac{1}{2}}, E_i]^\dagger = X^{\frac{1}{2}} \mathcal{E}(I) X^{\frac{1}{2}} - X^{\frac{1}{2}} \mathcal{E}(X^{\frac{1}{2}}) - \mathcal{E}(X^{\frac{1}{2}}) X^{\frac{1}{2}} + \mathcal{E}(X).$$

Clearly, the r.h.s. is zero if and only if the l.h.s. is zero. But the l.h.s. is zero if and only if $[X^{\frac{1}{2}}, E_i] = 0$ for every i, or equivalently, if and only if $[X, E_i] = 0$ for every i.

Problem 2.29

The instrument has POVM

$$A_0^\dagger A_0 = \sum_{n=1}^{\infty} |2n\rangle\langle 2n| + \tfrac{1}{2}|0\rangle\langle 0|, \quad A_1^\dagger A_1 = \sum_{n=1}^{\infty} |2n+1\rangle\langle 2n+1| + \tfrac{1}{2}|0\rangle\langle 0|,$$

which is evidently non-orthogonal. Moreover, one can see that $A_0 A_1 = A_1 A_0 = 0$, and

$$A_0^2 = \sum_{n=1}^{\infty} |8n\rangle\langle 2n| + \tfrac{1}{\sqrt{2}}|4\rangle\langle 0|, \quad A_1^2 = \sum_{n=1}^{\infty} |8n+1\rangle\langle 2n+1| + \tfrac{1}{\sqrt{2}}|5\rangle\langle 0|,$$

from which it follows that $(A_0^\dagger)^2 A_0^2 = A_0^\dagger A_0$ and $(A_1^\dagger)^2 A_1^2 = A_1^\dagger A_1$. It follows that the instrument satisfies Eqs. (2.121) and (2.121) for the repeatability.

Problem 2.30

We write the two purifications as in Eq. (2.66), namely

$$|\Psi_\rho\rangle_{\mathrm{AB}} = (I_{\mathrm{A}} \otimes V)|\rho^{\frac{1}{2}}\rangle\!\rangle, \quad |\Psi_\rho\rangle_{\mathrm{AB}'} = (I_{\mathrm{A}} \otimes W)|\rho^{\frac{1}{2}}\rangle\!\rangle,$$

---

[42] D'Ariano and Lo Presti (2001).

with $V \in \mathsf{Lin}(\mathcal{H}_A \to \mathcal{H}_B)$ and $W \in \mathsf{Lin}(\mathcal{H}_A \to \mathcal{H}_{B'})$ partial isometries with support on $\mathsf{Supp}\,\rho$. Then one has

$$|\Psi_\rho\rangle_{AB'} = (I_A \otimes WV^\dagger)|\Psi_\rho\rangle_{AB},$$

and $Z = WV^\dagger \in \mathsf{Lin}(\mathcal{H}_B \to \mathcal{H}_{B'})$ is a partial isometry with

$$Z^\dagger Z = (WV^\dagger)^\dagger WV^\dagger = VP_\rho V^\dagger = VV^\dagger = P'_\rho,$$

where $P_\rho = \mathsf{Proj}\,\mathsf{Supp}\,\rho$ and P'_ρ is the orthogonal projector on a subspace of $\mathcal{H}_B$ isomorphic to $\mathsf{Supp}\,\rho$. Therefore we have an atomic transformation connecting the two purifications with orthogonal quantum effect, which can be completed to a quantum channel by adding another atomic transformation with the Kraus operator being a partial isometry $X \in \mathsf{Lin}(\mathcal{H}_B \to \mathcal{H}_{B'})$ such that $XX^\dagger = I_B - P'_\rho$.

Problem 2.31

Most generally the system A can undergo a probabilistic transformation. However, if the transformation is probabilistic, there will be alternate transformations with overall coarse-graining occurring with certainty, namely the transformation is an element of a test. Being the system isolated, in principle there is no access to the outcome of the test, and the actual transformation is the deterministic one corresponding to the coarse-graining. Since such transformation cannot be non-trivially purified otherwise the system would not be isolated (it would interact with an environment), it must be a deterministic pure transformation. According to Corollary 2.10 such transformation must be isometric, hence unitary (the input and output system are the same finite-dimensional system). This corresponds to the usual assumption in thermodynamics that an isolated system must undergo a reversible transformation (see Corollary 2.12). The transformation is generally time-dependent, and we will denote by $U(t)$ the corresponding unitary operator (unique apart from a phase). Since the system is isolated, the t-dependence must be homogeneous, since there is no external influence that can determine a different evolution depending on t. Therefore we have $U(t_3 - t_2)U(t_2 - t_1) = U(t_3 - t_1)$, namely $U(t_2)U(t_1) = U(t_1 + t_2)$, and $U(t)$ must be a *one-parameter Abelian unitary group*. Therefore, it must be of the form $U(t) = \exp(-iHt)$ with $H \in \mathsf{Herm}(\mathcal{H}_A)$. Being obviously differentiable, the quantum Liouville equation (2.112) and the Schröedinger equation (2.113) follow immediately.

Problem 2.32[43]

In quantum theory a deterministic evolution depending on t is a quantum channel $\mathcal{C}_t$ parameterized by $t \geq 0$ with $\mathcal{C}_0 = \mathcal{I}_A$. Being t-homogeneous one must have (see Problem 2.31), hence

$$\mathcal{C}_{t_2}\mathcal{C}_{t_1} = \mathcal{C}_{t_1+t_2}. \tag{2.143}$$

Equation (2.143) is also called the *Markov condition*, since the t-dependent channel $\mathcal{C}_t$ belongs to the class of memory-less evolutions, namely $\rho_{t'}$ conditional on $\rho_{t''}$ with $t'' \leq t'$ is independent on ρ_t for $t \leq t''$. Therefore $\mathcal{C}_t$ for $t \geq 0$ is a *one-parameter Abelian channel semigroup*, and one has

---

[43] Alicki and Lendi (1987).

$$C_t =: \exp(t\mathcal{L}),$$

where $\mathcal{L}$ is a linear map over $\mathsf{Lin}(\mathcal{H}_A)$. One has

$$\mathcal{L} = \frac{d}{dt}\bigg|_{0^+} C_t \equiv \lim_{\epsilon \to 0^+} \frac{1}{\epsilon}(C_t - \mathcal{I}_A).$$

Introduce a linear basis F_k, $k = 0, 1, \ldots, d^2 - 1$ in $\mathsf{Lin}(\mathcal{H}_A)$, with $F_0 = I_A$ where $d := \dim(\mathcal{H}_A)$ (according to Exercise 2.43 such basis exists for every $d < \infty$). Expand the semigroup as follows:

$$C_t\rho = \sum_{kl=0}^{d^2-1} c_{kl}(t) F_k \rho F_l^\dagger,$$

where according to Exercise 2.46 $\{c_{kl}(t)\}$ is a positive-definite matrix. One has

$$\mathcal{L}\rho = \lim_{\epsilon \to 0^+} \frac{1}{\epsilon}(C_t\rho - \rho)$$

$$= \lim_{\epsilon \to 0^+} \left(\frac{c_{00}(\epsilon) - 1}{\epsilon}\rho + \sum_{k=1}^{d^2-1} \frac{c_{k0}(\epsilon)}{\epsilon} F_k\rho + \rho \sum_{k=1}^{d^2-1} \frac{c_{k0}^*(\epsilon)}{\epsilon} F_k^\dagger + \sum_{k,l=1}^{d^2-1} \frac{c_{kl}(\epsilon)}{\epsilon} F_k\rho F_l^\dagger \right)$$

$$= A\rho + \rho A^\dagger + \sum_{kl=1}^{d^2-1} a_{kl} F_k\rho F_l^\dagger,$$

where $\{a_{kl}\}$ $k, l = 1, \ldots, d^2 - 1$ is a positive-definite matrix (the submatrix of a positive matrix with a row and a column cancelled is still positive), and the term with c_{00} has been split into the two terms with A. Since the last sum represents a CP map, we can write it using a Kraus decomposition as follows:

$$\mathcal{L}\rho = A\rho + \rho A^\dagger + \sum_j V_j \rho V_j^\dagger.$$

The trace-preserving condition $\mathrm{Tr}(\mathcal{L}\rho) = 0$ corresponds to

$$A + A^\dagger = -\sum_j V_j^\dagger V_j,$$

and, upon denoting the imaginary part of A by $-iH$ with H self-adjoint, one has

$$A = -iH - \frac{1}{2}\sum_k V_k^\dagger V_k,$$

which leads to the general form

$$\mathcal{L}\rho = -i[H, \rho] + \frac{1}{2}\sum_k (2V_k\rho_t V_k^\dagger - V_k^\dagger V_k\rho_t - \rho_t V_k^\dagger V_k).$$

Notice that in absence of the V_k terms, the master equation reduces to the usual Schrödinger equation, with H playing the role of the Hamiltonian. The commutator with H thus describes the "coherent" part of the evolution, whereas the second term is the *dissipative evolution*.

PART II

THE INFORMATIONAL APPROACH

The Framework 3

> ..., all the sciences be only unconscious applications of the calculus of probabilities. To condemn this calculus would be to condemn the whole of science.
>
> Henri Poincaré, *La science et l'hypothèse.*

In this chapter we will introduce the framework of *operational probabilistic theories*. The framework is provided by first introducing the operational language that expresses the possible connections between events, and then by dressing the elements of the language with a probabilistic structure.

The operational language allows one to describe the physical processes and their mutual relations involved in an experimental setting. For example, one can specify whether two events $\mathcal{A}$ and $\mathcal{B}$ occur in sequence or in parallel in a given context. However, it is only the probabilistic structure that promotes the operational language from a merely descriptive tool to a framework for predictions, the predictive power being the crucial requirement for any scientific theory and for its testability – the essence of science itself. Different OPTs will have different rules for assigning the joint probabilities of events.

The OPT is an extension of probability theory, which in turn can be regarded as an extension of logic.[1] Therefore, the OPT can be viewed as an extension of logic. To the set of joint probabilities of probabilistic theory, the OPT adds the connectivity among events. Whereas the probabilistic aspect is only a facet of quantum theory, the event connectivity is its special trait, and calls for a thorough formulation. This is precisely the purpose of the present chapter: to provide the framework and the rules of such connectivity. In the chapters which will follow these will allow us to formulate principles for OPTs, and in particular, those of quantum theory.

3.1 The Operational Language

The primitive notions of any operational theory are those of *test, event,* and *system.* In addition to comprising a collection of *events*, the notion of test carries also the event-connectivity of the theory that is achieved by the systems. These can represent the *input* and the *output* of the test. The resulting representation of a test is the following diagram:

$$\overset{A}{\underline{\quad}}\boxed{\{\mathcal{E}_x\}_{x\in X}}\overset{B}{\underline{\quad}},$$

---

[1] Cox (1961); Jaynes (2003).

where $\{\mathcal{E}_x\}_{x\in X}$ denotes the collection of events $\mathcal{E}_x$ of the test, each labeled by the *outcome* x belonging to the *probability space* (or *outcome space*) of X. The wire on the left labeled as A represents the input system, whereas the wire on the right labeled as B represents the output system. The same diagrammatic representation is also used for any of the events, namely for $x \in X$

$$\begin{array}{c} A \ \boxed{\mathcal{E}_x} \ B \end{array}.$$

In the following, the systems will be denoted by capital Roman letters $A, B, \ldots, Z$, whereas the events by capital calligraphic letters $\mathcal{A}, \mathcal{B}, \ldots, \mathcal{Z}$.

A test can represent a single use of a physical device or a measuring apparatus, or else it epitomizes the single occurrence of a physical process. Examples are:

- A Stern–Gerlach apparatus: both input and output systems are a particle spin. The events are the two possible transformations corresponding to the particle passing through the upper or through the lower pinhole. The outcomes are "up" and "down."
- A photocounter: the input system is a mode of the electromagnetic field, whereas there is no output system (in the following we will also say that the output is the *trivial system*). The outcomes are the possible numbers of photons.
- Electron–proton scattering: both the input and the output systems are an electron–proton pair; the test contains only one event corresponding to the two-particle interaction; the outcome space is the singleton.

In the special case of the singleton test the event is *deterministic*, and we will denote the test by the event itself, e.g. $\mathcal{U}$. When there is more than a single outcome, the events of the test are probabilistic.

In some experimental protocols, it is useful to emphasize that there are *operationally equivalent systems*, meaning that two systems, being different in some sense (e.g. they are physically different or just have different spatial location) nevertheless perform in the same way and can be converted into each other perfectly and reversibly. This is the case e.g. of Alice's and Bob's photon polarization in a quantum teleportation experiment, or else the polarization of a photon and the spin of an electron, which both correspond to the same quantum system, i.e. the *qubit*. A formal definition of the notion of operationally equivalent systems will be given in the following.

Different tests can be combined in a *circuit*, which is a directed acyclic graph where the links are the systems (oriented from left to right, namely from input to output) and the nodes are the boxes of the tests. The same graph can be built up for a single test instance, namely with the network nodes being events instead of tests, corresponding to a joint outcome for all tests.

The circuit graph is obtained precisely by using the following rules.

Sequential Composition of Tests   When the output system of test $\{\mathcal{C}_x\}_{x\in X}$ and the input system of test $\{\mathcal{D}_y\}_{y\in Y}$ coincide, the two tests can be composed in sequence as follows:

$$\begin{array}{c} A \ \boxed{\{\mathcal{C}_x\}_{x\in X}} \ B \ \boxed{\{\mathcal{D}_y\}_{y\in Y}} \ C \end{array} \ =: \ \begin{array}{c} A \ \boxed{\{\mathcal{E}_{(x,y)}\}_{(x,y)\in X\times Y}} \ C \end{array},$$

resulting in the test $\{\mathcal{E}_{(x,y)}\}_{(x,y)\in X\times Y}$ called *sequential composition* of $\{\mathcal{C}_x\}_{i\in X}$ and $\{\mathcal{D}_y\}_{y\in Y}$. In formulas we will also write $\mathcal{E}_{(x,y)} := \mathcal{D}_y\mathcal{C}_x$.

Identity Test   For every system A, one can perform the *identity test* (shortly *identity*) that "leaves the system alone." Formally, this is the deterministic test $\mathcal{I}_A$ with the property

$$\xrightarrow{\text{A}}\boxed{\mathcal{I}_A}\xrightarrow{\text{A}}\boxed{\mathcal{C}}\xrightarrow{\text{B}} \quad = \quad \xrightarrow{\text{A}}\boxed{\mathcal{C}}\xrightarrow{\text{B}},$$

$$\xrightarrow{\text{B}}\boxed{\mathcal{D}}\xrightarrow{\text{A}}\boxed{\mathcal{I}_A}\xrightarrow{\text{A}} \quad = \quad \xrightarrow{\text{B}}\boxed{\mathcal{D}}\xrightarrow{\text{A}},$$

where the above identities must hold for any event $\xrightarrow{\text{A}}\boxed{\mathcal{C}}\xrightarrow{\text{B}}$ and $\xrightarrow{\text{B}}\boxed{\mathcal{D}}\xrightarrow{\text{A}}$, respectively. The sub-index A will be dropped from $\mathcal{I}_A$ where there is no ambiguity.

Operationally Equivalent Systems   We say that two systems A and A$'$ are *operationally equivalent* – denoted as $A' \simeq A$ or just $A' = A$ – if there exist two deterministic events $\xrightarrow{\text{A}}\boxed{\mathcal{U}}\xrightarrow{\text{A}'}$ and $\xrightarrow{\text{A}'}\boxed{\mathcal{V}}\xrightarrow{\text{A}}$ such that

$$\xrightarrow{\text{A}}\boxed{\mathcal{U}}\xrightarrow{\text{A}'}\boxed{\mathcal{V}}\xrightarrow{\text{A}} \quad = \quad \xrightarrow{\text{A}}\boxed{\mathcal{I}}\xrightarrow{\text{A}},$$

$$\xrightarrow{\text{A}'}\boxed{\mathcal{V}}\xrightarrow{\text{A}}\boxed{\mathcal{U}}\xrightarrow{\text{A}'} \quad = \quad \xrightarrow{\text{A}'}\boxed{\mathcal{I}}\xrightarrow{\text{A}'}.$$

Accordingly, if $\{\mathcal{C}\}_{i\in X}$ is any test for system A, *performing an equivalent test on system* A$'$ means performing the test $\{\mathcal{C}'_x\}_{x\in X}$ defined as

$$\xrightarrow{\text{A}'}\boxed{\mathcal{C}'_x}\xrightarrow{\text{A}'} \quad = \quad \xrightarrow{\text{A}'}\boxed{\mathcal{V}}\xrightarrow{\text{A}}\boxed{\mathcal{C}_x}\xrightarrow{\text{A}}\boxed{\mathcal{U}}\xrightarrow{\text{A}'}.$$

Composite Systems   Given two systems A and B, one can join them into the single *composite system* AB. As a rule, the system AB is operationally equivalent to the system BA, and we will identify them in the following. This means that the system composition is *commutative*,[2]

$$AB = BA.$$

We will call a system *trivial system*, reserving for it the letter I, if it corresponds to the identity in the system composition, namely

$$AI = IA = A.$$

The trivial system corresponds to having no system, namely I carries no information.

Finally we require the composition of systems to be associative, namely

$$A(BC) = (AB)C.$$

In other words, if we iterate composition on many systems we always end up with a composite system that only depends on the components, and not on the particular

---

[2] In order to accommodate a theory of *anyons* one might generalize composition of systems from Abelian to braided, requiring the *Yang–Baxter identity* for the "swap."

composition sequence according to which they have been composed. Systems then make an *Abelian monoid*. A test with input system AB and output system CD represents an *interaction* process (see the parallel composition of tests in the following).

Parallel Composition of Tests   Any two tests $\xrightarrow{A}\boxed{\{C_x\}_{x\in X}}\xrightarrow{B}$ $\xrightarrow{C}\boxed{\{D_y\}_{y\in Y}}\xrightarrow{D}$ can be composed in parallel as follows:

$$
\begin{array}{c}\xrightarrow{A}\boxed{\{C_x\}_{x\in X}}\xrightarrow{B}\\[4pt]\xrightarrow{C}\boxed{\{D_y\}_{y\in Y}}\xrightarrow{D}\end{array}\quad=:\quad \xrightarrow{AC}\boxed{\{\mathcal{F}_{(x,y)}\}_{(x,y)\in X\times Y}}\xrightarrow{BD}\ .
$$

The test $\xrightarrow{AC}\boxed{\{\mathcal{F}_{(x,y)}\}_{(x,y)\in X\times Y}}\xrightarrow{BD}$ is the *parallel composition* of tests $\xrightarrow{A}\boxed{\{C_x\}_{x\in X}}\xrightarrow{B}$ and $\xrightarrow{C}\boxed{\{D_y\}_{y\in Y}}\xrightarrow{D}$. Parallel and sequential composition of tests commute, namely one has

$$
\begin{array}{c}\xrightarrow{A}\boxed{C_z}\xrightarrow{B}\boxed{A_x}\xrightarrow{C}\\[6pt]\xrightarrow{D}\boxed{B_y}\xrightarrow{E}\boxed{D_w}\xrightarrow{F}\end{array}\quad=\quad\begin{array}{c}\xrightarrow{A}\boxed{C_z}\xrightarrow{B}\boxed{A_x}\xrightarrow{C}\\[6pt]\xrightarrow{D}\boxed{B_y}\xrightarrow{E}\boxed{D_w}\xrightarrow{F}\end{array}\quad .\tag{3.1}
$$

When one of the two operations is the identity, we will omit the identity box and draw only a straight line:

$$
\begin{array}{c}\xrightarrow{A}\boxed{C_x}\xrightarrow{B}\\[4pt]\xrightarrow{\hspace{2cm}C\hspace{2cm}}\end{array}\quad .
$$

Therefore, as a consequence of commutation between sequential and parallel composition, we have the following identity:

$$
\begin{array}{c}\xrightarrow{A}\boxed{C_x}\xrightarrow{\hspace{1.5cm}B}\\[4pt]\xrightarrow{C\hspace{1cm}}\boxed{D_y}\xrightarrow{D}\end{array}\quad=\quad\begin{array}{c}\xrightarrow{A\hspace{1cm}}\boxed{C_x}\xrightarrow{B}\\[4pt]\xrightarrow{C}\boxed{D_y}\xrightarrow{\hspace{1.5cm}D}\end{array}\quad .
$$

Preparation Tests and Observation Tests   Tests with a trivial input system are called *preparation tests*, and tests with a trivial output system are called *observation tests*. They will be represented as follows:

$$
\boxed{\{\rho_x\}_{x\in X}}\!\!\!\xrightarrow{B}\quad:=\quad\xrightarrow{I}\boxed{\{\rho_x\}_{x\in X}}\xrightarrow{B}\ ,
$$

$$
\xrightarrow{A}\boxed{\{a_y\}_{y\in Y}}\quad:=\quad\xrightarrow{A}\boxed{\{a_y\}_{y\in Y}}\xrightarrow{I}\ .
$$

The corresponding events will be called *preparation events* and *observation events*. In formulas we will also write $|\rho_i)_A$ to denote a preparation event and $(a_j|_A$ to denote an observation event.

Closed Circuits   Using the above rules we can build up *closed circuits*, i.e. circuits with no input and no output system. An example is given by the following circuit:

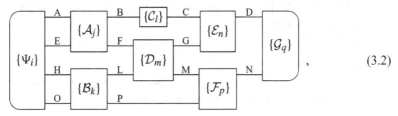

$$(3.2)$$

where we omitted the probability spaces of each test.

Independent Systems   For any (generally open) circuit constructed according to the above rules we call a set of systems *independent* if for each couple of systems in the set the two are not connected by a unidirected path (i.e. following the arrow from the input to the output). For example, in Eq. (3.2) the sets {A, E}, {H, O}, {A, E, H, O}, {A, L}, {A, E, L, P} are independent, whereas e.g. the sets {A, M}, {A, B}, {A, E, N} are not. A maximal set of independent systems is called a *slice*.

3.2 Operational Probabilistic Theory

The general purpose of an operational probabilistic theory is that of predicting and accounting for the joint probability of events corresponding to a particular circuit of connections. We are left with just a joint probability distribution if the circuit is closed, as in Eq. (3.2). Therefore, to a closed circuit of events as the following:

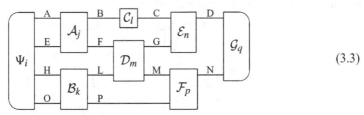

$$(3.3)$$

we will associate a joint probability $p(i, j, k, l, m, n, p, q)$, which we will consider as *parametrically dependent* on the circuit, namely, for a different choice of events and/or different connections we will have a different joint probability.

Since we are interested only in the joint probabilities and their corresponding circuits, we will build up probabilistic equivalence classes, and define:

> *Two events from system* A *to system* B *are equivalent if they occur with the same joint probability with the other events within the same circuit.*

We will call *transformation from* A *to* B – denoted as $\mathcal{A} \in \mathsf{Transf}(A \to B)$ – the equivalence class of events from A to B that are equivalent in the above sense. Likewise we will call *instrument* an equivalence class of tests, *state* an equivalence class of preparation events, and *effect* an equivalence class of observation events. We will denote the set of states of system A as $\mathsf{St}(A)$, and the set of its effects as $\mathsf{Eff}(A)$. Clearly, the input systems belonging to two different elements of an equivalence class will be operationally equivalent, and likewise for output systems.

We now can define an *operational probabilistic theory* as follows:

> *An operational probabilistic theory (OPT) is a collection of systems and transformations, along with rules for composition of systems and parallel and sequential composition of transformations. The OPT assigns a joint probability to each closed circuit.*

Therefore, in an OPT every test from the trivial system I to itself is a probability distribution $\{p_i\}_{i \in X}$ for the set of joint outcomes X, with $p(i) := p_i \in [0,1]$ and $\sum_{i \in X} p(i) = 1$. Compound events from the trivial system to itself are *independent*, namely their joint probability is given by the product of the respective probabilities for both the parallel and the sequential composition, namely

$$
\begin{array}{c}
\boxed{\rho_{i_1}} \!\!-\!\! A \!\!-\!\! \boxed{a_{i_2}} \\
\boxed{\sigma_{j_1}} \!\!-\!\! B \!\!-\!\! \boxed{b_{j_2}}
\end{array}
= \left(\boxed{\rho_{i_1}} \!\!-\!\! A \!\!-\!\! \boxed{a_{i_2}}\right)\!\!\left(\boxed{\sigma_{j_1}} \!\!-\!\! B \!\!-\!\! \boxed{b_{j_2}}\right) = p(i_1, i_2)\, q(j_1, j_2).
$$

A special case of OPT is the *deterministic OPT*, where all probabilities are 0 or 1.

3.3 States and Effects

Using the parallel and the sequential composition of transformations, we can regard every closed circuit as the composition of a state with an effect. This can be always done by cutting the circuit along a *slice*, i.e. a maximal set of independent systems. For example, the circuit in Eq. (3.3) can be split as follows:

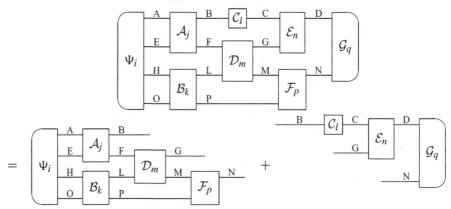

and thus is equivalent to the following state-effect circuit:

$$
\left(\Psi_i, \mathcal{A}_j, \mathcal{B}_k, \mathcal{D}_m, \mathcal{F}_p\right) \!\!-\!\! \text{BGN} \!\!-\!\! \left(\mathcal{C}_l, \mathcal{E}_n, \mathcal{G}_q\right)
$$

This also shows that *a state or an effect can be only defined for the systems of a slice.*

From the observations above it follows that any closed circuit can be regarded as the composition of a preparation event and an observation event, and a state is an equivalence class of preparation events that occur with the same probability with all possible effects, and vice versa. Therefore, a state $\rho \in \mathsf{St}(A)$ is a functional over effects $\mathsf{Eff}(A)$, the functional being denoted with the pairing $(a|\rho)$ with $a \in \mathsf{Eff}(A)$. With a similar argument we also conclude that an effect $a \in \mathsf{Eff}(A)$ is a functional over states. By taking linear combinations of functionals we see that $\mathsf{St}_{\mathbb{R}}(A) := \mathsf{Span}_{\mathbb{R}}[\mathsf{St}(A)]$ and $\mathsf{Eff}_{\mathbb{R}}(A) := \mathsf{Span}_{\mathbb{R}}[\mathsf{Eff}(A)]$ are dual spaces, and states are positive linear functionals over effects, and effects are positive linear functional over states.[3] Throughout the whole book, if not otherwise stated, the vector spaces $\mathsf{St}_{\mathbb{R}}(A)$ and $\mathsf{Eff}_{\mathbb{R}}(A)$ will be assumed *finite dimensional*, and we will denote as $D_A := \dim \mathsf{St}_{\mathbb{R}}(A) \equiv \mathsf{Eff}_{\mathbb{R}}(A)$, also called the *size of system* A. In the following we will also denote by $\mathsf{St}_1(A)$ and $\mathsf{Eff}_1(A)$ the sets of deterministic states and effects, respectively, and by $\mathsf{St}_+(A)$ and $\mathsf{Eff}_+(A)$ the cones containing all *conic* combinations – namely combinations with non-negative coefficients – of elements of $\mathsf{St}(A)$ and $\mathsf{Eff}(A)$, respectively.

According to the above definition, two states are different if and only if there exists an effect which occurrs on them with different joint probabilities. We also have that two effects are different if and only if there exists a state on which they have different probabilities. Therefore we conclude that:

States are separating for effects and effects are separating for states.

Exercise 3.1[4]   Let $(a|\rho) \neq (a|\sigma)$. Find $0 \leq q \leq 1$ and $0 \leq q' \leq 1$ such that

$$q(a|\rho) + q'[1 - (a|\rho)] > \frac{1}{2},$$

$$(1 - q)(a|\sigma) + (1 - q')[1 - (a|\sigma)] > \frac{1}{2}. \tag{3.4}$$

Under these circumstances there exists a strategy that allows one to discriminate between two states:

Lemma 3.1 (Discriminability of States)   *In any convex OPT if two states $\rho^{(0)}, \rho^{(1)} \in \mathsf{St}(A)$ are distinct (i.e. $\rho^{(0)} \neq \rho^{(1)}$), then one can discriminate them with error probability strictly smaller than $\frac{1}{2}$.*

Proof   A discrimination strategy for a pair of events $\rho^{(0)}, \rho^{(1)}$ must generally take into account that the events occur non-deterministically within two different tests $\{\rho_0^{(m)}, \rho_1^{(m)}\}$ with $m = 0, 1$ and $\rho_0^{(m)} = \rho^{(m)}$. This means that one can extract information for the discrimination even upon occurrence of the complementary event $\rho_1^{(m)}$. In other words, the event-discrimination becomes a test-discrimination. The strategy is then provided by an observation test $\{a_i\}_{i \in X}$ ($X = 0, 1$) along with a post-processing function $p(m|i, j)$ that

---

[3] Notice that in general the notion of dual vector space depends on the topological structure of the space, and not only on its linear structure. For example, the only spaces that coincide with their double dual are Hilbert spaces. However, we will consider here only finite-dimensional vector spaces, which are all Hilbert spaces when equipped with the Euclidean scalar product.

[4] A similar construction can be found in Chiribella *et al.* (2010c).

corresponds to probability of inferring the mth preparation test conditional on outcome (i,j) for the test $\{(a_i|\rho_j^{(m)})\}_{i,j\in X}$. The probability of estimating the value l, given that the test is the mth, is then given by

$$p(l|m) = \sum_{i,j\in X} p(l|i,j)(a_i|\rho_j^{(m)}). \tag{3.5}$$

Since effects are separating for states there exists an effect a such that $(a|\rho_0^{(0)}) \neq (a|\rho_0^{(1)})$. One can then choose a discrimination strategy based on the observation test $\{a_0, a_1\}$ with $a_0 = a$, along with conditional probabilities $p(0|0,0) = q$, $p(0|i,j) = q'$ for $(i,j) \neq (0,0)$, and $p(1|0,0) = (1-q)$, $p(1|i,j) = (1-q')$ for $(i,j) \neq (0,0)$. The success probabilities for $m = 0, 1$ calculated by Eq. (3.5) are then

$$p(0|0) = q(a|\rho_0^{(0)}) + q'[1 - (a|\rho_0^{(0)})], \tag{3.6}$$

$$p(1|1) = (1-q)(a|\rho_0^{(1)}) + (1-q')[1 - (a|\rho_0^{(1)})]. \tag{3.7}$$

One can now use the result of Exercise 3.1 to conclude the proof.   □

It is obvious that a similar lemma holds for effects, namely

Lemma 3.2 (Discriminability of Effects)   *In any convex OPT if two effects $a^{(0)}, a^{(1)} \in \mathsf{Eff}(A)$ are distinct (i.e. $a^{(0)} \neq b^{(1)}$), then one can discriminate them with error probability strictly smaller than $\frac{1}{2}$.*

The convexity of the sets of states $\mathsf{St}(A)$ and effects $\mathsf{Eff}(A)$ for every system A selects a relevant class of theories, which are called *convex theories*. Classical and quantum theory are both convex. Notice that if a theory is convex then one can consistently extend all the sets of transformations $\mathsf{Transf}(A \to B)$ to their convex hull $\mathsf{Co}(\mathsf{Transf}(A \to B))$. In the following, when referring to a convex theory we mean a theory where all the sets of transformations are convex.

3.4 Transformations

From what we said before, the following circuit is a state of system BFHO:

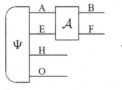

This means that any transformation connected to some output systems of a state maps the state into another state of generally different systems. Thus, while states and effects are linear functionals over each other, we can always regard a transformation as a map between states. In particular, a transformation $\mathcal{T} \in \mathsf{Transf}(A \to B)$ is always associated to a map $\hat{\mathcal{T}}$ from $\mathsf{St}(A)$ to $\mathsf{St}(B)$, uniquely defined as

$$\hat{\mathcal{T}} : |\rho) \in \mathsf{St}(A) \mapsto \hat{\mathcal{T}}|\rho) = |\mathcal{T}\rho) \in \mathsf{St}(B).$$

Similarly the transformation can be associated to a map from $\mathsf{Eff}(A)$ to $\mathsf{Eff}(B)$. The map $\hat{\mathcal{T}}$ can be linearly extended to a map from $\mathsf{St}_{\mathbb{R}}(A)$ to $\mathsf{St}_{\mathbb{R}}(B)$. Notice that the linear extension of $\mathcal{T}$ (which we will denote by the same symbol) is well defined. In fact, a linear combination of states of A is null – in formula $\sum_i c_i |\rho_i) = 0$ – if and only if $\sum_i c_i(a|\rho_i) = 0$ for every $a \in \mathsf{Eff}(A)$, and since for every $b \in \mathsf{Eff}(B)$ we have $(b|\mathcal{T} \in \mathsf{Eff}(A)$, then $(b|\mathcal{T} \left(\sum_i c_i |\rho_i) \right) = \sum_i c_i(b|\mathcal{T}|\rho_i) = (b| \sum_i c_i \mathcal{T}|\rho_i) = 0$, and finally $\sum_i c_i \mathcal{T}|\rho_i) = 0$.

We want to stress that if two transformations $\mathcal{T}, \mathcal{T}' \in \mathsf{Transf}(A \to B)$ correspond to the same map $\hat{\mathcal{T}}$ from $\mathsf{St}(A)$ to $\mathsf{St}(B)$, this does not mean that the two transformations are the same, since as an equivalence class, they must occur with the same joint probability in all possible circuits. In terms of state mappings, the same definition of the transformation as equivalence class corresponds to say that $\mathcal{T}, \mathcal{T}' \in \mathsf{Transf}(A \to B)$ as maps from states of AR to states of BR are the same for all possible systems R of the theory, namely $\mathcal{T} = \mathcal{T}' \in \mathsf{Transf}(A \to B)$ if and only if

$$\forall R, \forall \Psi \in \mathsf{St}(AR) \quad \Psi \begin{array}{c} A \; \boxed{\mathcal{T}} \; B \\ R \end{array} = \Psi \begin{array}{c} A \; \boxed{\mathcal{T}'} \; B \\ R \end{array} . \quad (3.8)$$

Indeed, as we will see in Chapter 6, there exist cases of OPT where there are transformations $\mathcal{T}, \mathcal{T}' \in \mathsf{Transf}(A \to B)$ corresponding to the same map when applied to $\mathsf{St}(A)$ and not when applied to $\mathsf{St}(AR)$ for some system R.

Since we can take linear combinations of linear transformations, $\mathsf{Transf}(A \to B)$ can be embedded in the vector space $\mathsf{Transf}_{\mathbb{R}}(A \to B)$. The deterministic transformations, whose set will be denoted as $\mathsf{Transf}_1(A \to B)$, will be also called *channels*. The conic span of elements of $\mathsf{Transf}(A \to B)$ will be denoted as $\mathsf{Transf}_+(A \to B)$.

Finally, a transformation $\mathcal{U} \in \mathsf{Transf}(A \to B)$ is *reversible* if there exists another transformation $\mathcal{U}^{-1} \in \mathsf{Transf}(B \to A)$ such that $\mathcal{U}^{-1}\mathcal{U} = \mathcal{I}_A$ and $\mathcal{U}\mathcal{U}^{-1} = \mathcal{I}_B$. The set of reversible transformations from A to B will be denoted by $\mathsf{RevTransf}(A \to B)$. When $A \equiv B$ the set of reversible transformations $\mathsf{RevTransf}(A)$ is actually a group, which will also be denoted by $\mathbf{G}_A$.

3.5 Coarse-graining and Refinement

When dealing with probabilistic events, a natural notion is that of *coarse-graining*, corresponding to merging events into a single event. According to probability theory, the probability of a *coarse-grained event* $S \subseteq X$ subset of the outcome space X is the sum of probabilities of the elements of S, namely $p(S) = \sum_{i \in S} p(i)$. We then correspondingly have that the coarse-grained event $\mathcal{T}_S$ of a test $\{\mathcal{T}_i\}_{i \in X}$ will be given by

$$\mathcal{T}_S = \sum_{i \in S} \mathcal{T}_i. \quad (3.9)$$

We stress that the equal sign in Eq. (3.9) is to be meant in the sense of equation (3.8). In addition to the notion of coarse-grained event we have also that of *coarse-grained test*, corresponding to the collection of coarse-grained events $\{\mathcal{T}_{X_l}\}_{l \in \mathbb{Z}}$ from a partition $X = \cup_{l \in \mathbb{Z}} X_l$ of the outcome space X, with $X_i \cap X_j = \emptyset$ for $i \neq j$.

The converse procedure of coarse-graining is what we call *refinement*. If $\mathcal{T}_S$ the coarse-graining in Eq. (3.9), we call any sum $\sum_{i \in S'} \mathcal{T}_i$ with $S' \subseteq S$ a *refinement* of $\mathcal{T}_S$. The same notion can be analogously considered for a test. Intuitively, a test that refines another is a test that extracts more detailed information, namely it is a test with better "resolving power."

The notion of refinement is translated to transformations (hence also to states, and effects), as equivalence classes of events. Refinement and coarse-graining define a partial ordering in the set of transformations $\mathsf{Transf}(A \to B)$ and in the cone $\mathsf{Transf}_+(A \to B)$, writing $\mathcal{D} \prec \mathcal{C}$ if $\mathcal{D}$ is a refinement of $\mathcal{C}$. This ordering corresponds to the ordering induced by the cone $\mathsf{Transf}_+(A \to B)$, namely $\mathcal{D} \prec \mathcal{C}$ if $\mathcal{C} - \mathcal{D} \in \mathsf{Transf}_+(A \to B)$. A transformation $\mathcal{C}$ is *atomic* if it has only trivial refinement, namely $\mathcal{C}_i$ refines $\mathcal{C}$ implies that $\mathcal{C}_i = p\mathcal{C}$ for some probability $p \geq 0$. A test that consists of atomic transformations is a test whose "resolving power" cannot be further improved.

It is often useful to refer to the set of all possible refinements of a given event $\mathcal{C}$. This set is called *refinement set* of the event $\mathcal{C} \in \mathsf{Transf}(A \to B)$, and is denoted by $\mathsf{RefSet}\,(\mathcal{C})$. In formula, $\mathsf{RefSet}\,(\mathcal{C}) := \{\mathcal{D} \in \mathsf{Transf}(A \to B) | \mathcal{D} \prec \mathcal{C}\}$.

In the special case of states, we will use the word *pure* as a synonym of atomic. A pure state describes an event providing maximal knowledge about the system's preparation, namely a knowledge that cannot be further refined.

As usual, a state that is not pure will be called *mixed*. An important notion is that of *internal state*. A state is called internal when any other state can refine it: precisely, $\omega \in \mathsf{St}(A)$ is internal if for every $\rho \in \mathsf{St}(A)$ there is a non-zero probability $p > 0$ such that $p\rho$ is a refinement of ω, i.e. $p\rho \in \mathsf{RefSet}\,(\omega)$. The adjective "internal" has a precise geometric connotation, since the state cannot belong to the border of $\mathsf{St}(A)$. An internal state describes a situation in which there is no definite knowledge about the system preparation, namely a priori we cannot in principle exclude any possible preparation.

Exercise 3.2   Show that for parallel composition $\mathcal{A} \boxtimes \mathcal{B}$ of transformation $\mathcal{A} \in \mathsf{Transf}(A \to C)$ and $\mathcal{B} \in \mathsf{Transf}(B \to D)$ one has

$$\mathsf{RefSet}_{\mathbb{R}}(\mathcal{A}) \otimes \mathsf{RefSet}_{\mathbb{R}}(\mathcal{B}) \subseteq \mathsf{RefSet}_{\mathbb{R}}(\mathcal{A} \boxtimes \mathcal{B}), \tag{3.10}$$

where $\mathsf{RefSet}_{\mathbb{R}} := \mathsf{Span}_{\mathbb{R}} \mathsf{RefSet}$, and we have introduced the temporary symbol $\boxtimes$ to denote the parallel composition of events.

3.6 Operational Distance Between States

The vector space $\mathsf{St}_{\mathbb{R}}(A)$ can be equipped with a natural norm, related to the optimal discrimination scheme for pairs of states ρ_0, ρ_1 making a binary preparation test. A special case of such a test is that of a couple of deterministic states with given prior probabilities

p and $1 - p$, which is the canonical case of state discrimination in quantum estimation theory.[5]

Given a binary preparation test $\{\rho_0, \rho_1\} \subseteq \mathsf{St}(A)$, consider the discrimination strategy for states ρ_0 and ρ_1 consisting in performing a binary observation test $\{a_0, a_1\}$. Upon defining $a := a_0 + a_1$, the success probability in the *discrimination* of ρ_0 and ρ_1 is given by

$$
\begin{aligned}
p_{\text{succ}} &= p(0,0) + p(1,1) = (a_0|\rho_0) + (a_1|\rho_1) \\
&= (a|\rho_0) + (a_1|\rho_1 - \rho_0) \\
&= (a|\rho_1) + (a_0|\rho_0 - \rho_1) \\
&= \frac{1}{2}[1 + (a_1 - a_0|\rho_1 - \rho_0)].
\end{aligned}
$$

If we now optimize over the possible strategies, classified by the observation tests $\{a_0, a_1\}$, we obtain

$$
p_{\text{succ}}^{(\text{opt})} = \frac{1}{2}[1 + \|\rho_1 - \rho_0\|], \tag{3.11}
$$

where the *operational norm* $\|\delta\|$ of an element $\delta \in \mathsf{St}_{\mathbb{R}}(A)$ is given by

$$
\|\delta\| := \sup_{\{a_0, a_1\}} (a_0 - a_1|\delta), \tag{3.12}
$$

with $\{a_0, a_1\}$ ranging over all two-outcome observation tests. For the proof that the functional $\|\cdot\|$ is actually a norm, see Exercise 3.5. In the special case where $A = I$, the operational norm coincides with the absolute value: indeed $\delta \in \mathsf{St}_{\mathbb{R}}(I)$ just means $\delta \in \mathbb{R}$, and since $0, 1 \in \mathsf{Eff}(I) \subseteq [0, 1]$ clearly one has

$$
\sup_{p \in [0,1]} (2p - 1)\delta = |\delta|, \tag{3.13}
$$

corresponding to the choice $p = 0$ if $\delta < 0$ and $p = 1$ if $\delta \geq 0$.

One can easily prove that

$$
\|\delta\| \leq \sup_{a_0 \in \mathsf{Eff}(A)} (a_0|\delta) - \inf_{a_1 \in \mathsf{Eff}(A)} (a_1|\delta).
$$

Moreover, whenever $(f|\delta) = (f'|\delta)$ for every pair $f, f' \in \mathsf{Eff}_1(A)$ of deterministic effects, also the converse bound holds (see Exercise 3.6). There are two relevant situations where the latter condition holds: (1) δ belongs to the linear span of deterministic states, which is the case for $\delta = \rho_0 - \rho_1$ for ρ_0, ρ_1 both proportional to deterministic states; (2) theories where the deterministic effect e_A is *unique* (this is a very relevant class of theories, called *causal*; see Chapter 5 for more details). In both cases the following identity holds:

$$
\|\delta\| = \sup_{a_0 \in \mathsf{Eff}(A)} (a_0|\delta) - \inf_{a_1 \in \mathsf{Eff}(A)} (a_1|\delta). \tag{3.14}
$$

Introducing a norm entails the existence of Cauchy sequences, $(\rho_n)_{n \in \mathbb{N}}$, as well as their equivalence relation $(\rho_n)_{n \in \mathbb{N}} \sim (\sigma_n)_{n \in \mathbb{N}}$ if $\lim_{n \to \infty}(\rho_n - \sigma_n) = 0$. In what follows, we will always take the set of states $\mathsf{St}(A)$ to be closed in the operational norm, i. e. a Banach

---

[5] Helstrom (1976).

space. This is a very natural assumption: the fact that there is a sequence of states $(\rho_n)_{n \in \mathbb{N}}$ that converges to $\rho \in \mathsf{St}_\mathbb{R}(A)$ means that there is a procedure to prepare ρ with arbitrary precision, and hence ρ can be considered as an ideal state.

Exercise 3.3 Show that in a non-deterministic operational probabilistic theory it is possible to generate every finite probability distribution.

Exercise 3.4 Prove that in a non-deterministic operational probabilistic theory for every deterministic effect $f \in \mathsf{Eff}_1(A)$ there exist all the observation tests $\{p_1 f, p_2 f, \ldots, p_N f\}$ where $\{p_1, p_2, \ldots, p_N\}$ is an arbitrary probability distribution.

Exercise 3.5 Prove that the functional $\| \cdot \|$ on $\mathsf{St}_\mathbb{R}(A)$ defined in Eq. (3.12) is actually a norm, namely it satisfies:

1 $\|\delta\| \geq 0$, with $\|\delta\| = 0$ if and only if $\delta = 0$;
2 $\|\delta + \theta\| \leq \|\delta\| + \|\theta\|$;
3 $\|k\delta\| = |k| \, \|\delta\|$ for $k \in \mathbb{R}$.

Exercise 3.6 Prove that whenever $(f|\delta) = (f'|\delta)$ for every pair of deterministic effects $f, f' \in \mathsf{Eff}(A)$, then

$$\|\delta\| = \sup_{a_0 \in \mathsf{Eff}(A)} (a_0|\delta) - \inf_{a_1 \in \mathsf{Eff}(A)} (a_1|\delta).$$

Exercise 3.7 Prove that in the case of quantum theory the operational norm over $\mathsf{St}_\mathbb{R}(A)$ coincides with the *trace-norm* $\|\delta\|_1 := \mathrm{Tr}|\delta|$, where $|\delta|$ denotes the absolute value of the operator $\delta \in \mathsf{Lin}(\mathcal{H}_A)$ of system A.

In addition to the usual properties of a norm, the operational norm also satisfies the following monotonicity property.

Lemma 3.3 (Monotonicity of the Operational Norm) *If $\mathcal{C} \in \mathsf{Transf}_1(A \to B)$ is a deterministic transformation, then for every $\delta \in \mathsf{St}_\mathbb{R}(A)$ one has*

$$\|\mathcal{C}\delta\|_B \leq \|\delta\|_A.$$

For reversible $\mathcal{C}$ equality holds.

Proof By definition, $\|\mathcal{C}\delta\|_B = \sup_{b_1 \in \mathsf{Eff}(B)} (b_1|_B \mathcal{C}|\delta)_A - \inf_{b_0 \in \mathsf{Eff}(B)} (b_0|_B \mathcal{C}|\delta)_A$. Since $(b_1|_B \mathcal{C}$ and $(b_0|_B \mathcal{C}$ are effects on system A, one has $\|\mathcal{C}\delta\|_B \leq \sup_{a_1 \in \mathsf{Eff}(A)} (a_1|_B|\delta)_A - \inf_{a_0 \in \mathsf{Eff}(A)} (a_0|_A|\delta)_A = \|\delta\|_A$. Clearly, if $\mathcal{C}$ is reversible one has the converse bound $\|\delta\|_A = \|\mathcal{C}^{-1}\mathcal{C}\delta\|_A \leq \|\mathcal{C}\delta\|_B$, thus proving the equality $\|\delta\|_A = \|\mathcal{C}\delta\|_B$. □

3.7 Operational Distances for Transformations and Effects

As for the vector space $\mathsf{St}_\mathbb{R}(A)$, also the vector space $\mathsf{Transf}_\mathbb{R}(A \to B)$ can be equipped with a natural norm related to the optimal discrimination scheme for pairs of transformations $\mathcal{A}_0$ and $\mathcal{A}_1$ making a binary test. Once we have a discrimination scheme for transformations, then we have also a discrimination scheme for effects $\mathsf{Eff}_\mathbb{R}(A) = \mathsf{Transf}_\mathbb{R}(A \to I)$ as special cases. Again, as for states, a special case of binary test is that of a couple of deterministic transformations (i.e. channels) with given prior probabilities p and $1 - p$, which is the canonical case considered in the quantum estimation theory literature.[6]

Given a binary test $\{\mathcal{A}_0, \mathcal{A}_1\}$, the optimal strategy for discriminating $\mathcal{A}_0$ and $\mathcal{A}_1$ is to apply both to a deterministic state $\rho \in \mathsf{St}_1(AC)$ with additional system C and to use the optimal strategy for discriminating the corresponding output states of the binary preparation test $\{\mathcal{A}_0\rho, \mathcal{A}_1\rho\}$ and finally optimizing over all possible states $\rho \in \mathsf{St}_1(AC)$ and over all possible systems C. Since the optimal strategy for state discrimination is already optimized over effects in $\mathsf{Eff}(BC)$, this scheme corresponds to optimize over all possible closed circuits containing the transformation $\mathcal{A}_i$, $i = 0, 1$. The scheme leads to the optimal probability of success

$$p_{succ}^{(opt)} = \frac{1}{2}(1 + \|\mathcal{A}_1 - \mathcal{A}_0\|_{A,B}), \tag{3.15}$$

where the operational norm over $\mathsf{Transf}_\mathbb{R}(A \to B)$ is defined by

$$\|\Delta\|_{A,B} := \sup_C \sup_{\rho \in \mathsf{St}_1(AC)} \|\Delta\rho\|_{BC}, \qquad \Delta \in \mathsf{Transf}_\mathbb{R}(A \to B). \tag{3.16}$$

As shown in the following exercise, in quantum theory the operational norm (3.16) reduces to the *diamond norm* in Schrödinger picture,[7] or equivalently, to the *completely bounded norm* (CB norm) in Heisenberg picture.[8]

Exercise 3.8   Prove that in the case of quantum theory the operational norm for transformations coincides with the completely bounded norm for operator maps.

In the case of trivial input system $A = I$, Eq. (3.16) gives back the norm for states, whereas for trivial output system $B = I$, it provides an operational norm for effects, given by

$$\|\delta\|_{A,I} = \sup_C \sup_{\rho \in \mathsf{St}_1(AC)} \|\delta\rho\|_C \qquad \delta \in \mathsf{Eff}_\mathbb{R}(A). \tag{3.17}$$

In fact, the extension with the ancillary system C is not needed in this case, as proved in the following exercise.

Exercise 3.9   Show that the operational norm over $\mathsf{Eff}_\mathbb{R}(A)$ is given by the expression

$$\|\delta\|_{A,I} = \sup_{\rho \in \mathsf{St}_1(A)} |(\delta|\rho)_A|. \tag{3.18}$$

As shown in Exercise 3.10, in quantum theory the effect norm corresponds to the usual operator norm, which, for $\mathsf{Herm}(\mathcal{H}_A)$ corresponding to $\mathsf{Eff}_\mathbb{R}(A)$ is just the maximum absolute value of the eigenvalues of the operator.

Exercise 3.10   Prove that in the case of quantum theory the operational norm for effects coincides with the usual operator norm.

---

[6] Belavkin *et al.* (2005).
[7] Aharonov *et al.* (1998).
[8] Paulsen (1986).

We conclude by mentioning a general monotonicity property of the operational norm for transformations.

Lemma 3.4 (Monotonicity of the Operational Norm for Transformations) *If* $C \in \mathsf{Transf}_1(A \to B)$ *and* $\mathcal{E} \in \mathsf{Transf}_1(C \to D)$ *are two deterministic transformations, then for every* $\Delta \in \mathsf{Transf}_\mathbb{R}(B \to C)$ *one has*

$$\|\mathcal{E}\Delta C\|_{A,D} \leq \|\Delta\|_{B,C}.$$

For both C and $\mathcal{E}$ reversible equality holds.

Proof Let R be an ancillary system, and $\rho \in \mathsf{St}_1(AR)$ be a normalized state of AR. Then, since $|\sigma)_{BR} = C|\rho)_{AR}$ is a normalized state of BR, we have $\|\mathcal{E}\Delta C\|_{A,D} = \sup_R \sup_{\rho \in \mathsf{St}_1(AR)} \|\mathcal{E}\Delta C\rho\|_{DR} \leq \sup_R \sup_{\sigma \in \mathsf{St}_1(BR)} \|\mathcal{E}\Delta\sigma\|_{DR}$. Now, using Lemma 3.3 we obtain $\|\mathcal{E}\Delta\sigma\|_{DR} \leq \|\Delta\sigma\|_{CR}$. Hence, $\|\mathcal{E}\Delta C\|_{A,D} \leq \sup_R \sup_{\sigma \in \mathsf{St}_1(BR)} \|\Delta\sigma\|_{CR} = \|\Delta\|_{B,C}$. Clearly, if both C and $\mathcal{E}$ are reversible, one also has the converse bound $\|\Delta\|_{B,C} = \|\mathcal{E}^{-1}(\mathcal{E}\Delta C)C^{-1}\|_{B,C} \leq \|\mathcal{E}\Delta C\|_{A,D}$, thus proving the equality. □

3.8 Summary

In this chapter we introduced the general framework of operational probabilistic theories. Operational probabilistic theories are characterized by two main ingredients: the operational language and the probabilistic structure. The former provides the connectivity rules for events resorting to general information-processing circuits. The latter sets the rules for evaluating the joint probabilities of events. The two notions together constitute the background for the formulation of the principles of quantum theory. We have also seen how the main objects of the theory – states, effects, and transformations – are also naturally endowed with a metric structure, which directly follows from their operational meaning.

Notes

Operational Probabilistic Theories The present formulation of the OPT has been first presented in Chiribella *et al.* (2010b). The circuit framework very closely resembles the notation used in quantum circuits. The mathematical formulation corresponds to that of a symmetric monoidal category theory (Joyal and Street, 1991), of which the circuit framework is a faithful interpretation. For a book on category theory see Ref. Mac Lane (1978). For an extended discussion on graphical calculus an interesting lecture is Penrose (1971). We also suggest the beautiful introductions in Coecke (2006) and Selinger (2011).

About the Trivial System An event with trivial input system should not be interpreted as a "creation of information out of nothing," but instead as knowledge of a preparation of systems, corresponding to forgetting previous evolutions. Similarly, an event with trivial output system should not be interpreted as a "destruction of systems," but as the simple fact that events occurring at its output are not considered. In short, the OPT is an input–output description. This is the core of a scientific theory: making predictions on the basis of prior knowledge.

In quantum theory the trivial system I is represented by the one-dimensional Hilbert space $\mathcal{H}_I = \mathbb{C}$, corresponding to having zero qubits. In the classical theory the trivial system, corresponding to zero bits, is associated to the linear space $\mathbb{R}$, with convex set of states $S_1^1 \sim S^0$ (see Section 2.9), which is a single point. In a general OPT the trivial system I has $\mathsf{St}_\mathbb{R}(I) = \mathbb{R}$, corresponding to a single deterministic state and to a segment of probabilistic states $p \in (0, 1]$. Even in the special case of a deterministic theory, corresponding to $\mathsf{St}(I) = \{1\}$, one has $\mathsf{St}_\mathbb{R}(I) = \mathbb{R}$. For any OPT the states of I are just probabilities.

Probabilities and Closed Circuits   According to the definition of probabilistic theory, only *closed circuits* – i.e. circuits with input and output system both trivial – correspond to a probability distribution. Considering classical and quantum theory, one would have expected a definition allowing for probabilities defined only in the presence of a preparation, namely without the need to specify the final observation, as it happens in the most common scenarios. However, in the next chapter we will see that such feature follows from a crucial property of the theory: causality. In theories without causality, instead, it is impossible to associate probabilities to events in a circuit if the output system is not trivial.

Trace-norm in Quantum Theory   In quantum theory the operational norm is the usual *trace-norm* $\|\cdot\|_1$. Indeed, quantum theory has a unique deterministic effect for every system A, represented by the identity operator I_A on $\mathcal{H}_A$, and identity (3.14) holds. Upon denoting by δ_+ and δ_- the positive and negative parts of the Hermitian operator δ, we have $\|\delta\| = \mathrm{Tr}[\delta_+] - \mathrm{Tr}[\delta_-] = \|\delta\|_1$ (see Exercise 3.7).

Discrimination and Success Probability   According to (3.11) the success probability is always larger than $\frac{1}{2}$, apart from the case where $\|\rho_1 - \rho_0\| = 0$, corresponding to $\rho_1 = \rho_0$. Having a success probability strictly greater than $\frac{1}{2}$ is indeed a crucial feature of any reasonable definition of discrimination. Notice that one can always make the success probability not smaller than $\frac{1}{2}$ by swapping the inference strategy, namely upon relabeling the outcomes $a_0' := a_1$, and $a_1' := a_0$. The case of success probability exactly equal to $\frac{1}{2}$ corresponds to a situation where one cannot do better than just random guessing, which is indeed the case only for $\rho_0 = \rho_1$.

About Operational Norms   The operational norms introduced in the chapter for states, effects, and transformations, restrict to the case when the pair of discriminated objects make a test. The general case of two objects each belonging to a different test, as in Lemma 3.1, is still an open problem (Lemma 3.1 is anyway needed for local discriminability in Chapter 6). As already mentioned, for states or transformations that are proportional to deterministic ones one can always introduce a test with the states or transformations rescaled by prior probabilities.

Problems

3.1 Prove Lemma 3.1 in the easiest case of deterministic states. More precisely, show that in any convex theory, for any two different deterministic states $\rho_0 \neq \rho_1 \in St_1(A)$ there exists a binary test $\{a_0, a_1\}$ with probabilities of error strictly smaller than $1/2$, namely

$$p(1|0) = p(0|1) < \frac{1}{2},$$

with $p(i|j) = p(i,j) / \sum_l p(l,j)$ conditioned probabilities, and $p(i,j) = (a_i|\rho_j)$.

Solutions to Selected Problems and Exercises

Exercise 3.2

By definition, for $\mathcal{X} \in RefSet\,\mathcal{A}$ and $\mathcal{Y} \in RefSet\,\mathcal{B}$, there exist $\mathcal{X}' \in RefSet\,\mathcal{A}$ and $\mathcal{Y}' \in RefSet\,\mathcal{B}$ such that $\mathcal{X} + \mathcal{X}' = \mathcal{A}$ and $\mathcal{Y} + \mathcal{Y}' = \mathcal{B}$. Then, if we define $\mathcal{W} := \mathcal{X} \boxtimes \mathcal{Y}' + \mathcal{X}' \boxtimes \mathcal{Y} + \mathcal{X}' \boxtimes \mathcal{Y}'$, we clearly have $\mathcal{X} \boxtimes \mathcal{Y} + \mathcal{W} = \mathcal{A} \boxtimes \mathcal{B}$. This proves that all the transformations $\mathcal{X} \boxtimes \mathcal{Y}$ belong to $RefSet\,(\mathcal{A} \boxtimes \mathcal{B})$. Thus, $Span_\mathbb{R}\,(RefSet\,\mathcal{A} \boxtimes RefSet\,\mathcal{B}) \subseteq Span_\mathbb{R}RefSet\,(\mathcal{A} \boxtimes \mathcal{B})$. Finally, by the linearity of parallel composition of transformations we have $Span_\mathbb{R}\,(RefSet\,\mathcal{A} \boxtimes RefSet\,\mathcal{B}) \equiv Span_\mathbb{R}RefSet\,(\mathcal{A}) \otimes Span_\mathbb{R}RefSet\,(\mathcal{B})$. We thus obtained Eq. (3.10). In particular, for states $\rho \otimes \sigma \in St(AB)$

$$RefSet_\mathbb{R}\rho \otimes RefSet_\mathbb{R}\sigma \subseteq RefSet_\mathbb{R}(\rho \boxtimes \sigma). \tag{3.19}$$

Exercise 3.3

First, we observe that since the theory is not deterministic there exists at least one transformation $p_0 = p$ of the trivial system such that $0 < p < 1$. Its complement is then $p_1 = 1 - p$. One can now show that an arbitrary test $\{p_0, p_1\}$ with $0 < p_0 < 1$ can be used to produce a distribution $\left\{\frac{1}{2}, \frac{1}{2}\right\}$ as follows: take the parallel composition test $\{p_s\}_{s \in \{0,1\}^N}$ where s are binary strings. The probability of obtaining either the string $s = 00 \ldots 0$ or $s = 11 \ldots 1$ is $2\varepsilon_N := p_0^N + p_1^N$, and decays exponentially in N. The remaining strings have then at least a couple of opposite adjacent bits, either $s_j = 0$ and $s_{j+1} = 1$ or vice versa, where s_j denotes the jth bit. For every such string we can then take the minimum j such that $s_j \neq s_{j+1}$, and we divide the strings in two sets S_0^N, S_1^N depending on whether $s_j = 0$ or $s_j = 1$. One can then take the coarse-graining with events $\{2\varepsilon_N, p_{S_0^N}, p_{S_1^N}\}$, and since for every string in S_0^N one can obtain a string in S_1^N having the same probability of the original string, just by switching $s'_j := s_{j+1}$ and $s'_{j+1} := s_j$, one concludes that $p_{S_0} = p_{S_1} = \frac{1}{2} - \varepsilon_N$. Since, by hypothesis, the set $St(A)$ is closed, and we just showed how to construct a sequence of events $p_{S_0^N}$ of the system I with $\lim_{N \to \infty} \|\frac{1}{2} - p_{S_0^N}\| = 0$, we conclude that the event $\frac{1}{2}$ is a state of I. Finally, once we have the preparation test $\{p_0, p_1\}$ with $p_0 = p_1 = \frac{1}{2}$, by taking its parallel composition N times, and with a suitable coarse-graining $\{q_N, 1 - q_N\}$, we can approximate every preparation test $\{q, (1 - q)\}$, with $\|q - q_N\| < \frac{1}{2^N}$. Using again the closure of $\mathsf{St}(\mathrm{I})$, we then have all the preparation tests $\{q, 1 - q\}$ with $q \in [0, 1]$. Finally any preparation test $\{p_1, p_2, \ldots, p_N\}$ of I can be obtained by parallel composition and coarse-graining from binary distributions.

Exercise 3.4

Let $\rho \in \mathsf{St}(A)$. In order to achieve the observation test $\{p_1 f, p_2 f, \ldots, p_N f\}$ it is sufficient to take the parallel composition $p_i \otimes \rho$ and then apply the deterministic observation test $1 \otimes f$, obtaining $(1|p_i)(f|\rho) = p_i(f|\rho) = (p_i f|\rho)$.

Exercise 3.5

Taking $a_0 = a_1 = \frac{1}{2} f$ for any deterministic effect f provides the lower bound $\|\delta\| \geq 0$. Moreover, if $\|\delta\| = 0$, then for every binary observation test $\{a, b\}$ one has $(a|\delta) \leq (b|\delta)$, but exchanging a and b gives the converse bound, hence $(a|\delta) = (b|\delta)$. This implies that $(a|\delta) = \frac{1}{2}(f|\delta)$ for the deterministic effect $f := a + b$. Now, taking the observation tests $\{pf, (1-p)f\}$ with $f = a + b$ and $0 \leq p \leq 1$ gives $(2p - 1)(f|\delta) = 0$ for every $0 \leq p \leq 1$, and finally $(f|\delta) = 0$, which implies $(a|\delta) = 0$ for every effect a, namely $\delta = 0$. The triangle inequality easily follows from the properties of sup as follows:

$$\|\delta + \theta\| = \sup_{\{a_0, a_1\}} (a_0 - a_1|\delta + \theta) = \sup_{\{a_0, a_1\}} [(a_0 - a_1|\delta) + (a_0 - a_1|\theta)]$$

$$\leq \sup_{\{a_0, a_1\}} (a_0 - a_1|\delta) + \sup_{\{a_0, a_1\}} (a_0 - a_1|\theta) = \|\delta\| + \|\theta\|.$$

Finally, for $k \in \mathbb{R}$ we have

$$\sup_{\{a_0, a_1\}} (a_0 - a_1|k\delta) = \begin{cases} k \sup_{\{a_0, a_1\}} (a_0 - a_1|\delta) & \text{for } k \geq 0, \\ -k \sup_{\{a_0, a_1\}} (a_1 - a_0|\delta) & \text{for } k \leq 0, \end{cases}$$

hence the result.

Exercise 3.6

Let $s := \sup_{a_0 \in \mathsf{Eff}(A)} (a_0|\delta)$ and $i := \inf_{a_1 \in \mathsf{Eff}(A)} (a_1|\delta)$. Then $i = (f|\delta) - s$. Indeed, if $i < (f|\delta) - s$, then by definition of infimum there exist $\varepsilon > 0$ and $b_1 \in \mathsf{Eff}(A)$ such that $i \leq (b_1|\delta) < i + \varepsilon < (f|\delta) - s$. Now, let $b_0 + b_1 = f' \in \mathsf{Eff}_1(A)$. By hypothesis $(b_0 + b_1|\delta) - s = (f'|\delta) - s = (f|\delta) - s > (b_1|\delta)$, and this implies $(b_0|\delta) > s$, contrarily to the hypothesis.

Exercise 3.7

The element of $\delta \in \mathsf{St}_{\mathbb{R}}(A)$ is represented by a self-adjoint trace-class operator $\delta \in \mathsf{Herm}(\mathcal{H}_A)$. Using the Jordan decomposition of a self-adjoint operator $\delta = \delta_+ - \delta_-$, with $\delta_\pm$ denoting the positive (negative) part of δ, one has

$$\|\delta\| = \sup_{a \in \mathsf{Eff}(A)} (a|\delta)_A - \inf_{a \in \mathsf{Eff}(A)} (a|\delta)_A$$

$$= \sup_{0 \leq E_a \leq I_A} \mathrm{Tr}[E_a \delta] - \inf_{0 \leq E_a \leq I_A} \mathrm{Tr}[E_a \delta]$$

$$= \mathrm{Tr}[P_+ \delta] + \mathrm{Tr}[P_- \delta] = \mathrm{Tr}|\delta| =: \|\delta\|_1,$$

$E_a \in \mathsf{Lin}_+(\mathcal{H}_A)$ denoting the positive operator corresponding to the effect a, and $P_\pm$ the orthogonal projector over the linear space corresponding to positive (negative) eigenvalues of δ. This result extends also to infinite-dimensions, with δ trace-class, as a consequence of the identity

$$\|\delta\|_1 = \sup_{\|Y\| \le 1} |\mathrm{Tr}[Y\delta]|, \quad Y \in \mathsf{Lin}(\mathcal{H}_A). \tag{3.20}$$

Exercise 3.8

The definition (3.16) corresponds *mutatis mutandis* to the definition of *CB norm* for a *completely bounded map*[9] $\mathcal{A} \in \mathsf{Lin}(\mathsf{Lin}(\mathcal{H}), \mathsf{Lin}(\mathcal{K}))$

$$\|\mathcal{A}\|_{CB} = \sup_{\mathcal{E}} \sup_{\|\rho\|_1 \le 1} \|\mathcal{A} \otimes \mathcal{I}_{\mathcal{E}}(\rho)\|_1, \quad \rho \in \mathsf{Lin}(\mathcal{H} \otimes \mathcal{E}). \tag{3.21}$$

In the Heisenberg picture instead one has the *diamond norm*[10]

$$\|\mathcal{A}^\dagger\|_\diamond = \sup_{\mathcal{E}} \sup_{\|X\| \le 1} \|\mathcal{A}^\dagger \otimes \mathcal{I}_{\mathcal{E}}(X)\|, \quad X \in \mathsf{Lin}(\mathcal{K} \otimes \mathcal{E}). \tag{3.22}$$

Exercise 3.9

Taking $C = I$ in Eq. (3.16) yields $\|\delta\|_{A,I} \ge \sup_{\rho \in \mathsf{St}_1(A)} \|(\delta|\rho)_A\|_1 = \sup_{\rho \in \mathsf{St}_1(A)} |(\delta|\rho)_A|$, where we used the fact that the norm of a real number $x \in \mathbb{R} \equiv \mathsf{St}_\mathbb{R}(I)$ is given by its modulus: $\|x\|_I = |x|$. To prove the equality of Eq. (3.18) we now prove that $\|\delta\|_{A,I} \le \sup_{\rho \in \mathsf{St}_1(A)} |(\delta|\rho)_A|$. By the definition of the operational norm for states in Eq. (3.14), for every $\sigma \in \mathsf{St}_1(AC)$ we have

$$\begin{aligned}
\|\delta\sigma\|_C &= \sup_{c_1 \in \mathsf{Eff}(A)} (\delta|_A (c_1|_C|\sigma)_{AC} - \inf_{c_0 \in \mathsf{Eff}(A)} (\delta|_A (c_0|_C|\sigma)_{AC} \\
&= \sup_{\{c_0, c_1\}} (\delta|_A (c_1 - c_0|_C|\sigma)_{AC},
\end{aligned}$$

where the optimization in the last equation is over all possible binary tests $\{c_0, c_1\}$ for system C. Now, applying the observation test $\{c_0, c_1\}$ to the bipartite state $|\sigma)_{AC}$ we obtain a preparation test $\{\rho_0, \rho_1\}$ for system A, defined by $|\rho_i)_A = (c_i|_C|\sigma)_{AC}, i = 0, 1$. Defining the probabilities $p_i = (e|\rho_i)_A$ and the normalized states $\bar{\rho}_i = \rho_i/(e|\rho_i)_A$ we then have

$$\begin{aligned}
(\delta|_A (c_1 - c_0|_C|\sigma)_{AC} &= p_1(\delta|\bar{\rho}_1)_A - p_0(\delta|\bar{\rho}_0)_A \\
&\le \sup\{(\delta|\bar{\rho}_1)_A, -(\delta|\bar{\rho}_0)_A\} \\
&\le \sup_{\rho \in \mathsf{St}_1(A)} |(\delta|\rho)_A|.
\end{aligned}$$

Exercise 3.10

The elements of $\mathsf{Eff}_\mathbb{R}(A)$ in quantum theory are self-adjoint operators. The definition in Eq. (3.18) corresponds to the following supremum for self-adjoint E

---

[9] Paulsen (1986).
[10] Aharonov *et al.* (1998).

$$E \in \mathsf{Herm}(\mathcal{H}_A), \quad \|E\|_A := \sup_{\rho \in \mathsf{St}(A)} |\mathrm{Tr}[E\rho]| = \sup_{\rho \geq 0, \|\rho\|_1 \leq 1} |\mathrm{Tr}[E\rho]| \quad (3.23)$$

$$\equiv \sup_{\|\rho\|_1 \leq 1} |\mathrm{Tr}[E\rho]|.$$

Using the bound $|\mathrm{Tr}[E\rho]| \leq \|E\| \|\rho\|_1$ we see that the supremum is achieved for ρ eigenstate of E corresponding to its eigenvalue with maximal modulus. Notice that the last identity in Eq. (3.23) extends to infinite dimensions.

Problem 3.1

Since the states are distinct there exists at least an effect a such that $(a|\rho_0) > (a|\rho_1)$. Moreover, since the theory is convex we can choose without loss of generality $(a|\rho_1) \geq 1/2$ (if a does not meet this condition, we can replace it with the convex combination $a' = 1/2(a + e)$). Now define the binary test $\{a_0, a_1\}$ as follows:

$$\begin{cases} a_0 = qa \\ a_1 = e - a_0 \end{cases} \quad q = \frac{1}{(a|\rho_0) + (a|\rho_1)} < 1,$$

For this test one has $p(1|0) = p(0|1) = (a|\rho_1)/[(a|\rho_0) + (a|\rho_1)] < 1/2$.

The New Principles

In this chapter we provide an overview of the new principles used for reconstructing quantum theory. We have already seen the principles in Chapter 2, where they have been derived as theorems of the quantum theory. In Part IV of this book we will actually do the reverse process, namely to derive quantum theory from the principles. All features of quantum theory – ranging from the superposition principle, entanglement, no cloning, teleportation, Bell's inequalities violation, quantum cryptography – can be understood and proved using only the principles, without using Hilbert spaces, and this is indeed the aim of Part III of the book, whereas Part IV contains the derivation of the theory from the principles.

All the six principles are operational, in that they stipulate whether or not certain tasks can be accomplished: they set the rules of the game for all the experiments and all the protocols that can be carried out in the theory. They also provide a great insight into the worldview at which quantum theory hints.

We review the list of the principles:

1. Atomicity of composition
2. Perfect discriminability
3. Ideal compression
4. Causality
5. Local discriminability
6. Purification

All six principles, with the exception of purification, express standard features that are shared by both classical and quantum theory. The principle of purification picks up uniquely quantum theory among the theories allowed by the first five, partly explaining the magic of quantum information.

The principles of causality, local discriminability, and purification will be thoroughly analyzed in Chapters 5, 6, and 7, respectively. The other three principles, atomicity of composition, perfect discriminability, ideal compression, are examined in the remainder of the present chapter.

4.1 Atomicity of Composition

In the general framework we encountered the notions of coarse-grained and atomic operation. A coarse-grained operation is obtained by joining together outcomes of a test, corresponding to neglect some information. The inverse process of coarse-graining is that of refining. An atomic operation is one where no information has been neglected, namely an operation that cannot be refined. When the operation is atomic, the experimenter has maximal knowledge of what's happening in the lab. A test consisting of atomic operations represents the highest level of control achievable according to our theory. But, is it possible to maintain such a level of control throughout a sequence of experiments?

The principle of atomicity of composition answers in the affirmative, stating precisely what follows:

Axiom 1 (Atomicity of Composition)   The sequence of two atomic operations is an atomic operation.

Atomicity of composition is a very primitive rule about how maximal information propagates in time. Think of a world where the principle does not hold. In that world, Alice performs an operation $\mathcal{A}_x$, with such degree of control that she could not possibly know better what happened to her system. Immediately after, Bob performs another operation $\mathcal{B}_y$ on Alice's system, and he also has a maximal knowledge of what he is doing. The resulting operation $\mathcal{B}_y\mathcal{A}_x$ is coarse-grained, i.e. it is an operation that can be simulated by a third party – Charlie – by performing one test $\{\mathcal{C}_z\}_{z\in\mathsf{Z}}$ and joining together the outcomes in a suitable subset $\mathsf{S}_{xy} \subset \mathsf{Z}$

$$\begin{array}{c} \text{A}\!-\!\boxed{\mathcal{A}_x}\!-\!\text{A}\!-\!\boxed{\mathcal{B}_y}\!-\!\text{A} \end{array} = \sum_{z\in\mathsf{S}_{xy}} \begin{array}{c}\text{A}\!-\!\boxed{\mathcal{C}_z}\!-\!\text{A}\end{array}. \tag{4.1}$$

Although this scenario is logically conceivable, it raises the questions: What is the extra information about? Which physical parameters correspond to the outcome z? Surely it is not about what happened in the first step, because Alice already had maximal knowledge about this. Nor it is about what happened in the second step, because Bob has maximal information about that. The outcome z has to specify a feature of how the two time steps interacted together – in a sense, a kind of information that is *non-local in time*. Quantum theory is non-local, but not in such an extreme way! Indeed, atomic operations in quantum theory are described by completely positive maps with a single Kraus operator, i.e. of the form $\mathcal{A}_x(\cdot) = A_x \cdot A_x^\dagger$ and $\mathcal{B}_y(\cdot) = B_y \cdot B_y^\dagger$, and clearly the composition of two atomic operations is still atomic: $\mathcal{B}_y\mathcal{A}_x(\cdot) = (B_yA_x)\cdot(B_yA_x)^\dagger$. Atomicity of composition guarantees this property at the level of first principles.

In a world that violates the atomicity of composition, another odd feature could arise. Suppose that Alice prepares system A in the pure state α and then Bob prepares system B in the pure state β. If atomicity did not hold, the resulting state $\alpha \otimes \beta$ could be mixed: having maximal knowledge of how the individual systems A and B have been prepared would not be enough to guarantee maximal knowledge about the preparation of the composite system AB. Again, this is not the case in quantum theory, where the tensor product of two pure states with wavevectors $|\alpha\rangle$ and $|\beta\rangle$ is the pure state with wavevector $|\alpha\rangle \otimes |\beta\rangle$. In a few paragraphs we will preview the local discriminability principle, which, like atomicity of composition, leads to the consequence that the product of two pure states must be a pure state. It is worth stressing, however, that the content of the two principles is different and logically independent. The reason why they both lead to the same consequence for the product of pure states is that for preparation processes the sequential composition (to which atomicity of composition refers) collapses into the parallel composition (to which local discriminability refers).

4.2 Perfect Discriminability

Both in classical and quantum theory, every non-trivial system has at least two perfectly discriminable states. For example, the states 0 and 1 for a classical bit, or the states $|0\rangle$ and $|1\rangle$ for a quantum bit. In our general setting, two deterministic states ρ_0 and ρ_1 are perfectly discriminable if there exists a measurement $\{m_y\}_{y\in\{0,1\}}$ such that

$$(m_y \mid \rho_x) = \delta_{xy} \qquad \forall x, y \in \{0, 1\}.$$

The existence of perfectly discriminable states is important, because these states can be used to communicate classical information without errors. In a communication protocol, the sender can encode the value of a bit x into the state ρ_x and then transmit the system to the receiver, who can decode the value of the bit using the measurement $\{m_y\}_{y\in\{0,1\}}$.

In principle, one can imagine theories where there are no perfectly discriminable states at all. For example, one can consider a noisy version of classical theory, where the possible states of a bit are described by probability distributions $p(x)$ such that $p(x) \geq \epsilon$ for every $x \in \{0, 1\}$. The noisy bit has two (and only two) pure states, corresponding to the probability distributions $p_0(x) = (1 - \epsilon)\delta_{x,0} + \epsilon\delta_{x,1}$ and $p_1(x) = (1 - \epsilon)\delta_{x,1} + \epsilon\delta_{x,0}$. All the other states are mixtures of these two. It is easy to see that no states of the noisy bit can be discriminated perfectly (cf. Exercise 4.2). Noisy theories, like the one we just described, may be useful models for realistic experiments, where in practice there is always some imperfection that prevents us from discriminating states perfectly. However, if we want to describe nature at the most fundamental level, noisy theories seem rather unappealing.

The perfect discriminability axiom ensures that our ability to discriminate states is as sharp as it could possibly be: except for trivial cases, every state can be perfectly discriminated from some other state. The "trivial cases" are those states that cannot be discriminated from anything else because they contain every other state in their convex decomposition. Geometrically, these states are those in the interior of the convex set of states. We can call them *internal*, or *completely mixed*.

Exercise 4.1 Suppose that the deterministic states of system A form a convex set. Show that a state $\rho \in \mathsf{St}_1(A)$ is completely mixed if and only if it is in the interior of the convex set.

Axiom 2 (Perfect Discriminability) Every deterministic state that is not completely mixed is perfectly discriminable from some other state.

Note that both classical and quantum theory have this feature, as highlighted by Exercises 4.3 and 4.4.

As anticipated, the perfect discriminability axiom guarantees that every non-trivial system has at least two perfectly discriminable states:

Proposition 4.1 *In a theory satisfying perfect discriminability, every physical system has at least two perfectly discriminable states, unless the system is trivial (i.e. it has only one deterministic state).*

Proof Pick a pure state $\alpha \in \mathsf{St}(A)$. If α is not internal, then perfect discriminability guarantees that α is perfectly discriminable from some other state α', hence A has two perfectly discriminable states. If α is internal every pure state belongs to its refinement set. Moreover, since it is also pure, i.e. extremal, one has that every other deterministic state $\rho_1 \in \mathsf{St}_1(A)$ must be equal to α, i.e. A has only one deterministic state.   □

An easy consequence of this result is that the theory can describe noiseless classical communication: if Alice wants to communicate a bit string $\mathbf{x} = (x_1, \ldots, x_k)$ to Bob, she can encode each bit into a pair of perfectly discriminable states – say ρ_0 and ρ_1 – of her system and transmit to him a sequence of identical systems prepared in the state $\rho_{x_1} \otimes \rho_{x_2} \otimes \cdots \otimes \rho_{x_k}$.

Exercise 4.2 Let $q_0(x)$ and $q_1(x)$ be two probability distributions on $\{0, 1\}$, describing two states of the noisy classical bit. Show that the total variation distance between the two probability distributions, defined as $\|q_0 - q_1\|_1 := \sum_x |q_0(x) - q_1(x)|$ is upper bounded by

$$\|q_0 - q_1\|_1 \leq 2(1 - 2\epsilon).$$

Exercise 4.3 Let $p(x)$ and $q(x)$ be two probability distributions over a finite set X.

1. Show that $p(x)$ is compatible with $q(x)$ if and only if $\mathsf{Supp}(p) \supseteq \mathsf{Supp}(q)$, where $\mathsf{Supp}(f) = \{x \in \mathsf{X} \mid f(x) \neq 0\}$ is the support of the function f.
2. Show that p is completely mixed if and only if it has full support.
3. Show that classical probability theory satisfies the perfect discriminability axiom.

Exercise 4.4 Let ρ and σ be two density matrices on a finite-dimensional Hilbert space $\mathcal{H}$.

1. Show that ρ is compatible with σ if and only if $\mathsf{Supp}(\rho) \supseteq \mathsf{Supp}(\sigma)$, where $\mathsf{Supp}(A) = \mathsf{Ker}(A)^\perp$ is the support of the matrix A.
2. Show that ρ is completely mixed if and only if it has full rank.
3. Show that quantum theory satisfies the perfect discriminability axiom.

4.3 Ideal Compression

In a trivial sense, every physical theory is a theory of information – the state of a physical system at a certain time just describes the information available at that time about the results of future experiments. What makes the expression "theory of information" less trivial is that in some theories one can talk about "information" independently of the particular physical support that carries it. This is the case for classical theory, where information can be transferred faithfully from a newspaper or a strand of DNA to a laptop or a USB stick, and it is also the case for quantum theory, where, at least in principle, information can be transferred faithfully e.g. from the polarization of a photon to the electronic levels of an atom.

Not every physical theory has the property that information can be transferred faithfully from one system to another. In principle, one can conceive theories where every transformation from an input system A to a different output system B leads to an irreversible loss of information. Let us make this idea more precise: suppose that Alice has a preparation device, which prepares system A in some state α. Alice does not know the state α, but she knows that on average the device prepares the deterministic state $\rho \in \mathrm{St}_1(A)$. Now, suppose Alice wants to transfer the state of her system to Bob's laboratory, but unfortunately she cannot send system A directly. Instead, she has to encode the state α into the state of another system B, by applying a suitable deterministic operation $\mathcal{E}$ (the *encoding*), which transforms the state α into the state

$$\beta := \mathcal{E}\alpha. \tag{4.2}$$

We say that the encoding is *lossless for the state* ρ iff there exists another deterministic operation $\mathcal{D}$ (the *decoding*) such that

$$\mathcal{D}\mathcal{E}\alpha = \alpha \qquad \forall \alpha \in F_\rho \tag{4.3}$$

where F_ρ is the refinement set of ρ, which is made of the set of all states α that are compatible with ρ (on the convex set of states this would be the face to which ρ belongs).

Our fourth axiom establishes the possibility of a particular type of lossless encoding, called *ideal compression*. Intuitively, information compression is the task of encoding information from a larger system to a smaller system, e.g. from the memory of a laptop to a USB stick. The ultimate limit to the lossless compression of a given state ρ is reached when every state of the encoding system B is a codeword for some state in F_ρ, namely when every state $\beta \in \mathrm{St}(B)$ is of the form $\mathcal{E}\alpha$ for some $\alpha \in F_\rho$. When this is the case, we say that the compression is *efficient*, and we call the triple $(B, \mathcal{E}, \mathcal{D})$ an *ideal compression protocol*.

Our fourth axiom ensures that such kind of protocols always exists:

Axiom 3 (Ideal Compression)   Every state can be compressed in a lossless and efficient way.

It is easy to see that both classical and quantum theory satisfy the ideal compression axiom (cf. Exercises 4.6 and 4.7). Moreover, it is easy to see that, unless the state ρ is completely mixed, the system B used for ideal compression must be strictly smaller than system A:

Proposition 4.2   *Let* $(B, \mathcal{E}, \mathcal{D})$ *be an ideal compression protocol for the state* $\rho \in \mathrm{St}_1(A)$. *If* ρ *is not a completely mixed state, then* $B \neq A$.

Proof   It is easy to prove the contrapositive: if $B = A$, then ρ must be a completely mixed state. Indeed, when $B = A$ the mapping $\mathcal{E} : F_\rho \to \mathrm{St}_1(A)$ is both injective (because the compression is lossless), and surjective (because the compression is efficient). Hence, we have $F_\rho \equiv \mathsf{St}_1(\mathrm{A})$, meaning that ρ is compatible with every other state of system A.   □

Exercise 4.5   Let $\mathcal{E} \in \mathsf{Transf}(\mathrm{A} \to \mathrm{B})$ be a lossless encoding for the state $\rho \in \mathsf{St}_1(\mathrm{A})$.

1. Show that the dimension of the vector space $\mathsf{Span}_\mathbb{R}\{F_\rho\}$ must be smaller or equal than the dimension D_B of the vector space $\mathsf{St}_\mathbb{R}(\mathrm{B})$.
2. Show that the two dimensions are equal when $\mathcal{E}$ is an ideal compression.

Exercise 4.6   Let $p(x)$ be a probability distribution on some finite set X and let $\mathsf{Supp}\,(p)$ be the support of p. Show that there exists an ideal compression protocol for $p(x)$ where the states of the encoding system are the probability distributions over a set Y of cardinality $|\mathsf{Y}| = |\mathsf{Supp}\,(p)|$.

Exercise 4.7   Let ρ be a density matrix on a finite-dimensional Hilbert space and let r be the rank of ρ. Show that there exists an ideal compression protocol for ρ where the encoding system is an r-dimensional quantum system.

4.4 A Preview of the Three Main Principles

In this section we give a very short preview of the other three principles – causality, local discriminability, and purification – which will be used to derive quantum theory in the last part of this book, and will be thoroughly analyzed in Chapters 5, 6, and 7.

4.4.1 Causality

In Chapter 3 we described a framework that can accommodate a large number of physical theories. So large that in it we can find theories with fairly odd features, like signaling from the future or instantaneous signaling across space. Quantum theory does not allow such kinds of signaling, as a consequence of the following principle.

Axiom 4 (Causality)   The probability of the outcome of a preparation test is independent of the choice of observation tests connected at its output.

The causality principle identifies the input–output ordering of a circuit with the direction along which information flows, identifying such ordering with a proper-time arrow, corresponding to the request that future choices cannot influence the present.

The causality axiom has a number of consequences, which will be thoroughly analyzed in Chapter 5. It shapes the convex structure of the theory in a very special way.

The assumption of causality is implicitly embedded in the framework of most works in the tradition of generalized probabilistic theories. It is, however, conceptually very important to recognize it as a separate principle, since it is possible to formulate a version of quantum theory where causality is not obeyed. We will provide some ideas in Chapter 5; however, in this book we will content ourselves with the reconstruction of the ordinary quantum theory.

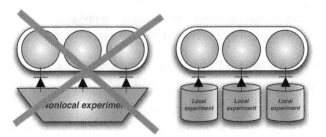

Fig. 4.1 **Local discriminability principle (also called local tomography).** Alice can reconstruct the state of compound systems using only local measurements on the components. A world where this property did not hold would contain global information that cannot be accessed by local experiments.

4.4.2 Local Discriminability

Axiom 5 (Local Discriminability)   It is possible to discriminate any pair of states of composite systems using only local measurements.

As we will see in Chapter 6 devoted to the axiom, the statement is also equivalent to state that every state of a composite system can be reconstructed from the statistics of local measurements on the components, which is also referred to as the *local tomography* axiom (see Fig. 4.1). As explained in Chapter 6, the local discriminability axiom has crucial consequences in the possibility of checking experimentally a physical law by testing it only on a given input system, instead of being forced to consider all possible systems, as e.g. testing the Maxwell equations on all possible radiation modes in the universe. The principle is trivially satisfied by classical theory, where the states of every composite system are joint probability distributions for the outcomes of two local measurements. In Chapter 2 we have seen that the principle holds for quantum theory, whereas it doesn't hold for quantum theory on real Hilbert spaces, where there exist bipartite states that cannot be discriminated by local measurements.

The reader may have probably noticed that the local discriminability principle has a similar flavor with that of atomicity of composition (both of them exclude the existence of some inaccessible global information), but the two principles are actually very different. The fact that the two principles are indeed logically independent is proved by the simple fact that quantum theory on real Hilbert spaces satisfies the atomicity of composition axiom, but not the local discriminability axiom.

4.4.3 Purification

The five principles presented so far define a family of theories of information that can be regarded as standard. If it were for these principles only, the world may very well be classical. What is special about quantum theory, compared to all other theories, is that it always admits a description with maximal knowledge, and such a description is compatible with any other description with generally limited knowledge. Knowledge about physical systems is provided by the state of the systems: maximal knowledge corresponds to having the state pure, partial knowledge corresponds to having it mixed. On the other hand, partial knowledge also means that we are ignoring some systems, including their correlations with the systems under observation. This means that the partial knowledge is compatible with the perfect knowledge when this means to marginalize some of the systems involved in the perfect knowledge. In other words, the mixed-state description must be the marginal of the pure-state description. We conclude that the compatibility requirement of the two kinds of knowledges corresponds to assuming the existence of a pure state Ψ_ρ of which ρ is marginal. The state Ψ_ρ is what we call a *purification* of the state ρ.

Now, an important question is whether or not the perfect-knowledge description is unique: suppose that two theoreticians are asked to explain why system A is in the mixed state ρ. Will they give the same explanation? In a good theory, the answer should be *Yes*, up to trivial differences. By "trivial differences" we mean the following:

1. $\Psi'_\rho := \Psi_\rho \otimes \phi \in \mathsf{PurSt}(AEF)$ is a purification of ρ, for every pure state ϕ of every system F.
2. $\Psi'_\rho := (\mathcal{I}_A \otimes \mathcal{U}_E)\Psi_\rho$ is a purification of ρ for every reversible transformation $\mathcal{U}_E$ acting on the environment.

If every two purifications of the same state are equivalent up to these trivial differences, we say that the purification is *essentially unique*. The purification axiom is then phrased as follows:

Axiom 6 (Purification) *Every state has an essentially unique purification.*

Purification rules out classical probability theory, because there is no way to obtain a generic probability distribution for a variable X as the marginal of a probability distribution for two variables X and Y, which is concentrated on a single value $(x_0, y_0) \in X \times Y$. In fact, purification not only rules out classical theory, but also provides the seed for most of the non-classical features and of the advantages of quantum theory. A large number of quantum features are derived in Chapter 7, where we elaborate in depth on the significance of the purification principle.

4.5 Summary

In this chapter we gave a bird's eye view of the six principles used to reconstruct quantum theory. Five principles – causality, local discriminability, atomicity of composition, perfect discriminability, and ideal compression – define a standard class of theories of information, which includes both quantum theory and classical probability theory. The sixth principle – purification – brings in the quantum nature of information, by requiring that the ignorance about the preparation of a physical system can be always explained out, in an essentially unique way, by including the environment in the description. For finite systems (systems whose state is determined by a finite number of outcome probabilities) the six principles presented above are equivalent to quantum theory. In Chapter 2 we have derived these principles as theorems from the theory, in the rest of the book we will derive quantum theory from the six principles. Complex Hilbert spaces, superposition principle, Heisenberg's uncertainty relations, entanglement, no cloning, teleportation, violation of Bell's inequalities, quantum cryptography – every quantum feature is already zipped into the six principles; the rest of this book will be devoted to the unzipping.

Notes

About Atomicity of Composition   In the context of our axioms, atomicity of composition is equivalent to the axiom of pure conditioning, proposed by Wilce (2010). Pure conditioning is the requirement that when a bipartite system is in a pure state, an atomic measurement performed on one part induced pure states on the other part.

About Ideal Compression and the Usual Compression in Information Theory   It is worth noting that the notion of ideal compression in Section 4.3 is different from the one used in Shannon's compression in classical information and the Schumacher's compression in quantum information. The difference is twofold: first, in our case we consider a single use of the source (i.e. a single use of Alice's preparation device), while Shannon's and Schumacher's theorems refer to the asymptotic scenario where the same source is used a large number of times. Second, we require the compression protocol to be perfectly lossless (i.e. to have zero error), while the framework of Shannon's and Schumacher's compression allows for compression protocols that have an error, provided that the error vanishes when the number of uses of the source tends to infinity.

Causal Theories   5

In this chapter we introduce the postulate of causality, and derive the mathematical structure of a causal OPT. We show how the present definition of causality is equivalent to the Einsteinian one, and argue that it is the only possible notion once it is stripped from its spurious deterministic connotation. In addition, the notion of causality used here is compliant with the historical philosophical concept since David Hume, and with the idea used in common reasoning, inference, and modeling in human sciences.[1]

In brief, the present notion of causality can be stated as the requirement of *no signaling from the future*. This is also the only requisite for an OPT in order to satisfy the principle of *no signaling without interaction*. In mathematical terms in an OPT the causality postulate is equivalent to the uniqueness of the deterministic effect, which in turn is equivalent to the ability to normalize states. The notion of causality is so naturally embedded in our understanding of reality that it often remains unconsciously implicit in foundational research, even in recent axiomatizations.[2]

5.1 Causality: From Cinderella to Principle

Causality has been the object of a very extensive literature encompassing hundreds of books and technical articles in a wide spectrum of disciplines, ranging from pure philosophy to economics, law, natural sciences, and obviously, physics. Also due to its involvement in such heterogeneous blend of disciplines, causality has long been a vexed notion. Perhaps the most natural connection between concepts of causality in different branches of knowledge is the one at the borderline between physics and philosophy, since the early work of Aristotle, up to the cornerstone of Renée Descartes, who broke the ground for the modern view of David Hume and Immanuel Kant, through to contemporary works on physical causation.[3]

Causality has always remained in the realm of philosophy, staying only in the background of physics, without the status of a "law" or the rank of a "principle." Most of the time causality creeps into physical theories in the form of *ad hoc* assumptions based on empirical evidences, as when we discard advanced potentials in electrodynamics or when

---

[1] Pearl (2012).

[2] Hardy (2001).

[3] The most relevant works are those of Salmon (1998) and Dowe (2007).

we motivate the Kramers–Kronig relations. In other cases, causality is embodied in the "interpretation" of the theory, as for the Special Relativity of Einstein. In the framework of OPTs causality often remains hidden in the theoretical framework, as in the recent axiomatization of Hardy.[4]

One of the main sources of confusions has been the traditional connection between causality and the notion of determinism,[5] the latter being so deeply entangled with physical causation to the extent that the two notions are often merged into that of *causal determinism*. An example of this connection can be found in the quotation of the founding father of quantum theory, Max Planck: "An event is causally determined if it can be predicted with certainty."[6] Such a confusion between causality and determinism is the main source of the common misinterpretation of quantum correlations as "spooky action at a distance," namely the commonplace situation of EPR correlations interpreted as a sort of causation.

The property of causality as considered here is trivialized in the classical mechanical context by the identification of "observation" with "preparation": there is only one classical measurement, and the outcome is the pure classical state revealed by the measurement without disturbing it. Complementarity in quantum theory breaks the classical identification between observation and preparation (measurement and state), due to the possibility of choosing among different incompatible measurements. Causality in quantum theory is then the assumption of independence of the probability of preparation from the choice of observation. This position distills all the intuitive guises in which causality appears in physics, and has an intimate relation with the Einsteinian notion. In this formulation it is the first axiom of quantum theory in the derivation of the present book.[7] The two notions of causality and determinism thus become completely disentangled in quantum theory (and, more generally, in an OPT), while classical theory is only the degenerate case of conceptual overlapping between determinism and causality, corresponding to the peculiar feature of identification of the state with the measurement outcome. In a probabilistic context determinism is identified with the tautological property of a theory of having all probabilities of physical events being equal to either zero or one, which is clearly a definition with no causal connotation.

The connectivity between events is a crucial feature of the operational framework in order to define causality properly, i.e. in the presence of plurality of "causes" and "effects," which, otherwise, would make the notion of causality vague, as remarked in the critique of Russel.[8] Such connectivity gives rise to the same network construction of the methodological causal approach of Pearl,[9] and endows the causality notion with the structure of a partial ordering. Such an ordering allows us to build up foliations over the operational circuits, ultimately leading to the equivalence with the Einsteinian notion.[10]

---

[4] Hardy (2001).

[5] The logical independence between the two notions of causality and determinism is proved by the existence of a causal OPT that is not deterministic (quantum theory, of course), and vice versa of a deterministic theory that is not causal (D'Ariano *et al.*, 2014a).

[6] Planck (1941).

[7] Chiribella *et al.* (2011).

[8] Russel (1912).

[9] Pearl (2012).

[10] See e.g. D'Ariano and Tosini (2013).

5.2 No Signaling from the Future

The causality axiom will ultimately lead us to interpret the input–output connections between tests as causal links, understanding their sequential composition as a series of tests performed in cascade on the same system. Let us now review the statement of the axiom.

Causality Axiom *The probability of preparations is independent of the choice of observations.*

Let us analyze what the causality axiom says precisely. Consider the joint test consisting of a preparation test $\mathcal{X} = \{\rho_i\}_{i\in X} \subset \mathsf{St}(A)$ followed by the observation test $\mathcal{Y} = \{a_j\}_{j\in Y} \subset \mathsf{Eff}(A)$ performed on system A:

$$\boxed{\mathcal{X}} \!\!-\!\!\!-^{A}\!\!\!-\!\! \boxed{\mathcal{Y}}$$

The joint probability of preparation ρ_i and observation a_j is given by

$$p(i,j|\mathcal{X},\mathcal{Y}) := (a_j|\rho_i) \equiv \boxed{\rho_i} \!\!-\!\!^{A}\!\!-\!\! \boxed{a_j}$$

The marginal probability of the preparation alone does not depend on the outcome j. Yet, it generally depends on which observation test $\mathcal{Y}$ is performed, namely

$$\sum_{a_j\in\mathcal{Y}}(a_j|\rho_i) =: p(i|\mathcal{X},\mathcal{Y}).$$

The marginal probability of preparation ρ_i is then generally conditioned on the choice of the observation test $\mathcal{Y}$. What the causality axiom states is that $p(i|\mathcal{X},\mathcal{Y})$ is actually independent of $\mathcal{Y}$, namely for any two different observation tests $\mathcal{Y} = \{a_j\}_{j\in Y}$ and $\mathcal{Z} = \{b_k\}_{k\in Z}$ one has

$$p(i|\mathcal{X},\mathcal{Y}) = p(i|\mathcal{X},\mathcal{Z}) = p(i|\mathcal{X}).$$

The causality postulate is not just a restriction to probability distributions of circuits made only of two tests – preparation and observation. It actually regulates the joint probability distribution of any closed circuit made of multiple systems and tests, since any closed circuit can be always regarded as the composition of a preparation and an observation test by cutting the circuit along a slice of systems, as explained in Section 3.3. For example, if we cut the circuit

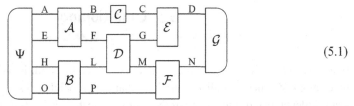

$$(5.1)$$

along the slice BGN any test within the circuit either pertains to preparation $\mathcal{X} = \mathcal{F}\mathcal{D}(\mathcal{A} \otimes \mathcal{B})|\Psi)$, or to observation $\mathcal{Y} = (\mathcal{G}|\mathcal{E}\mathcal{C})$. If we consider the test $\mathcal{F} \equiv \{\mathcal{F}_f\}_{f\in F}$, and take the marginal probability distribution of its outcomes f, by causality we have

$$\forall f \in F: \quad p(f|\mathcal{X}\mathcal{Y}) = p(f|\mathcal{X}) = p(f|\Psi, \mathcal{A}, \mathcal{B}, \mathcal{D}, \mathcal{F}).$$

Causality means that the probability distribution of a test within a circuit is not conditioned by tests not connected to its inputs. In the following, we will also say that a test $\mathcal{A}$ *precedes* a test $\mathcal{B}$ if some output system of $\mathcal{A}$ is connected to some input system of $\mathcal{B}$, and extend the notion transitively, namely a test $\mathcal{A}$ *precedes* a test $\mathcal{B}$ if there is a chain of tests $\mathcal{C}_1, \mathcal{C}_2, \ldots, \mathcal{C}_{n+1}$ with $\mathcal{C}_1 = \mathcal{A}$ and $\mathcal{C}_{n+1} = \mathcal{B}$ and $\mathcal{C}_i$ precedes $\mathcal{C}_{i+1}$ for $i = 1, \ldots, n - 1$. We will also call the chain $\mathcal{C}_1, \mathcal{C}_2, \ldots, \mathcal{C}_{n+1}$ a *causal chain*. Equivalently, we will say that $\mathcal{B}$ *follows* $\mathcal{A}$. The above notions are trivially extended to events.

Causal OPT An OPT satisfying the causality postulate for all closed circuits is called *causal*.

The causality axiom is also equivalent to the following statement.

No Signaling From the Future An OPT is *causal*, if for any test $\mathcal{A}$ that does not follow a test $\mathcal{B}$, one has that the marginal probability distribution of test $\mathcal{A}$ is independent of the choice of test $\mathcal{B}$.

The above statement implies the causality axiom, since the axiom is a special case of it. The reverse implication is also true, since as we have seen, any closed circuit can be always split into a preparation and an observation test, with the preparation containing test $\mathcal{A}$ and the observation containing test $\mathcal{B}$.

A parametric dependence of the probability distribution $p(j|\mathcal{A}, \mathcal{B})$ of the test $\mathcal{B} := \{\mathcal{B}_j\}_{j \in Y}$ on the choice of a preceding test $\mathcal{A} := \{\mathcal{A}_i\}_{i \in X}$ describes a causal relation between the two tests. Indeed, in such a case we can communicate information to test $\mathcal{B}$ by changing test $\mathcal{A}$. The input test $\mathcal{A}$ can be regarded as a "cause" for the outcome of test $\mathcal{B}$, which in turn plays the role of an "effect." Causality allows one to choose the input–output direction as the same direction from past to future, i.e. the arrow of time, and the naming "from the future" corresponds to such a choice. In practice, in a physics lab this corresponds to assert that the probability distribution of a test performed on a system (e.g. an electron or an optical beam) cannot depend on which test will be executed later in time on the same system (namely on the electron or the optical beam at the output of the previous test). On the other hand, the probability distribution of a test on a system generally depends on which test has been performed on the same system, which can then be regarded as contributing to the preparation of the system itself.

5.3 Conditioning

In a causal OPT the choice of a test on a system can be conditioned on the outcomes of a preceding test, since causality guarantees that the probability distribution of the preceding test is independent of the choice of the following test. This leads us to introduce the notion of *conditioned test*.

Conditioned test *If $\{\mathcal{A}_i\}_{i \in X}$ is a test from* A *to* B, *and $\left\{\mathcal{B}_j^{(i)}\right\}_{j \in Y_i}$ is a test from* B *to* C *for every* $i \in X$, *then the conditioned test is a test from* A *to* C, *with outcomes*

$(i,j) \in Z := \bigcup_i \{\{i\} \times Y_i\}$, *and events* $\{\mathcal{B}_j^{(i)} \circ \mathcal{A}_i\}_{(i,j) \in Z}$. *Diagrammatically, the events* $\mathcal{B}_j^{(i)} \circ \mathcal{A}_i$ *are represented as follows:*

$$
\begin{array}{c}
\xrightarrow{\text{A}} \boxed{\mathcal{A}_i} \xrightarrow{\text{B}} \boxed{\mathcal{B}_j^{(i)}} \xrightarrow{\text{C}}
\end{array} . \tag{5.2}
$$

The notion of conditioned test generalizes both notions of sequential composition and randomization of tests. In fact, composition of test is just a special case of conditioned test for $\{\mathcal{B}_j^{(i)}\}_{j \in Y_i} \equiv \{\mathcal{B}_j\}_{j \in Y}$ all equal independently of i. On the other hand, a randomized test is just a test conditioned by a test from the trivial system to the trivial system, namely a probability.

Among conditioned test, a special role is played by the *observe-and-prepare test*,[11] where the "connecting" system is the null system I. They are thus made of a preparation test conditioned by an observation test, as follows:

$$
\begin{array}{c}
\xrightarrow{\text{A}} \boxed{l_i} \quad \left(\!\omega^{(i)}\right) \xrightarrow{\text{C}}
\end{array} ,
$$

which can be also represented as $\left\{ |\omega^{(i)})(l_i| \right\}_{i \in X}$.

Remark The requirement of causality for introducing conditioned test is dictated from the fact that in a theory without causality it is impossible to include *every* conceivable conditioned test, since there exist conditioned tests that would lead to violations of the basic rules of probability. As an example, consider a non-causal theory and take two deterministic effects $e^{(0)} \neq e^{(1)}$ for system A, namely there exists a state $\tau_0 \in \text{St}(A)$ such that $(e^{(0)}|\tau_0) \neq (e^{(1)}|\tau_0)$. The state τ_0 can be complemented to a complete preparation test by a state τ_1 to the make the binary test $\{\tau_i\}_{i=0,1}$. Construct now the conditioned test in which the preparation τ_i conditions which deterministic effect $e^{(i)}$ is chosen. Equation (5.2) can be rewritten

$$
\begin{array}{c}
\xrightarrow{\text{I}} \boxed{\tau_i} \xrightarrow{\text{B}} \boxed{e^{(i)}} \xrightarrow{\text{I}} \equiv \left(\tau_i \xrightarrow{\text{A}} \boxed{e^{(i)}} \right) = (e^{(i)}|\tau_i).
\end{array} \tag{5.3}
$$

The test $\{(e^{(i)}|\tau_i)\}_{i=0,1} \subseteq \text{Transf}(I \to I)$ then violates the rules of probability.

Exercise 5.1 Prove that the test $\{(e^{(i)}|\tau_i)\}_{i=0,1} \subseteq \text{Transf}(I \to I)$ violates the probability rules.

5.4 A Unique Wastebasket

Causal theories can be characterized as follows:

Lemma 5.1 *An OPT is causal if and only if for every system* A *there is a unique deterministic effect.*

---

[11] Often in the literature the observe-and-prepare tests are named "measure-and-reprepare." In the early literature about quantum measurements the measure-and-reprepare quantum instruments were also called "Gordon–Louisell Measurements" (Gordon and Louisell, 1966) or "demolitive measurements."

Proof We will prove the two directions separately, namely: (1) if the probability of preparation of states is independent of the observation test, then the deterministic effect is unique; (2) vice versa. (1) The probability of the preparation ρ is given by the marginal of the joint probability with the observation, namely $p(\rho) = \sum_{i \in X}(a_i|\rho)$. Upon denoting the deterministic effects of two different tests as $a = \sum_{i \in X} a_i$ and $b = \sum_{j \in Y} b_j$, the statement that the preparation probability is independent of the observation tests translates to $(a|\rho) = (b|\rho)$ for every preparation $\rho \in \mathsf{St}(A)$, which implies that $a = b$, since the set of states is separating for events. (2) Uniqueness of the deterministic effect implies that the preparation probability of each state is independent of the test, since the effect $a = \sum_{i \in X} a_i$ for any test $\{a_i\}_{i \in X}$ is deterministic, and $(a|\rho)$ for any deterministic effect $a \in \mathsf{Eff}(A)$ is the probability of preparation ρ. $\qquad\square$

We will denote the unique deterministic effect for system A as e_A, and the subindex will be dropped when no confusion can arise.

In the following we will use the notation $\leq$ to denote the partial ordering between effects, defined as follows:

$$a, b \in \mathsf{Eff}(A), \ a \leq b \quad \Leftrightarrow \quad (a|\rho) \leq (b|\rho), \ \forall \rho \in \mathsf{St}(A).$$

It is immediate to show that the causality condition of Lemma 5.1 spawns the following lemmas.

Lemma 5.2 *Causality is equivalent to the following statements regarding tests:*

1. **Completeness of observation tests:** *For any system* A *and for every observation test* $\{a_i\}_{i \in X}$ *one has*

$$\sum_{i \in X} a_i = e_A. \tag{5.4}$$

2. **Completeness of tests:** *For any systems* A, B *and for every test* $\{\mathcal{C}_i\}_{i \in X}$ *from* A *to* B *one has*

$$\sum_{i \in X} (e_B|\mathcal{C}_i = (e_A|.$$

3. **Domination of transformations:** *For any systems* A, B *a transformation* $\mathcal{C} \in \mathsf{Transf}(A \to B)$ *satisfies the condition*

$$(e_B|\mathcal{C} \leq (e_A|, \tag{5.5}$$

with the equality if and only if $\mathcal{C}$ *is a channel, i.e. a deterministic transformation corresponding to a single-outcome test.*

4. **Domination of effects:** *For any system* A *all effects are dominated by a unique effect* e_A *which is deterministic*

$$\forall a \in \mathsf{Eff}(A), \qquad 0 \leq a \leq e_A. \tag{5.6}$$

An immediate consequence of uniqueness of the deterministic effect is the identification of all transformations of the form

$$\forall \mathcal{B} \in \mathsf{Transf}(B \to C), \quad \sum_{i \in X} a_j \otimes \mathcal{B} = e_A \otimes \mathcal{B},$$

for any observation test $\{a_i\}_{i \in X}$ of system A. In particular, we have the factorization of the deterministic effect of composite systems

$$e_{AB} = e_A \otimes e_B.$$

The causality condition greatly simplifies the evaluation of probabilities of events. Indeed, since the probability of an event in a test is independent of the choice of tests following it, we can substitute the network with another one in which all tests following the event of interest are substituted by any arbitrary deterministic test. For example, in order to evaluate the marginal probability $p(i, j | \Psi, \mathcal{A})$ of events $\mathcal{A}_j$, Ψ_i in the following circuit:

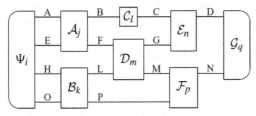

we just need the simplified circuit

$$p(\mathcal{A}_j, \Psi_i) = \begin{array}{c} \Psi_i \end{array}$$

with e denoting the deterministic effect of the corresponding input system. Therefore, the deterministic effect works as a *wastebasket*, since it erases all the unwanted information (Fig. 5.1).

5.4.1 The Normalization of States

In a causal theory we can achieve any state deterministically by a *conditional preparation*, also called *post-selection*, a procedure made as follows. A preparation test $\{\rho_i\}_{i \in X}$ including the state $\rho \equiv \rho_{i_0}$ is performed. If the test gives outcome i_0 one knows for sure *a posteriori* that the normalized state $\bar{\rho} = \rho/(e|\rho)$ has been prepared. This is possible in a causal theory, since the probability of preparing a state is independent of the choice of the observation test. We will therefore assume that the state $\bar{\rho} = \rho/(e|\rho)$ is a deterministic state of the theory.

We now prove the equivalent relation for causality of an OPT.

Lemma 5.3 *An OPT is causal if and only if every state is proportional to a deterministic one.*

The deterministic effect in a causal theory works as a wastebasket, where information can be dumped.

Fig. 5.1

Proof We first prove that a theory where every state is proportional to a deterministic one is causal. Let $\rho \in \mathsf{St}(A)$ be an arbitrary state and $e_A \neq e'_A$ be two deterministic effects. By hypothesis there exists $k > 0$ such that $\rho = k\bar{\rho}$ with $\bar{\rho}$ deterministic. This implies $(e|\rho)_A = k = (e'|\rho)_A$, and, since ρ is arbitrary $e = e'$, and by Lemma 5.1, this implies that the theory is causal. The converse assertion is trivial, since for any state $\rho \in \mathsf{St}(A)$ in a causal theory the state $\rho/(e|\rho)$ is deterministic.   □

We emphasize that in a causal theory the state $\rho/(e|\rho)$ can be actually prepared deterministically by a post-selection procedure. In a non-causal theory it is not generally possible to achieve a given state deterministically. Indeed, by negating Lemma 5.3 we deduce that if a theory is non-causal there must exist probabilistic states that cannot be prepared deterministically. Therefore, in a sense, the lack of causality in an OPT corresponds to a limitation of preparations.

The procedure connecting a probabilistic state with its deterministic version

$$\bar{\rho} := \frac{\rho}{(e|\rho)_A} \tag{5.7}$$

is called *normalization*. Causality is so natural an assumption that we use it all the time without even knowing, and this is the case when we normalize quantum states.

5.4.2 The Cone Structure of a Non-deterministic Causal Theory

For a convex OPT, for every system A the sets $\mathsf{St}(A)$ of states, $\mathsf{St}_1(A)$ of normalized states, and $\mathsf{Eff}(A)$ of effects are all convex. In Section 5.4.4 we will see that a non-deterministic OPT is necessarily convex, due to operational closure of OPTs. Operationally, convex combinations of states are performed by randomization of preparations, e.g. the state $\rho_p := p\rho_1 + (1-p)\rho_2$ can be prepared by performing a binary test with outcome probabilities $p_1 = p$ and $p_2 = 1 - p$, and prepare the state ρ_i for outcome i, thus realizing the preparation test $\{p_i\rho_i\}_{i=1,2}$, and finally by coarse-graining the test, thus obtaining $\rho_p = \rho_1 + (1-p)\rho_2$. A conceptually relevant point is that the step of preparation of a state conditionally on the binary test is an observe-and-prepare test, which is granted to be possible when the theory is causal.

From Lemma 5.3 it follows that the convex set of deterministic states $\mathsf{St}_1(A)$ of a causal OPT is the intersection of the pointed convex cone $\mathsf{St}_+(A)$ with the normalization (affine) hyperplane $\{\rho \in \mathsf{St}_{\mathbb{R}}(A)|(e|\rho) = 1\}$. The set $\mathsf{St}(A)$ of probabilistic states is then the truncated cone contained between the normalization hyperplane and the cone-vertex $\rho = 0$. This geometric construction is depicted in Fig. 5.2. The deterministic states make a *base* for the cone.[12] It follows that if a state is atomic (i.e. it belongs to extremal ray of the cone) its deterministic state is extremal in $\mathsf{St}_1(A)$.

On the other hand, according to point 4 of Lemma 5.2, the set of effects $\mathsf{Eff}(A)$ is the intersection of the pointed cone $\mathsf{Eff}_+(A)$ with the pointed cone $\{a \in \mathsf{Eff}_{\mathbb{R}}(A)|a \leq e_A\}$ of domination by the unique deterministic effect e_A.

---

[12] We recall that a subset $\mathsf{B} \subset \mathsf{C}$ of a cone C is called *base* if $0 \notin \mathsf{B}$ and for every point $0 \neq u \in \mathsf{C}$, there is a unique representation $u = \lambda v$, with $v \in \mathsf{B}$ and $\lambda > 0$ (Barvinok, 2002).

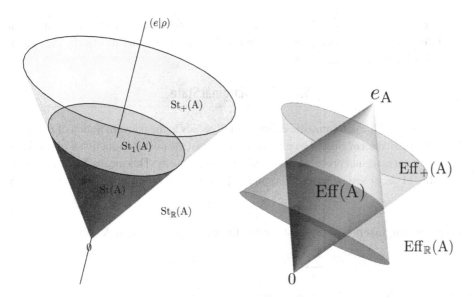

Illustration of the convex structure of a causal theory. On the left we have the state cone $\mathsf{St}_+(A)$ of system A
embedded in the linear space $\mathsf{St}_\mathbb{R}(A)$. This is a pointed convex cone with the convex set of deterministic states
$\mathsf{St}_1(A)$ as a base. The latter is the intersection of the cone with the (affine) hyperplane of normalization
condition $(e|\rho) = 1$. The set $\mathsf{St}(A)$ of probabilistic states is the truncated cone contained between such a
plane and the vertex $\rho = 0$. On the right we have the set of effects $\mathsf{Eff}(A)$ embedded in the linear space
$\mathsf{Eff}_\mathbb{R}(A)$. This is the spindle obtained as the intersection of the pointed cone $\mathsf{Eff}_+(A)$ with the pointed cone
$\{a \in \mathsf{Eff}_\mathbb{R}(A) | a \le e_A\}$ of domination by the unique deterministic effect e_A. (We recall that the specific
geometric picture in the figure corresponds to a special embedding in a Euclidean space, e.g. the orthogonality of
the normalization hyperplane with the cone axis.)

Fig. 5.2

Exercise 5.2 Prove the equivalence of conditions in Lemma 5.2.

Exercise 5.3 For causal theories we will often make use of the *renormalized refinement set*
defined as follows:

$$\mathsf{RefSet}_1\omega := \{\sigma \in \mathsf{St}_1(A) : \exists \alpha \in (0, 1] | \alpha\sigma \in \mathsf{RefSet}\,\omega\}.$$

Show that these sets for varying state $\omega \in \mathsf{St}_1(A)$ are the faces[13] of the convex set
of deterministic states $\mathsf{St}_1(A)$.

Exercise 5.4 Show that in quantum theory the partial trace of density operator ϱ of the state
$\mathrm{Tr}_A\,\varrho$ over the Hilbert space $\mathcal{H}_A$ of system A corresponds to the deterministic effect
e_A of the theory.

Exercise 5.5 Show that in quantum theory instruments are tests, POVMs are observation tests, whereas the von Neumann measurement is an observe-and-prepare test.

Exercise 5.6 [Existence of minimal informationally complete observation test] In Chapter
2 we defined as *informationally complete* an observation test that is state-separating.
We call the informationally complete test *minimal* if the effects of the tests are

---

[13] We recall that a *face* F of a convex set A is a non-empty subset of A with the property that if $x, y \in A$,
$\theta \in (0, 1)$, and $\theta x + (1 - \theta)y \in F$, then $x, y \in F$. A face F that is strictly smaller than A is called a *proper face*.

linearly independent. Show that it is always possible to construct a minimal informationally complete observation test for system A using its effects. [See again the solution of Problem 2.21.]

5.4.3 The Marginal State

The uniqueness of the deterministic effect naturally leads to the relevant notion of *marginal state* or also called *local state*. The marginal state is just the positive functional over effects giving the right marginal probabilities for all observation tests. This means that for the joint probability of the observation tests $\{a_i\}$ and $\{b_j\}$ on systems A and B, respectively, namely

$$p(i,j|a,b,\sigma) = (a_i|_A(b_j|_B|\sigma)_{AB},$$

one has the marginal probability distribution for local tests on system A only

$$p_i^{(A)} := \sum_j p(i,j|a,b,\sigma) = \sum_j (a_i|_A(b_j|_B|\sigma)_{AB}.$$

Using the identity (5.4), this can be written as follows:

$$p_i^{(A)} = (a_i|_A(e|_B|\sigma)_{AB} =: (a_i|_A|\rho)_A. \tag{5.8}$$

Equation (5.8) defines the marginal state $|\rho)_A$ of system A of the joint state $|\sigma)_{AB}$. Therefore, in summary:

Marginal state   *The marginal state of $|\sigma)_{AB}$ on system A is the state*

$$|\rho)_A := (e|_B|\sigma)_{AB}.$$

represented by the diagram

5.4.4 For Causal OPTs Closure Means Convexity

A theory having $\mathsf{St}(A)$ which is closed with respect to the operational norm will contain all the states that can be approximated arbitrarily well by states of the theory. Since probabilities are just states $\mathsf{St}(I)$ of the trivial system I, if an OPT is operationally closed, then also the set of possible values of probabilities is closed. Now, if the only available values for the probability are just $p = 0, 1$ – i.e. $\mathsf{St}(I) = \{0, 1\}$ – then the probabilistic theory will be deterministic. We will say that the theory is a *deterministic OPT*, considering deterministic theories as a special case of probabilistic theories. Now, a relevant fact is that if the OPT contains at least a non-deterministic test, then the operational closure of the OPT automatically guarantees that the whole interval $[0, 1]$ of probabilities is available. In equations

$$0 < p < 1, p \in \mathsf{St}(I) + \text{operational closure} \implies \mathsf{St}(I) = [0, 1].$$

Indeed the availability of a non-deterministic test means that at least a binary test with $0 < p < 1$ is available. We can then use it as a biased coin which can be tossed many times, and by randomness extraction we can approximate any coin bias $p \in [0, 1]$. Hence the available probabilities are a dense set in $[0, 1]$, and closure of the set $\mathsf{St}(I)$ implies that $\mathsf{St}(I) \equiv [0, 1]$, namely the whole interval of probabilities is available (see Exercise 3.3).

Now, if the theory is causal, the availability of a non-deterministic test along with the possibility of conditioning will provide any possible convex combination of events, as stated in the following lemma.

Theorem 5.4 (Approximation of Convex Combinations) *In a causal OPT containing at least one non-deterministic test any convex combination of events is allowed.*

Proof Let $q \in [0, 1]$ an arbitrary probability. Then by Exercise 3.3 $q \in \mathsf{St}(I)$. By conditioning two tests $\{\mathcal{C}_i\}_{i \in X}$ and $\{\mathcal{D}_j\}_{j \in Y}$ on the preparation $\{q, 1 - q\}$ of system I we get the test $\{q\mathcal{C}_i\}_{i \in X} \cup \{(1 - q)\mathcal{D}_j\}_{j \in Y}$, and then, by coarse-graining we obtain the convex combination $q\mathcal{C}_i + (1 - q)\mathcal{D}_j$. □

As a simple consequence of the above lemma we have the following corollary.

Corollary 5.5 *An operationally closed causal OPT containing a non-deterministic test is convex.*

Closed OPT We will name an OPT complete if it satisfies the no-restriction hypothesis and is operationally closed.

Exercise 5.7 (Alternative Definition of *State* for Causal OPT)[14] For a causal theory a state $\omega \in \mathsf{St}(A)$ for system A is the *probability rule* $\omega(\mathcal{A})$ for any event $\mathcal{A} \in \mathsf{Transf}(A \to B)$. The rule is given by $\omega(\mathcal{A}) = {}_B(e|\mathcal{A}|\omega)_A$. The *conditional state* $\omega_{\mathcal{A}}$ provides the probability rule for any event $\mathcal{B} \in \mathsf{Transf}(B \to C)$ under the condition that event $\mathcal{A} \in \mathsf{Transf}(A \to B)$ occurred with system A prepared in the state ω. Show that

$$\omega_{\mathcal{A}}(\mathcal{B}) := \frac{\omega(\mathcal{B}\mathcal{A})}{\omega(\mathcal{A})} = \frac{(\mathcal{A}\omega)(\mathcal{B})}{\omega(\mathcal{A})}.$$

Show that $\mathcal{A}\omega$ is the transformed probability rule after occurrence of event $\mathcal{A}$ on the state ω. In this way the evolution of state under a transformation can be regarded as a *conditioning*.

Exercise 5.8 Show that the notion of conditional state in a causal theory leads to the two following equivalence relations between transformations:

1. *Conditioning equivalence:* the transformations $\mathcal{A}_1, \mathcal{A}_2 \in \mathsf{Transf}(A \to B)$ are *conditioning-equivalent* when $\omega_{\mathcal{A}_1} = \omega_{\mathcal{A}_2} \, \forall \omega \in \mathsf{St}(A)$, namely $\mathcal{A}_1$ and $\mathcal{A}_2$ produce the same conditional state for all prior states ω.

[14] D'Ariano (2006b).

2. *Probabilistic equivalence:* the transformations $\mathcal{A}_1, \mathcal{A}_2 \in \mathsf{Transf}(\mathrm{A} \to \mathrm{B})$ are *probabilistically-equivalent* when $\omega(\mathcal{A}_1) = \omega(\mathcal{A}_2) \; \forall \omega \in \mathsf{St}(\mathrm{A})$, namely $\mathcal{A}_1$ and $\mathcal{A}_2$ occur with the same probability for all prior states ω.

Show that the two equivalence classes completely specify the transformation, and two transformations that are conditionally equivalent are necessarily multiples of each other.

Exercise 5.9[15] Show that the effects $a \in \mathsf{Eff}(\mathrm{A})$ for system A are in one-to-one correspondence with the probabilistic equivalence classes of transformations in $\mathsf{Transf}(\mathrm{A} \to \mathrm{B})$ for all systems B.

Exercise 5.10 Show that the effect of $\mathcal{A}_1 + \mathcal{A}_2$ is the sum of their respective effects, whereas the corresponding conditional state is given by

$$\omega_{\mathcal{A}_1 + \mathcal{A}_2} = \frac{\omega(\mathcal{A}_1)}{\omega(\mathcal{A}_1 + \mathcal{A}_2)} \omega_{\mathcal{A}_1} + \frac{\omega(\mathcal{A}_2)}{\omega(\mathcal{A}_1 + \mathcal{A}_2)} \omega_{\mathcal{A}_2}, \quad \forall \omega \in \mathsf{St}(\mathrm{A}).$$

5.5 No Signaling at a Distance

The "no signaling from the future," i.e. the causality requirement, implies another kind of no signaling, namely the impossibility of signaling without interaction, i.e. by just performing local tests. This is precisely expressed by the following theorem.

Theorem 5.6 (No Signaling Without Interaction) *In a causal OPT it is impossible to send signals by performing only local tests.*

Proof Suppose the general situation in which two "distant" parties Alice and Bob share a bipartite state $|\Psi)_{\mathrm{AB}}$ of systems A and B. Alice performs her local test $\{\mathcal{A}_i\}_{i \in \mathrm{X}}$ on system A and similarly Bob performs his local test $\{\mathcal{B}_j\}_{j \in \mathrm{Y}}$ on system B. The joint probability of their outcomes is

$$p_{ij} = (e|_{\mathrm{AB}}(\mathcal{A}_i \otimes \mathcal{B}_j)|\Psi)_{\mathrm{AB}}.$$

The marginal probabilities p_i^{A} at Alice and p_j^{B} at Bob are given by

$$p_i^{\mathrm{A}} := \sum_j p_{ij}, \quad p_j^{\mathrm{B}} := \sum_i p_{ij}.$$

Alice's marginal does not depend on the choice of test $\{\mathcal{B}_j\}$ of Bob, since

$$p_i^{(\mathrm{A})} = \sum_j (e|_{\mathrm{A}}(e|_{\mathrm{B}}(\mathcal{A}_i \otimes \mathcal{B}_j)|\Psi)_{\mathrm{AB}} = (e|_{\mathrm{A}} \left(\mathcal{A}_i \otimes \left[\sum_j (e|_{\mathrm{B}}\mathcal{B}_j \right] \right) |\Psi)_{\mathrm{AB}} \tag{5.9}$$

$$= (e|_{\mathrm{A}}\mathcal{A}_i|\rho)_{\mathrm{A}}, \quad |\rho)_{\mathrm{A}} := (e|_{\mathrm{B}}|\Psi)_{\mathrm{AB}},$$

where we used Eq. (3.1) and the normalization condition $\sum_j (e|_{\mathrm{B}}\mathcal{B}_j = (e|_{\mathrm{B}}$. The same argument holds for Bob's marginal. $\qquad \square$

---

[15] D'Ariano (2007b).

5.6 Causality and Space-Time

If we identify the input–output direction with the arrow of time, a causal chain can be also regarded as a timeline. Correspondingly, two events on the same causal chain have a definite time ordering. On the contrary, for two causally independent events, namely not belonging to a common causal chain, the choice of which of the two is in the past and which in the future is arbitrary. The no signaling from the future implies that there can be no signaling in both directions.

The same notion of causal dependence/independence can be used for systems in place of events, and two causal systems are causally connected if they belong to a common causal chain, whereas they are independent if they belong to a common slice. The no signaling coincides with the so-called *Einstein locality*, which states that if two physical systems do not interact (i.e. they remained isolated) for a time interval, then the evolution of the physical properties of one system cannot be affected by whatever operation is performed on the other system.[16] Indeed, in the Minkowskian view two systems that cannot interact belong to the same slice and vice versa.[17]

In a causal network we can build the causal cone for any event as the union of all causal chains starting from such event. Vice versa we can use such causal cone to define causal ordering, by stating that an event B causally follows event A if it belongs to the causal cone of A. Since Einstein causality is defined in terms of a causal cone (the Minkowskian causal cone), we see that our notion of causality is equivalent to the Einstein one.[18]

5.7 Theories without Causality

Non-causal theories look quite odd, because we are so familiar with causality that its negation results in behaviors that look strange or impossible to us. Upon negating the statement of Lemma 5.3, we can conclude that, for example:

> *In a theory that is not causal there exist states that cannot be prepared deterministically by any means.*

One may erroneously think that if a state is probabilistic, then in principle it could be obtained deterministically by *post-selection*, namely upon repeating the preparation test sufficiently many times, until the state is achieved. However, as the name says, post-selection presupposes a notion of causal order. On the other hand, upon negating the statement of Lemma 5.1 one can realize that in a non-causal theory there must exist more than one deterministic effect for each system, and the convex set of effects is no longer

---

[16] Einstein *et al.* (1935).

[17] In this fashion a foliation made of slices that cover all systems is the equivalent of a reference frame in relativity theory. See e.g. D'Ariano and Tosini (2013).

[18] For a close connection between the Minkowski space in 1+1 dimensions and the causal homogeneous network, see D'Ariano and Tosini (2013).

spindle-shaped. The existence of multiple deterministic effects implies that there exist preparations that are achieved with different probability depending on the observation test, since states are separating for effects.

It is natural now to ask if there exist probabilistic theories that are non-causal, and, in such case, what a non-causal input–output relation looks like. The affirmative answer comes from a simple procedure that allows us to build up a non-causal theory over a causal one.

Consider the horseshoe-shaped box

and regard it as a portion of a circuit to which one can connect tests according to our connectivity rules, i.e. attaching wires with the same label without making closed loops. The horseshoe-shaped box can now be regarded as a map that transforms a test in $\mathsf{Transf}(B\to C)$ to a new test in $\mathsf{Transf}(A\to D)$ by operating the following insertion:

We now regard the box insertion as the input–output connection of the new theory, the box $\Omega \in \mathsf{Transf}(B\to C)$ becomes the state $\Omega \in \mathsf{St}(B \triangleright C)$ of the new theory with output system denoted as $B \triangleright C$, whereas the horseshoe-shaped box becomes the transformation $\mathcal{T} \in \mathsf{Transf}(B\triangleright C\to A\triangleright D)$ of the new theory. In the new theory, diagrammatically we have

It is clear that since the preparation Ω is getting input from the horseshoe-shaped box $\mathcal{F}$, its probability distribution will depend on the choice of it, and the new theory is not causal. An example of a theory of this kind is provided by quantum theory itself when we consider transformations as preparations. Indeed, the cone of states of bipartite system BC coincides with the cone of transformations from C to B due to the Choi–Jamiołkowski isomorphism

$$\mathsf{Transf}_+(C\to B) \equiv \mathsf{St}_+(BC) \equiv \mathsf{Lin}_+(\mathcal{H}_B \otimes \mathcal{H}_C).$$

However, the deterministic condition for the two sets is very different, since for states $\rho \in \mathsf{St}_+(BC)$ it is given by $\mathrm{Tr}\,\rho = 1$, whereas for transformations $\mathcal{T} \in \mathsf{Transf}_+(B\to C)$ it is given by $\mathrm{Tr}_C[R_{\mathcal{T}}] = I_B$, where $R_{\mathcal{T}} = \mathcal{T} \otimes \mathcal{I}|I\rangle\!\rangle\langle\!\langle I| \in \mathsf{Lin}_+(BC)$ is the Choi operator of the CP map $\mathcal{T} \in \mathsf{Lin}(\mathcal{H}_B, \mathcal{H}_C)$ describing the transformation $\mathcal{T} \in \mathsf{Transf}_+(B\to C)$, and $|I\rangle\!\rangle =: \sum_{n=1}^{\dim(B)} |n\rangle|n\rangle \in \mathcal{H}_B^{\otimes 2}$. As a consequence, the convex structure of the theory is very different from that depicted in Fig. 5.2: for example, the set of states $\mathsf{St}(A \triangleright B)$ is not the cone $\mathsf{St}_+(A \triangleright B)$ truncated by an hyperplane of dimension $D_{A\triangleright B} - 1$ (see also Exercise 2.48), and analogously the set of effects $\mathsf{Eff}(A \triangleright B)$ is not spindle-shaped.

Exercise 5.11 Show that the deterministic condition for quantum states in $\mathsf{St}_+(BC)$ corresponds to a single hyperplane of dimensions $\dim(\mathcal{H}_B)^2 \dim(\mathcal{H}_C)^2 - 1$ in the space of Hermitian operators on $\mathcal{H}_B \otimes \mathcal{H}_C$, whereas for transformations in $\mathsf{Transf}(B\to C)$ is given by an hyperplane of dimensions $\dim(\mathcal{H}_B)^2[\dim(\mathcal{H}_C)^2 - 1]$ in the same space.

The construction of the non-causal theory proceeds upon introducing comb-shaped boxes, as the following one:

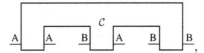

and then combs with increasing number of teeth, and then composes them with transformations as in the following:

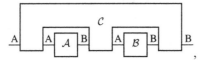

which in the non-causal theory would be represented as follows:

More generally we would have circuits as in the following:

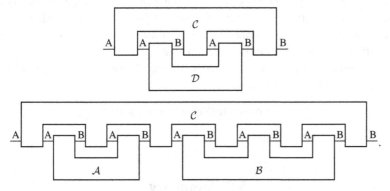

The above OPT is non-causal, but can be achieved by an underlying causal OPT, i.e. with the combs achieved by circuits from a causal OPT with missing tests in correspondence of the holes between the teeth of the comb. However, there exist also non-causal OPTs with no causal underlying OPT. One can just consider the case of the comb

where the resulting transformation can be either $\mathcal{A} \circ \mathcal{B}$ or $\mathcal{B} \circ \mathcal{A}$, depending on the input at C. It is easy to convince oneself that such a comb cannot be achieved by a circuit, since no circuit can exchange the ordering of composition of the two boxes.[19] This situation in which the causal relations cannot be pre-established is an example of genuinely non-causal OPT.[20]

---

[19] Chiribella *et al.* (2013a).

[20] The mentioned construction (Chiribella *et al.*, 2013a) of a non-causal OPT achieved by circuits from a causal OPT has spawned a full research line about a non-causal variation of quantum theory (Oreshkov *et al.*, 2012;

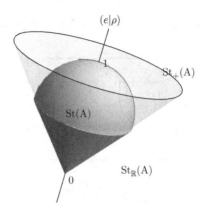

Fig. 5.3 A strange OPT ... or it is an OPT?

Exercise 5.12 Write the the representation of the last two circuits within the non-causal theory.

Exercise 5.13 Consider the *ice-cream cone* shaped convex set of states in Fig. 5.3. Is the theory causal? Is such a theory possible? What can be said about the theory? [Hint: there is a full spherical cap of states that are not dominated by any other state, hence it is made of deterministic states. Mixtures of deterministic states are deterministic. The affine span of deterministic is deterministic ...]

Exercise 5.14 [21] Show that in a finite-dimensional convex causal OPT, for every couple of states $\sigma, \rho \in \mathsf{St}_1(A)$ one has

$$\sigma \in \mathsf{Span}_+ \mathsf{RefSet}\, \rho \implies k\sigma \in \mathsf{RefSet}\, \rho,$$

for some non-zero probability k.

5.8 Summary

We have given a precise definition of causality in the generally non-deterministic context of OPTs, in terms of the independence of the probability of preparations of the choice of observations. We have derived the mathematical structure of a causal OPT, and shown that causality is equivalent to the uniqueness of the deterministic effect, which, in turn is equivalent to the ability to normalize states. Correspondingly one has a precise cone structure for the theory, as illustrated in Fig. 5.2. The causality postulate greatly simplifies the evaluation of probabilities in the theory, and allows one to introduce the notion of a conditioned test. Since causality is equivalent to the impossibility of signaling from the future, it also implies the impossibility of signaling without interaction, which is the Einstein notion of locality. We have seen how the present notion of causality corresponds to the Einsteinian notion, and finally, we have provided explicit examples of OPTs that do not satisfy causality.

Brukner, 2014a,b; Baumeler and Wolf, 2015), and even experimental tests have been devised (Procopio *et al.*, 2015; Rambo *et al.*, 2012).

[21] Chiribella *et al.* (2010a).

Notes

No Signaling from the Future   The need to consider no signaling from the future as a separate postulate was first noticed by M. Ozawa during a visit by G. M. D'Ariano in Sendai, March 2007, and appeared for the first time in the form of a causality postulate in D'Ariano (2010).

Causality and Conditioned Tests   It should be emphasized that causality guarantees the feasibility of the conditioned test, which depends on the knowledge of which outcome happened in a past test. In the absence of a causal connection between the conditioning and the conditioned tests, in principle one can find a reference frame where the time relation between the two tests is interchanged, with the conditioning test occurring after the conditioned one (see Section 5.6). In principle, one could also consider conditioning where the output system of each test $\left\{ \mathcal{D}_j^{(i)} \right\}$ is a system C_i that depends on the outcome i. In this case the output space of the conditioned test would be a "direct sum" system "$C := \bigoplus_{i \in X} C_i$." This would also require treating the outcome spaces X as a *classical system* that can be the input or the output of some classical information-processing device.

Is our Definition of *State* the Usual One?   In any scientific construction – mechanical, chemical, or whatsoever – the state is the collection of the variables of a system whose knowledge is sufficient to make all predictions. For example, in classical dynamics the state is the collection of positions and momenta of all particles. In equilibrium thermodynamics the state is the specification of the physical variables – e.g. (U, V, N), U energy, V volume, and N number of moles – sufficient to evaluate all thermodynamical quantities. In quantum theory the state allows one to evaluate all possible expectations. Analogously, in a causal OPT, the state allows one to predict the results of any test. In the logic of performing experiments to predict results of forthcoming experiments with the same preparation, the information gathered in the experiment must concern whatever is needed to make predictions, which is, by definition, the state itself of the input system. Such information is provided by the outcome of the preparation test, and the knowledge of the test itself.

About the Notion of *System*   Bohr would have said that there is no particle before we reveal it: only the outcome of the experiment is real. The *particle* is a system in quantum mechanics. In second quantization, the system is a *field mode*, and the particle becomes a state. The system is thus an intimately theoretical notion, not to be regarded as a concrete material object, whereas only the performed tests and their outcomes have the objectivity status. Ultimately, in a causal OPT, the system is the rule that determines the connection between events.

Alternative Notion of Effect as Equivalence Class of Events   The notion of effect as equivalence class of transformations occurring with the same probability for all possible states is the same notion as in Ludwig (1983), who also introduced the word "effect."

It is possible to rederive many results presented here, by using this definition of effect (D'Ariano, 2010).

Non-locality   Theorem 5.6 is crucial in clarifying the nature of the so-called *quantum non-locality*, a shorthand name for *quantum non-local realism*. Indeed, quantum theory, being a causal OPT, cannot violate Einstein locality. The correlations produced by quantum entangled states are "non-local," in the sense that they cannot be achieved by *local hidden variables*, but still they cannot be used for superluminal communications, thus not violating Einstein locality.

Solutions to Selected Problems and Exercises

Exercise 5.1

It is not restrictive to assume $(e^{(0)}|\tau) - (e^{(1)}|\tau) > 0$. Since $\tau_0 + \tau_1$ is a deterministic state and $e^{(i)}$ are both deterministic effects, we have $(e^{(i)}|\tau_0) + (e^{(i)}|\tau_1) = 1$ for $i = 0, 1$. If we now consider the conditioned test $\{(e^{(i)}|\tau_i)\}_{i=0,1} \subseteq \mathsf{Transf}(\mathrm{I} \to \mathrm{I})$, its coarse-graining is

$$(e^{(0)}|\tau_0) + (e^{(1)}|\tau_1) = (e^{(0)}|\tau_0) + 1 - (e^{(1)}|\tau_0)$$
$$= 1 + [(e^{(0)}|\tau_0) - (e^{(1)}|\tau_0)]$$
$$> 1.$$

Theories with Local Discriminability

6

In this chapter we introduce the principle of local discriminability, which stipulates the possibility of discriminating states of composite systems via local measurements on the component systems. The phenomenon of entanglement has relevant consequences on the accessibility of information in composite systems and on the complexity of characterization of transformations: local discriminability brings dramatic simplifications to the structure of the sets of states and transformations, reconciling the holism of entangled states with the reductionist scientific approach. In order to illustrate the structural relevance of the principle, we will briefly review some OPTs that do not satisfy it, contrasting the features of such theories with those of quantum and classical probabilistic theories, where the principle holds.

6.1 Entanglement and Holism

Holism is a widely known feature of quantum theory, related to the existence of entangled states. In quantum theory one has pure entangled states, whose marginal states are completely mixed. This situation corresponds to having maximal knowledge of the whole with minimal knowledge of the parts. The main manifestation of the holism of entanglement is the impossibility of reconciling the statistics of joint measurements on entangled states with *local realism*. Local realism is the assumption that any measurement on a system is just the reading of a pre-existing quantity encoded on the system. In other words, we cannot explain the statistics of measurements on the parts of a composite system as a manifestation of our ignorance of locally pre-existing values, as in classical OPT.

Does entanglement make information about composite systems locally unaccessible? This would pose severe practical issues on the possibility of acquiring information about physical systems. For example, suppose that two distant telescopes observe a source of entangled photons from a stellar emission. Holism would imply that some features of the emission process could never be detected, unless we are able to make the entangled photons interact in a joint measurement. On the other hand, if a reductionist approach is possible, local observations of the two photons at the sites of the two telescopes would be enough to characterize the emitting source. Thanks to the local discriminability principle, this is actually the case with entangled quantum sources.

Another consequence of holism is the fact that transformations generally cannot be characterized by local experiments, namely by applying the unknown transformation only to its input system. On the contrary, one may potentially need to run the transformation with its input system entangled with any possible additional system in the whole universe. Testing the validity of any physical law in such a situation would be practically impossible, e.g. one would need to check the Maxwell's equations not only on single modes, but on entangled states involving all multiple modes as well. However, thanks to the local discriminability of quantum theory, the reductionist approach is safe.

The reconciliation of holism with reductionism in quantum theory hinges on the local discriminability principle. The principle guarantees that local measurements on the component systems are sufficient to identify the state of any composite system. This means that, though two systems are prepared in an entangled state, the information extracted from them separately is sufficient to reconstruct the whole information about their joint preparation.

Entanglement and holism are not at all specific to quantum theory, but pertain to a general OPT. Entangled states are defined, by negation, as those states that are *not separable*. The separable states of a composite system AB are those of the form

$$\left(\Sigma \begin{array}{c} A \\ B \end{array}\right. = \sum_{i \in Y} p_i \begin{array}{c} \boxed{\alpha_i} \quad A \\ \boxed{\beta_i} \quad B \end{array} \quad , \tag{6.1}$$

where $\{p_i\}$ is a probability distribution and $\{\alpha_i\}$ and $\{\beta_i\}$ are states of system A and B, respectively. Operationally, the separable states are those that can be prepared by two distant parties, Alice and Bob, using only local preparation devices and exchanging classical communication. In the classical OPT all the states of composite systems are separable.[1]

Entangled states of an OPT exhibit the same puzzling features of entangled quantum states, and generally cannot be prepared by local operations and classical communication. As for the quantum case, the entangled states of an OPT exhibit the holistic feature of having marginals that are mixed, corresponding to having maximal knowledge of the whole without having a corresponding maximal knowledge of the parts. Interestingly, the converse also holds: maximal knowledge of the whole without maximal knowledge of the parts implies entanglement. Indeed, a pure state with a mixed marginal must be entangled, since all separable pure states are of the form $\alpha \otimes \beta$, with α and β pure, namely they have pure marginals. We can therefore conclude that the existence of entangled states is equivalent to the holism of the OPT.

6.1.1 Holism of Transformations

Entanglement has the disconcerting facet that, in order to identify a transformation acting on system A, one generally needs to test its action on arbitrary entangled states of AB for arbitrary B. This is not an issue for separable joint states, since the map is extended to AB via convex combinations through the following identity of states:

---

[1] An example of OPT having separable states only is that of quantum theory with the *minimal tensor product* of Barnum *et al.* (2009), corresponding to forbidden entangled states. Notice that for such a theory the convex set of joint effects would be larger than that of the usual quantum theory, since restricting a convex set generally increases its dual.

$$\Sigma \left[\begin{array}{c} A \\ \boxed{\mathcal{A}} \\ B \end{array} \right. \begin{array}{c} A' \\ \end{array} = \sum_{i \in Y} p_i \begin{array}{c} \boxed{\alpha_i} \quad A \quad \boxed{\mathcal{A}} \quad A' \\ \boxed{\beta_i} \quad\quad B \end{array} .$$

For a theory with entangled states, however, it generally happens that two transformations acting in the same way on all states of system A can give a different output on some entangled states of system AB. This would require to specify the transformation on all possible composite systems AB, and practically this means e.g. testing a physical law on a global experiment running on the whole universe.

6.1.2 The Reductionism of Quantum Theory

We have seen that in a general OPT entanglement can disrupt the reductionist approach to the extent that testing a physical law would be virtually impossible. But why does this not happen in quantum theory? The answer is that quantum theory is "holistic, but not too much." It has entangled states, but, at the same time, it enjoys a property that allows the reductionist approach: local discriminability. This property means that any two different states of a composite system AB have different joint probability distribution for some local measurement. In other words, by testing the correlations of local measurements on A and B, two distant observers can distinguish between any two joint states.

Local discriminability implies that the dimension of the state space of any composite system is the product of the dimensions of component systems. This rule plays a crucial role in selecting quantum theory on complex Hilbert spaces versus quantum theory on real or quaternionic Hilbert spaces.[2] The product rule lies at the basis of the tensor product structure for composite systems, which is the core of the reconciliation between holism and reductionism.[3] Indeed, the full information contained in an entangled state that belongs to a tensor product space rests in the correlations of local measurements. And, it is the tensor product structure of state spaces of composite systems that allows us to fully characterize a transformation by running it only on its input system, without considering input entangled states with any other additional system.

6.2 The Principle

Local Discriminability Axiom *It is possible to discriminate any pair of states of composite systems using only local measurements.*

Mathematically the axiom asserts that for every two joint states $\rho, \sigma \in \mathsf{St}(AB)$, with $\rho \neq \sigma$, there exist effects $a \in \mathsf{Eff}(A)$ and $b \in \mathsf{Eff}(B)$ such that the joint probabilities for the two states are different, namely, in circuits

---

[2] See e.g. Araki (1980).

[3] In D'Ariano (2006a) the principle was called *local tomography*. Indeed, we will see that local discriminability is equivalent to the possibility of performing a reconstruction (called tomography) of joint states by local measurements.

$$
\left(\rho \;\begin{array}{|c|} \hline A \\ \hline B \\ \hline \end{array}\right) \neq \left(\sigma \;\begin{array}{|c|} \hline A \\ \hline B \\ \hline \end{array}\right) \;\Rightarrow\; \left(\rho \;\begin{array}{|c|} \hline A \!-\! \boxed{a} \\ \hline B \!-\! \boxed{b} \\ \hline \end{array}\right) \neq \left(\sigma \;\begin{array}{|c|} \hline A \!-\! \boxed{a} \\ \hline B \!-\! \boxed{b} \\ \hline \end{array}\right). \tag{6.2}
$$

Exercise 6.1 proves that the two local effects can be actually achieved with a local test (namely made only of local effects), and in addition the error probability can be made $p_E < 1/2$, thus providing a local discrimination strategy. Therefore, the statement of the axiom is equivalent to Eq. (6.2).[4]

Exercise 6.1   Prove that Eq. (6.2) holds if and only if there exists a local discrimination strategy for the states ρ and σ, namely a local test giving error probability $p_E < 1/2$. [Hint: consider the two binary tests $\{a, \bar{a}\}$ and $\{b, \bar{b}\}$ completing the effects a and b in Eq. (6.2), take the local test given by their parallel composition, and obtain a coarse-grained binary local discriminating test. For the error probability use Exercise 3.1.]

In Chapter 2 we have shown that local discriminability holds for quantum theory. Here we raise it to a principle for general OPTs, and in Chapter 17 we will show how it can be used as an axiom to derive quantum theory.

The role played by local discriminability in the derivation of quantum theory may be blurred by the participation of the other axioms. However, there are two crucial consequences that can be traced back directly to local discriminability alone: (1) the state space of a composite system is the tensor product of state spaces of component systems; and (2) transformations are completely defined by their local action. Both assertions hold in quantum theory. The first is trivially true. The second is just the simple fact that the extension of a quantum operation $\mathcal{A} \in \mathsf{Transf}(A \to A')$ to $\mathcal{A} \in \mathsf{Transf}(AB \to A'B)$ containing an ancillary system B is achieved upon considering the map $\mathcal{A} \otimes \mathcal{I}_B$,[5] which is just the tensor product of map $\mathcal{A}$ with the identity map $\mathcal{I}_B$. The same is true also for the classical OPT, and more generally for every OPT enjoying local discriminability, as we will see in the following. The OPT properties (1) and (2) represent the essence of the reconciliation between holism and reductionism.

6.3   Reconciling Holism with Reductionism

In this section we provide a proof that the local discriminability axiom justifies the reductionist approach in the presence of holism. In the process, we will also illustrate the two main consequences of local discriminability mentioned in the previous section, namely the tensor-product structure of state space, and the extendibility of local transformations to composite input systems.

---

[4] We stress the conceptual relevance of achieving error probability $p_E < 1/2$, namely the discrimination test must be better than random guessing. We also emphasize that for local discriminability perfect discrimination is not required, i.e. corresponding to $p_E = 0$ (the two probabilities in Eq. (6.2) would be equal to 0 and 1, respectively).

[5] In the present context we are abusing the notation $\mathcal{A} \otimes \mathcal{B}$ to denote the parallel composition of $\mathcal{A}$ and $\mathcal{B}$. This coincides with the tensor product map only when local discriminability holds.

We can now prove the two main theorems following from the principle of local discriminability.

Theorem 6.1 (Product Rule for Composite Systems) *A theory satisfies local discriminability if and only if, for every composite system* AB, *one has*

$$D_{AB} = D_A D_B. \tag{6.3}$$

Proof By Eq. (6.2), a theory satisfies local discriminability if and only if local effects $a \otimes b \in \mathsf{Eff}(AB)$, with $a \in \mathsf{Eff}(A)$ and $b \in \mathsf{Eff}(B)$, are separating for joint states $\mathsf{St}(AB)$. Equivalently, the set $T := \{a \otimes b | a \in \mathsf{Eff}(A), \ b \in \mathsf{Eff}(B)\}$ is a spanning set for $\mathsf{Eff}_\mathbb{R}(AB)$. Since the dimension of $\mathsf{Span}_\mathbb{R}(T)$ is $D_A D_B$ and the spaces of states and effects have the same dimension, we have $D_{AB} = D_A D_B$. Conversely, if Eq. (6.3) holds, then the product effects are a spanning set for the vector space $\mathsf{Eff}_\mathbb{R}(AB)$, hence they are separating, and local discriminability holds. $\square$

Theorem 6.2 (Local Characterization of Transformations) *If local discriminability holds, then for any two transformations* $\mathcal{A}, \mathcal{A}' \in \mathsf{Transf}(A \to A')$, *the condition* $\mathcal{A}\rho = \mathcal{A}'\rho$ *for every* $\rho \in \mathsf{St}(A)$ *implies that* $\mathcal{A} = \mathcal{A}'$.

Proof Let B be a system and $\Psi \in \mathsf{St}(AB)$. Then, for every effect $a \in \mathsf{Eff}(A)$ and $b \in \mathsf{Eff}(B)$ we have

Now, suppose that $\mathcal{A}\rho = \mathcal{A}'\rho$ for all $\rho \in \mathsf{St}(A)$. This implies

$$\mathcal{A}\rho_b = \mathcal{A}'\rho_b \qquad \forall b \in \mathsf{Eff}(B), \ \forall B,$$

and, therefore $\forall a \in \mathsf{Eff}(A), \forall b \in \mathsf{Eff}(B)$, one has

By the local discriminability principle, we then conclude that $(\mathcal{A} \otimes \mathcal{I}_B)\Psi = (\mathcal{A}' \otimes \mathcal{I}_B)\Psi$, for every state $\Psi \in \mathsf{St}(AB)$ and for every system B. By definition, this means that $\mathcal{A}$ coincides with $\mathcal{A}'$. $\square$

Upon extending the notion of *separating set* from linear functionals to linear maps, we can restate Theorem 6.2 as follows.

Corollary 6.3 *Local input states are separating for transformations.*

Theorem 6.2 dramatically reduces the resources needed to characterize transformations, both in terms of number of tests (one has only to run tests on the input system A), and in terms of entanglement (entanglement is not necessary for distinguishing transformations). Entanglement, though, can help in reducing all input states to a single entangled state with a reference system B with the same Hilbert-space dimension of system A, as in the quantum ancilla-assisted tomography.[6] As we will see in Chapter 7, this is a consequence of the purification axiom.

To prove that the reductionist approach is justified by local discriminability, we now show that in a causal operational theory with local discriminability, maximal knowledge of the states of the component of a composite system implies maximal knowledge of the state of the component systems. First we need the result in the following exercise.

Exercise 6.2 Prove that in a causal theory with local discriminability, a state $\rho \in \mathsf{St}(AB)$ has pure marginal β on B if and only if it is the product state $\rho = \alpha \otimes \beta$.

Solution

Thanks to causality, without loss of generality we can take ρ as deterministic. Moreover, for every $a \in \mathsf{Eff}(A)$ one has

$$|\beta_a) := (a|_A|\rho) \le (e|_A|\rho) = |\beta), \qquad (6.4)$$

where the marginal state $|\beta)$ is pure, hence also the state $|\beta_a)$ is pure, and $|\beta_a) = \kappa_a|\beta)$, with $0 < \kappa_a \le 1$. One has $(a|\alpha) = \kappa_a$, where the state $|\alpha)$ is the marginal state of ρ for system A, namely

$$\kappa_a = \left(\rho \begin{array}{c} \underline{}^{A}\!\!-\!\!\boxed{a} \\ B\,-\!\!\boxed{e} \end{array} \right. = \boxed{\alpha}\!-^{A}\!\!\boxed{a}.$$

Thus, applying b on both sides of Eq. (6.4), one has

$$\left(\rho \begin{array}{c} \underline{}^{A}\!\!-\!\!\boxed{a} \\ B\,-\!\!\boxed{b} \end{array} \right. = (b|\beta_a)\kappa_a = (b|\beta) = \begin{array}{c} \boxed{\alpha}\!-^{A}\!\!\boxed{a} \\ \boxed{\beta}\!-_{B}\!\!\boxed{b} \end{array}, \quad \forall a \in \mathsf{Eff}(A), b \in \mathsf{Eff}(B).$$

Thanks to local discriminability, this implies

$$\rho = \alpha \otimes \beta.$$

Exercise 6.3 Prove that a product of two pure states is pure.

Solution

Consider that the state $\rho = \alpha \otimes \beta$, and take an arbitrary decomposition $\rho = \sum_{i \in X} \rho_i$. Since $(e|_A|\rho) = |\beta)$ is pure, then $(e|_A|\rho) = \sum_{i \in X}(e|_A|\rho_i) = |\beta)$, and then $(e|_A|\rho_i) = p_i|\beta)$ for every $i \in X$. Similarly, $(e|_B|\rho_i) = q_i|\alpha)$ is also pure, with $p_i = q_i = (e|_A(e|_B|\rho_i)/(e|_A(e|_B|\rho)$. Since the two marginals of ρ_i are pure for every $i \in X$, from Exercise 6.2 it follows that each ρ_i is the product state of its marginals, whence $\rho_i = p_i\rho$, namely any decomposition of ρ is trivial and ρ is pure.

We can now prove the following theorem.

Theorem 6.4 (Maximal Knowledge of the Parts Implies Maximal Knowledge of the Whole) *In a causal theory with local discriminability, a state with pure marginals is pure.*

Proof According to Exercise 6.2 a state has pure marginals iff it is the product state of its marginals. But according to Exercise 6.3 a product of two pure states is itself pure.   □

[6] Childs *et al.* (2001); D'Ariano and Lo Presti (2001).

Exercise 6.4 Prove that in a theory with local discriminability if the sets of states $\{\rho_i\}_{i \in X} \subseteq$
St(A) and $\{\sigma_j\}_{j \in Y} \subseteq$ St(B) span the linear spaces $St_{\mathbb{R}}(A)$ and $St_{\mathbb{R}}(B)$, respectively,
then $\{\rho_i \otimes \sigma_j\}_{(i,j) \in X \times Y}$ spans $St_{\mathbb{R}}(AB)$.

Exercise 6.5 Prove that if a set of states $\{\rho_i \otimes \sigma_j\}_{(i,j) \in X \times Y} \subseteq$ St(AB) spans $St_{\mathbb{R}}(AB)$, then
the probabilities of local effects are sufficient to characterize bipartite states. [Hint:
consider an observation test that is the product of informationally complete local
tests.]

6.4 Consequences of Local Discriminability

We now prove the main theorems that are consequence of local discriminability. These will
be very useful in the derivation of quantum theory.

The first theorem is the equivalence with the following property of an OPT:

Local Tomography Property of an OPT *For any state of a composite system, the joint
probabilities of all local effects completely identify the state.*

It is immediate to realize that the statement of the local tomography property is
equivalent to the requirement that local effects are separating for states of composite
systems. This immediately leads to the following lemma.

Lemma 6.5 *The local discriminability axiom is equivalent to the local tomography
property of an OPT.*

Proof Local effects are separating for St(AB), iff they span $Eff_{\mathbb{R}}(AB)$ (see Lemma (2.4)),
hence the statement of the local discriminability axiom and that of the local tomography
property are both equivalent to the requirement that local effects $a \otimes b \in$ Eff(AB) are
separating for St(AB).   □

We remark now that according to Eq. (3.19) we have the inclusion

$$\mathsf{Span}[\mathsf{RefSet}\,(\rho) \otimes \mathsf{RefSet}\,(\sigma)] \subseteq \mathsf{Span}[\mathsf{RefSet}\,(\rho \otimes \sigma)].$$

Actually, in the presence of local discriminability, there is a set of product states $\{\rho_i \otimes \sigma_j\}$
that is complete in $\mathsf{Span}[\mathsf{RefSet}\,(\rho \otimes \sigma)]$ (see Exercise 5.14). In other words,

$$\mathsf{Span}[\mathsf{RefSet}\,(\rho \otimes \sigma)] = \mathsf{Span}[\mathsf{RefSet}\,(\rho) \otimes \mathsf{RefSet}\,(\sigma)]. \tag{6.5}$$

6.5 Different Degrees of Holism

There exist principles weaker than local discriminability that can reconcile holism with
reductionism. Such principles are called *n-local tomography*.[7] They all reduce to consider
sets of a finite number of systems on which one needs to perform joint measurements

[7] Hardy and Wootters (2012).

for reconstructing joint states. An equivalent formulation of such principles is the following.

n-local Discriminability Axiom *If two m-partite states are different, with $m > n$, then they give different probabilities for at least one factorized effect involving k-partite effects, with $k \leq n$.*

The most popular example of theory where local discriminability fails is *real quantum theory* (RQT), which enjoys bi-local discriminability, instead.

6.5.1 Real Quantum Theory

Real quantum theory (RQT) can be formulated in the same way as quantum theory, apart from the requirement that systems correspond to Hilbert spaces on the real field rather than on the complex field. RQT can be viewed as a restriction of quantum theory, where states are constrained to satisfy the identity $\rho = \rho^T$, T denoting the transposition with respect to a given orthonormal basis. The states of the system A of the theory are symmetric density operator on the Hilbert space $\mathbb{R}^{d_A}$, corresponding to dimension

$$D_A = \frac{d_A(d_A + 1)}{2}. \tag{6.6}$$

Violation of local discriminability can be simply checked upon considering two *rebits*, namely two systems A and B with $d_A = d_B = 2$. Then, the composite system AB has $d_{AB} = 4$, and, according to Eq. (6.6), one has $D_A = D_B = 3$, whereas $D_{AB} = 10 > D_A D_B = 9$. One can see that the additional dimension is spanned by the operator $\sigma_y \otimes \sigma_y$, which is real, but does not belong to the span of tensor products of two-dimensional real matrices (which contains only I, σ_x and σ_z). It follows that two states $\rho_0, \rho_1 \in \mathsf{St}(AB)$ with $\rho_1 - \rho_0 \propto \sigma_y \otimes \sigma_y$ cannot be discriminated by local effects. As an example, take $\rho_0 := \frac{1}{4} I_4$ and $\rho_1 := \frac{1}{4}(I_4 + \sigma_y \otimes \sigma_y)$.

A consequence of violation of local discriminability in RQT is that there exist transformations that are not characterized only by their local action. For example, the rebit completely positive transformation $\mathcal{A}(\rho) := \frac{1}{2}(\sigma_y \rho \sigma_y + \rho^T)$ is locally indistinguishable from the transformation $\mathcal{B}(\rho) := \frac{1}{2}(\sigma_y \rho \sigma_y + \rho)$. However, upon applying $\mathcal{A} \otimes \mathcal{I}$ and $\mathcal{B} \otimes \mathcal{I}$ to the bipartite state $R = \frac{1}{2} |I\rangle\!\rangle \langle\!\langle I|$ of AB one obtains the two different states

$$R_\mathcal{A} = \frac{1}{4} I_4, \quad R_\mathcal{B} = \frac{1}{4}(|I\rangle\!\rangle \langle\!\langle I| + |\sigma_y\rangle\!\rangle \langle\!\langle \sigma_y|).$$

It is remarkable that RQT satisfies the following relevant restriction of local discriminability.

Local Discriminability Over Pure States Axiom *If the state R of a bipartite system AB is pure, then for every state $S \neq R$ of AB (generally mixed) there are local effects $a \in \mathsf{Eff}(A)$ and $b \in \mathsf{Eff}(B)$ such that*

We conjecture that RQT is the only theory satisfying local discriminability over pure states. RQT also enjoys causality and unique purification. These properties are directly inherited from QT.

6.6 Summary

In this chapter we have analyzed the principle of local discriminability, namely the possibility of discriminating states of composite systems by local observation tests. We have seen that such a property for an OPT allows the possibility of fully characterizing any transformation only by its action on local states, and therefore, for the above two reasons, the principle reconciles the holism of entanglement with reductionism. We explored the consequences of the principle, in particular the tensor-product structure of state-spaces for composite systems. We have then considered examples of OPTs that violate local discriminability, to gain more insight into the principle.

Notes

Local Realism   One of the features of the holism of quantum entanglement is the violation of Bell's inequalities corresponding to the breach of local realism, namely the impossibility of interpreting the correlations of local measurements on e.g. a singlet state in terms of reading of local pre-determined values. There are still few attempts at proving similar no-go theorems in OPTs. For a discussion of no-go theorems for local realism in OPTs see e.g. Brandenburger and Yanofsky (2008); Schumacher and Westmoreland (2012).

The Notion of Local Transformations   One would conjecture that, for a fixed set of states for each system, there is only a single theory with the largest class of transformations. However, in the absence of local discriminability it is possible to construct different "maximal" theories that differ in the action of local transformations on multipartite systems (see e.g. Barnum *et al.*, 2009; D'Ariano *et al.*, 2014b).

Entanglement and Local Discriminability   In the context of OPTs we have defined a state as entangled when it is non-separable. However, if we regard entanglement as a resource as in quantum information theory, an appropriate quantification of the entanglement of a state ρ is provided by an analysis of the asymptotic rate of conversion of $\rho^{\otimes n}$ from/to a maximally entangled state $\psi_E^{\otimes m}$ by an optimized local operation and classical conditioning (LOCC) protocol (Nielsen and Chuang, 1997). In this respect it may happen that in an OPT without local discriminability there exist non-separable states that are LOCC indistinguishable from a separable state, even asymptotically. This phenomenon is more radical than that of bound entanglement in quantum information theory, that exhibits no violation of any Bell inequality. A geometric interpretation for this fact is that in the presence of local discriminability any functional defining a separating hyperplane between the entangled state and the convex set of separable states can be expanded over local effects, corresponding to an *entanglement witnessing* made with local tests. On the contrary, in absence of local discriminability it may happen that, for some entangled state, the separating functional lies outside the span of local effects. Since Bell-like inequalities are a special case of local entanglement witness, such an entangled state would not allow for any violation of this kind.

Problems

6.1 Show that, by definition, for a theory with 2-local discriminability and without local discriminability the following inequalities must be satisfied

$$D_{AB} > D_A D_B,$$
$$D_{ABC} \leq D_A D_B D_C + \tilde{D}_{AB} D_C + \tilde{D}_{BC} D_A + \tilde{D}_{CA} D_B, \tag{6.7}$$

where $\tilde{D}_{AB} := (D_{AB} - D_A D_B)$. We call *maximally bi-local* a theory with 2-local discriminability for which the bound in Eq. (6.7) is saturated.

6.2 Provide a condition for 3-local discriminability analogous to that of Eq. (6.7).

6.3 Consider a theory where all systems A are composition of qubits with states and effects being quantum respectively, but both restricted to the cone of block-diagonal positive operators over $\mathbb{C}^{2N} = \mathcal{H}_0 \oplus \mathcal{H}_1$, where $\mathcal{H}_0$ ($\mathcal{H}_1$) is the parity space spanned by vectors with an even (odd) number of qubits up. Show that for such theory local discriminability is violated.

6.4 By dimensional counting, show that the theory described in Problem 6.3 satisfies bi-local discriminability.

6.5 Assuming that the transformations for the theory in Problem 6.3 are all CP-map that preserve the block-diagonal structure, find two transformations that cannot be discriminated on a single-system input–ouput.

6.6 Using the result of Exercise 5.14, prove that in an operational theory with local discriminability the parallel composition of completely mixed states is completely mixed.

6.7 Prove that if $\sigma \in \mathsf{Span}_+ \mathsf{RefSet}\, \rho$, then for some suitable $0 < k < 1$ one has $k\sigma \in \mathsf{RefSet}\, \rho$.

6.8 Using Exercise 5.14 and Problem 6.7, prove that in a theory with local discriminability, for any bipartite state $\rho \in \mathsf{St}(AB)$ there exists $0 < k < 1$ such that $k\rho \in \mathsf{RefSet}\,(\rho_A \otimes \rho_B)$, where ρ_A and ρ_B are the marginals of ρ.

Solutions to Selected Problems and Exercises

Exercise 6.1

Condition (6.2) is clearly necessary for the possibility of locally discriminating ρ from σ, which would otherwise be identical when tested with local observation tests. On the other hand, condition (6.2) corresponds to the following condition:

$$(a|(b| \,|\rho) \neq (a|(b| \,|\sigma).$$

The effect $(a|(b|$ can be completed to a local observation test $\{A_0, A_1\}$, with $A_0 := (a|(b|$, and $A_1 := (e_A|(e_B| - (a|(b|$, where $e_A \geq a$ and $e_B \geq b$ are deterministic local effects. The test $\{A_0, A_1\}$ is clearly local, since it is obtained by coarse-graining from the local test $\{(a|(b|, (a|(e_B - b|, (e_A - a|(b|, (e_A - a|(e_B - b|\}$. Now we can complete the proof using the result of Exercise 3.1.

7 The Purification Principle

What makes quantum theory so special? Everyone has an opinion on that: one can point at entanglement, non-locality, complementarity, uncertainty, and many other features. In a trivial sense, every answer is legitimate: every feature in which quantum theory *differs* from other familiar theories can be taken as *the* distinctive feature, because it allows you to tell quantum theory apart from what you already know. But this would not take you very far. The real question is not how to distinguish quantum theory from theories that you already know, but how to distinguish it from all theories that you could possibly imagine. Most features are not good at that – for example, physicists have invented many toy theories that exhibit entanglement, non-locality, and complementarity and still are not quantum theory. In this chapter we put forward our answer to the question of what makes quantum theory special: purification. In the following, we will explore some of the most fundamental facts about quantum theory, like entanglement and teleportation, connecting purification with a deep stream of thought that goes back to Schrödinger and von Neumann. The results presented in this chapter are also a first example of how one can *reason* on quantum theory starting from first principles.

7.1 A Distinctive and Fundamental Trait

The purification principle stipulates that, whenever you are ignorant about the state of a system A, you can always claim that your ignorance comes from the fact that A is a part of a larger system AB, of which you have full knowledge. When you do this, the pure state that you have to assign to the composite system AB is determined by the state of A in an essentially unique way.

The purification of mixed states is a peculiar feature – surely, not one that we experience in our everyday life. How can you claim that you know A *and* B if you do not know A alone? This counterintuitive feature had been noted already in the early days of quantum theory, when Erwin Schrödinger famously wrote: "Another way of expressing the peculiar situation is: *the best possible knowledge of a whole does not necessarily include the best possible knowledge of all its parts*."[1] And, in the same paper: "I would not call that *one* but rather *the* characteristic trait of quantum mechanics, the one that enforces its entire departure from classical lines of thought."

---

[1] Schrödinger (1935b) (italics added).

This is a bold statement, if you think that it was made in 1935! Nowadays, however, there is plenty of evidence supporting it. Here are three good arguments: first, we know that purification, combined with five rather basic principles, picks up quantum theory among all possible theories one can imagine. Second, physicists have been trying to fabricate toy theories that exhibit quantum-like features without being quantum theory. While they succeeded with many features, up to now purification resists: if you want to purify mixed states, then quantum theory seems to be the only option, up to minimal variations.[2] Third, many important features of quantum theory can be derived *directly* from purification, without the need of deriving quantum theory first. This fact strongly suggests that, when it comes to isolating what is specific of quantum theory, purification just hits the spot.

The purification of mixed states is specifically quantum. But why should we assume it as a fundamental principle of Nature? At first, it looks like a weird feature – and it must look so, because quantum theory itself is weird and if you squeeze it inside a principle, it is likely that the principle will look weird too. However, on second thought one realizes that purification is a fundamental requirement: essentially, it is the link between physics and information theory. Information theory would not make sense without the notions of probability and mixed state, for the whole point about information is that there are things that we do not know in advance. But in the world of classical physics of Newton and Laplace, every event is determined and there is no space for information at the fundamental level. In principle, it does not make sense to toss a coin or to play a game of chance, for the outcome is already determined and, with sufficient technology and computational power, can always be predicted. In contrast, purification tells us that "ignorance is physical." Every mixed state can be generated in a single shot by a reliable procedure, which consists in putting two systems in a pure state and discarding one of them. As a result of this procedure, the remaining system will be a physical token of our ignorance. This discussion suggests that, only if purification holds, information can aspire to a fundamental role in physics.

Here we emphasized the role of "information"and "ignorance." But if you are bothered by the incursion of these subjective notions into the temple of physics, you can just replace them by "randomness" and think that what is at stake here is whether randomness exists at the fundamental level. Indeed, a world with randomness is a world with mixed states (if you have a random number generator, you can always use it to prepare a mixed state). In such a world, it is natural to ask "Why do we have mixed states?" From the realist's point of view, if a mixed state cannot be purified, it is always possible that, in fact, the system is in a pure state – a pure state that, accidentally, we do not happen to know. Hence, purification is a necessary requirement if you want to guarantee that, at the most fundamental level, some states are mixed.

In the relation between purification and mixedness, there is also an interesting way to turn the table around: purification allows one to *eliminate* mixedness from the description, enabling us to replace mixed states with pure states of enlarged systems. Even more than that, purification allows us to replace *every* physical process with a *reversible* process acting on a larger system. Let us spell precisely what we mean. Suppose that system

---

[2] Purification is satisfied also by quantum theory on the real field. But quantum theory on the real and complex field are just two variants grown on the same theoretical core.

A undergoes an irreversible process, which transforms it into system A'. Then, purification tells us that we can simulate the process with the following recipe:

Pure and Reversible Simulation

1. Prepare a system B in a pure state β.
2. Evolve systems A and B together by a reversible transformation, which turns them into systems A' and B'.
3. Discard system B'.

The pure and reversible simulation is loaded with consequences:

1. **Conservation of information.** In this simulation, discarding system B' is the only irreversible operation in the picture – if we keep system B' the whole evolution is reversible. This is good news if you believe that, at the fundamental level, information cannot be destroyed. Among the experts, this belief is called *conservation of information*, and its validity is often upheld as one of the most fundamental laws of physics.
2. **Reversible computation.** The conservation of information is important not only to theoretical physicists, but also to computer scientists. All in all, a computation is a physical process, which transforms input data into output data. In general, the process does not have to be reversible – think for example of the computation of the function $f(x) = x^2$. But is it possible to realize every computation using only reversible gates? Bennett asked this question in the 1970s, giving a positive answer: every deterministic computation that can be realized in classical theory can be achieved using only reversible gates, by adding a system in the description of the process. Now, we know that this is true also in quantum theory, but it is easy to come up with toy theories where some computations cannot be done reversibly.[3]
3. **Simulation of randomized algorithms.** The pure and reversible simulation ensures that also probabilistic computations can be realized without pumping entropy from the outside – that is, without preparing system B in a mixed state. This means that even a randomized algorithm or a Monte Carlo simulation can be run without an external random number generator, starting off only with pure states.

The three points above provide good reasons to require the pure and reversible simulatability as a fundamental property of physical processes. Since purification gives this as a bonus, there are at least three good reasons to be happy about it.

But, do we *need* purification in order to have a pure and reversible simulation? The answer is "yes," because the preparation of a state is a special case of physical process – a process with no input. Hence, if you want the pure and reversible simulatability to hold for every process, then you also need purification as a special case.

In the following, we will delve deeper into the consequences of purification, giving a first illustration of how the high-level reasoning from first principles can reconstruct crucial quantum features.

---

[3] This is the case e.g. in the theory known as *boxworld*, where two elementary systems cannot interact through reversible transformations (Gross *et al.*, 2010). In this theory it is impossible to implement controlled operations in a reversible way.

7.2 The Purification Principle

Here is the precise statement of the purification principle:

Purification Axiom *For every system* A *and for every state* $\rho \in \mathsf{St}(A)$, *there exists a system* B *and a pure state* $\Psi \in \mathsf{PurSt}(AB)$ *such that*

$$\rho\!-\!\!A \;=\; \Psi \begin{array}{c} -A \\ -B-\!\!e \end{array} \ . \tag{7.1}$$

If two pure states Ψ *and* Ψ' *satisfy*

$$\Psi' \begin{array}{c} -A \\ -B-\!\!e \end{array} \;=\; \Psi \begin{array}{c} -A \\ -B-\!\!e \end{array} \ ,$$

then there exists a reversible transformation $\mathcal{U}$, *acting only on system* B, *such that*

$$\Psi' \begin{array}{c} -A \\ -B \end{array} \;=\; \Psi \begin{array}{c} -A \\ -B-\mathcal{U}-B \end{array} \ . \tag{7.2}$$

Here we say that Ψ is a *purification* of ρ and that B is the *purifying system*. Informally, Eq. (7.1) guarantees that you can always find a pure state of AB that is compatible with your limited knowledge of A alone. On top of this, Eq. (7.2) specifies that all the states of AB that are compatible with your knowledge of A are essentially the same, up to a reversible transformation on B. We will call this property the *uniqueness of purification*. Note that the two purifications in Eq. (7.2) have the same purifying system. It is easy to generalize the statement to the case where the purifying systems are different:

Proposition 7.1 *If two pure states* $\Psi \in \mathsf{PurSt}(AB)$ *and* $\Psi' \in \mathsf{PurSt}(AB')$ *are purifications of the same mixed state, then*

$$\Psi' \begin{array}{c} -A \\ -B' \end{array} \;=\; \Psi \begin{array}{c} -A \\ -B-\mathcal{C}-B' \end{array} \tag{7.3}$$

for some deterministic transformation $\mathcal{C}$ *transforming system* B *into system* B'.

Proof Pick two pure states $\beta \in \mathsf{PurSt}(B)$ and $\beta' \in \mathsf{PurSt}(B')$. Since $\Psi \otimes \beta'$ and $\Psi' \otimes \beta$ are purifications of the same state on A, the uniqueness of purification implies

$$\begin{array}{c} \Psi' \begin{array}{c} -A \\ -B' \end{array} \\ \beta -\!\!B \end{array} \;=\; \begin{array}{c} \Psi \begin{array}{c} -A \\ -B \\ \end{array} \\ \beta' -\!\!B' \end{array} \begin{array}{c} \\ \mathcal{U} \end{array} \begin{array}{c} -B' \\ -B \end{array}$$

for some reversible transformation $\mathcal{U}$ (we have also assumed local discriminability). Discarding system B on both sides we then obtain $\Psi' = (\mathcal{I}_A \otimes \mathcal{C})\Psi$, where $\mathcal{C}$ is the deterministic transformation defined by

$\square$

This is all you need to know about purification – now let us start deriving consequences.

7.3 Entanglement

The most obvious consequence of purification is entanglement.

Proposition 7.2 (Existence of Entangled States)   *The purification of a mixed state is an entangled state.*

Proof   Suppose that Ψ is a purification of ρ. Now, if Ψ is not entangled, by definition this means that Ψ is the product of two pure states. Hence, ρ must be pure. By contrapositive, if ρ is mixed, the only possibility is that Ψ is entangled. $\square$

Whenever we purify a mixed state, we get an entangled state. Interestingly, the converse is also true:

Proposition 7.3   *The marginal of a pure entangled state is a mixed state.*

Proof   Suppose that Ψ is a pure state of AB and that its marginal on system A is pure – call it α (for the existence of the marginal state we are implicitly assuming causality). Then, for every pure state β, the product state $\Psi' = \alpha \otimes \beta$ will be a purification of α. The uniqueness of purification, stated by Eq. (7.2), implies that $\Psi = \alpha \otimes \mathcal{U}\beta$ for some reversible transformation $\mathcal{U}$ acting only on B. This means that Ψ is a product state. Hence, if a pure state is entangled, then its marginal must be mixed. $\square$

Summarizing, we have proved that, in a theory satisfying our principles, a state is mixed if and only if its purification is entangled. By this observation, the only theories that satisfy purification and have no entanglement are the theories where there are no mixed states at all. In these theories no event can be random, because random events could be used to generate mixed states. In other words, for a causal theory we have proven the implication: *"purification + no entanglement $\implies$ determinism."* This is mostly a curiosity here, because in this book we will focus our attention to probabilistic theories where not all outcomes are determined in advance. In these theories, purification implies the existence of entanglement.

7.4 Reversible Transformations and Twirling

Purification implies not only that there are entangled states, but also that there are "enough reversible transformations" in our theory. For example, one has the following.

Proposition 7.4 *For every pair of normalized pure states ψ and ψ' of a generic system B there must be a reversible transformation $\mathcal{U}$ such that*

$$\left(\overline{\psi'}\,\frac{\mathrm{B}}{}\right. = \left(\overline{\psi}\,\frac{\mathrm{B}}{}\right\rfloor\mathcal{U}\left\lfloor\frac{\mathrm{B}}{}\right. . \tag{7.4}$$

Proof Easy corollary of the uniqueness of purification stated by Eq. (7.2): if we erase system A from the diagram (mathematically, if we set it to be the trivial system I), then the uniqueness condition reads "if $_\mathrm{I}(e|\psi') = _\mathrm{I}(e|\psi)$, then there exists a reversible transformation $\mathcal{U}$ such that $\psi' = \mathcal{U}\psi$." $\qquad\square$

The ability to transform any pure state into any other by means of reversible transformations will be called *transitivity*, meaning that the action of the set of reversible transformations is *transitive* on the set of pure states.

Transitivity, combined with the existence of entanglement, leads us straight to the existence of *entangling gates*, i.e. reversible gates that transform product states into entangled states. Another consequence of transitivity is every physical system has a unique *maximally mixed state*, i.e. a unique state that is invariant under the action of every reversible transformation. In order to see that, we need to make a small digression in the structure of the reversible transformations acting on a given physical system. First of all, observe that the reversible transformations acting on a given system form a group, i.e.:

1. the identity is a reversible transformation;
2. the inverse of a reversible transformation is a reversible transformation;
3. the composition of two reversible transformations is a reversible transformation.

For system A we will denote by $\mathbf{G}_\mathrm{A}$ the group of reversible transformations acting on A. When A is finite dimensional, it is possible to show that $\mathbf{G}_\mathrm{A}$ is a compact group. We postpone the proof of this fact to Section 7.12, because the proof becomes easy once the right techniques have been developed. For the moment, we just use the compactness of the group $\mathbf{G}_\mathrm{A}$ to construct a special channel, called the *twirling channel*. The twirling channel, $\mathcal{T}$, is defined as

$$\mathcal{T} := \int_{\mathbf{G}_\mathrm{A}} \mathrm{d}\mathcal{U}\,\mathcal{U}, \tag{7.5}$$

where $\mathrm{d}\mathcal{U}$ is the normalized Haar measure on $\mathbf{G}_\mathrm{A}$, the integral runs over the whole group. Assuming local discriminability, it is easy to see that the linear map defined here is a channel: indeed, since the transformations on A form a finite-dimensional space, the convex set generated by the reversible transformations in $\mathbf{G}_\mathrm{A}$ is a finite-dimensional convex set, and, since the map $\mathcal{T}$ is inside the convex set, Carathéodory's theorem implies that one can write it as a finite convex combination

$$\mathcal{T} := \sum_{x \in \mathsf{X}} p_x \mathcal{U}_x. \tag{7.6}$$

Clearly, a convex combination of channels is itself a channel.

Note that the action of the twirling channel does not change if we compose it with reversible transformations: as a consequence of the definition of $\mathcal{T}$ and thanks to the invariance of the Haar measure we have

$$\mathcal{W}\mathcal{T}\mathcal{V} = \mathcal{T} \qquad \forall \mathcal{V}, \mathcal{W} \in \mathbf{G}_A.$$

This fact leads us immediately to the conclusion that there is a unique invariant state. On the one hand, the output of the twirling must be an invariant state, since

$$(\mathcal{T}\rho) = \mathcal{W}(\mathcal{T}\rho), \qquad \forall \mathcal{W} \in \mathbf{G}_A, \forall \rho \in \mathsf{St}(A).$$

This means that there exists at least one *invariant state*, say

$$\chi := \mathcal{T}\rho_0 \tag{7.7}$$

for a fixed state ρ_0. On the other hand, it is easy to see that the above definition is independent of the choice of ρ_0, and that, in fact, there can be only *one* invariant state. The proof is an easy exercise:

Exercise 7.1 Show that:

1. assuming causality, for every pair of states ρ and ρ', one has $\mathcal{T}\rho \propto \mathcal{T}\rho'$;
2. if χ' is an invariant state, it must satisfy $\mathcal{T}\chi' = \chi'$.

Using the two points above, conclude that there is a unique invariant state, and, therefore

$$\underset{A}{—}\boxed{\mathcal{T}}\underset{A}{—} = \underset{A}{—}\boxed{e} \quad \underset{A}{(\chi}\boxed{}\underset{A}{—}. \tag{7.8}$$

Exercise 7.2 Assuming local discriminability, prove that the invariant state χ_{AB} of the composite system AB is the parallel composition of the invariant states χ_A and χ_B.

Like every state, the invariant state χ can be decomposed as a convex combination of pure states, say $\chi = \sum_{x \in \mathsf{X}} p_x \alpha_x$. One important thing is that we can always find a decomposition of this form where one of the states $\{\alpha_x\}$ is any desired pure state:

Exercise 7.3 Show that:

1. for every pure state $\alpha \in \mathsf{PurSt}(A)$ there exists a non-vanishing probability $p > 0$ such that $\chi_A = p\alpha + (1 - p)\rho$, where ρ is some other state of system A;
2. the state χ_A can be written as $\chi_A = \sum_x p_x\alpha_x$, where $\{\alpha_x\}$ is a set of pure states that spans $\mathsf{St}_{\mathbb{R}}(A)$ and $\{p_x\}$ is a probability distribution.

We have seen that every pure state is contained with non-zero probability in some convex decomposition of the invariant state. This is an important property with major consequences. To see why, we need to make ourselves acquainted with an important consequence of the purification principle: steering.

7.5 Steering

Suppose that you have a preparation device that prepares system A in a state α_x with probability p_x. If you ignore the value of x, this procedure results in the preparation of the mixed state

$$\rho = \sum_{x \in X} p_x \alpha_x,$$

where X is the set of the possible values of x. Now, suppose that, instead of ignoring x, you encode it in the state of another physical system – call it C – in such a way that everyone who measures this system can find out the value of x. This can be done by preparing system C in a state γ_x, with the property that the states $\{\gamma_x \mid x \in X\}$ are perfectly discriminable. As the result of this procedure, the composite system AC is in the state

$$\sigma = \sum_{x \in X} p_x \alpha_x \otimes \gamma_x, \tag{7.9}$$

(this is possibly granted by the perfect discriminability axiom). The state σ has two important properties: first, it is an *extension* of ρ, that is,

$$\left(\sigma \begin{array}{c} A \\ C \end{array} \!\!\!\! \underline{e} \right) = \left(\rho \begin{array}{c} A \end{array} \right). \tag{7.10}$$

Second, if one measures system C with the test that discriminates among the states $\{\gamma_x \mid x \in X\}$, one can simulate the original preparation device for system A: indeed, one has

$$\left(\sigma \begin{array}{c} A \\ C \end{array} \!\!\!\! \underline{c_x} \right) = p_x \left(\alpha_x \begin{array}{c} A \end{array} \right) \qquad \forall x \in X, \tag{7.11}$$

where $c := \{c_x\}$ is the observation test that discriminates among the states $\{\gamma_x \mid x \in X\}$. This is an interesting trick, because it allows us to replace the preparation of a random pure state with the preparation of a single state of a larger system AC, followed by a measurement on C.

Clearly, the trick that we showed here works for every ensemble decomposition of ρ: given an ensemble decomposition, we can always find a suitable system C, a state of AC, and a measurement on C such that Eq. (7.11) is satisfied. But can we find an extension that works *for every ensemble?* Thanks to purification, the answer is affirmative:

Proposition 7.5 (Steering) *Let $\Psi \in \mathrm{PurSt}(AB)$ be a purification of $\rho \in \mathrm{St}(A)$. Then, for every ensemble decomposition $\rho = \sum_x p_x \alpha_x$ there exists a measurement $b = \{b_x\}$, such that*

$$\left(\Psi \begin{array}{c} A \\ B \end{array} \!\!\!\! \underline{b_x} \right) = p_x \left(\alpha_x \begin{array}{c} A \end{array} \right) \qquad \forall x \in X. \tag{7.12}$$

Proof For every ensemble $\{p_x \alpha_x\}$, construct an extension $\sigma \in \mathrm{St}(AC)$ as in Eq. (7.9) and take a purification of it, say $\Psi' \in \mathrm{PurSt}(ACD)$. Since Ψ' and Ψ are two purifications of ρ, the uniqueness of purification implies that there must exist a channel $\mathcal{C} \in \mathsf{Transf}(\mathrm{B} \to \mathrm{CD})$ such that $\Psi' = (\mathcal{I}_A \otimes \mathcal{C})\Psi$ (cf. Proposition 7.1). Using Eq. (7.11) we then obtain

Defining the measurement $\{b_x\}$ by $b_x := (c_x \otimes e)\mathcal{C}$, we then have Eq. (7.12).   □

    By choosing different measurements on system B we can "steer" the ensemble decomposition of ρ, in the sense that we decide which particular ensemble we want to generate.[4] This feature is quite striking when the state ρ has more than one ensemble decomposition into pure states; in this case, we cannot say that the state before the measurement was in an unknown pure state, because even the set of alternative pure states in which the system could be depends on the choice of the measurement. This fact means that we don't have a *local realistic* interpretation in terms of the ensembles describing the state preparation.

7.6 Process Tomography

Purification, along with local discriminability, establishes an interesting correspondence between transformations and states. This is easy to see: let us take a set of states $\{\alpha_x \mid x \in \mathsf{X}\}$ that span the whole state space of system A and a set of positive probabilities $\{p_x\}_{x \in \mathsf{X}}$. Then, take a purification of the mixed state $\rho = \sum_x p_x \alpha_x$ – say $\Psi \in \mathsf{PurSt}(\mathrm{AB})$. Now, if two transformations $\mathcal{A}$ and $\mathcal{A}'$ satisfy

it is clear that $\mathcal{A}$ must be equal to $\mathcal{A}'$, namely the correspondence $\mathcal{A} \mapsto (\mathcal{A} \otimes \mathcal{I}_B)\Psi$ is injective.

    Indeed, using the steering property of Eq. (7.12) we obtain

Since the states $\{\alpha_x\}$ span the whole state space, this also means that $\mathcal{A}\rho = \mathcal{A}'\rho$ for every state ρ and – by local discriminability

[4] Note, however, that we cannot decide which particular state α_x is prepared – otherwise we would violate causality.

This result gives us a recipe to identify an unknown transformation:

1. prepare the entangled state Ψ;
2. apply the unknown transformation $\mathcal{A}$ on system A;
3. perform a complete measurement on the output state $(\mathcal{A} \otimes \mathcal{I}_B)\Psi$.

Repeating this procedure a large number of times, the statistics of the measurement outcomes can be used to identify the state $\Psi_{\mathcal{A}}$ and, therefore, to identify $\mathcal{A}$. This procedure is known as *process tomography*. An example of state that can be used for process tomography is a purification of the invariant state χ: indeed, by Exercise 7.3 we know that χ can be written as $\chi = \sum_x p_x \alpha_x$, where $\{\alpha_x\}$ is a set of pure states that spans the whole state space.

By itself, process tomography is not a big deal: if you just want an injective correspondence you do not need purification – instead, you can just take a correlated state, like the state σ in Eq. (7.9). The key point here is that the state Ψ is *pure*, a feature that leads directly to striking consequences, like the principle of *no information without disturbance*.

7.7 No Information Without Disturbance

Suppose that you want to extract some information about system A and, at the same time, you want to be "invisible," in the sense that you leave no trace of your operations. This is the situation if you want to eavesdrop a secret communication between Alice and Bob, without making them realize that you are listening to what they say. Mathematically, the condition that your test $\{\mathcal{A}_x\}$ does not introduce any disturbance is

$$\sum_x \mathcal{A}_x = \mathcal{I}_A, \tag{7.13}$$

which means that, for those who ignore the outcome, your test is indistinguishable from the identity on system A.

In the familiar classical world, it is easy to satisfy the no disturbance condition. For example, we can find out whether a coin landed on heads or tails without introducing any detectable change. In a world that satisfies purification, things are very different. If we take a pure state $\Psi \in \mathsf{PurSt}(AB)$ that can be used for process tomography, then the no disturbance condition implies $\sum_x (\mathcal{A}_x \otimes \mathcal{I}_B)\Psi = \Psi$. But Ψ is pure: hence, each unnormalized state $(\mathcal{A}_x \otimes \mathcal{I}_B)\Psi$ must be proportional to Ψ. Precisely, there must be a set of probabilities $\{p_x\}$ such that $(\mathcal{A}_x \otimes \mathcal{I}_B)\Psi = p_x\Psi$. Since the map $\mathcal{A} \mapsto (\mathcal{A} \otimes \mathcal{I}_B)\Psi$ is injective (see Section 7.6), we conclude that $\mathcal{A}_x = p_x\mathcal{I}_A$. In other words, our test is equivalent to the following procedure:

1. do nothing on system A;
2. output an outcome x chosen at random with probability p_x.

Clearly, this means that our test does not extract any information at all. In summary, we proved the implication "no disturbance $\Longrightarrow$ no information": in a world that satisfies purification, there is no way to extract information without leaving a trace behind. This fact has enormous implications for cryptography: when two parties are communicating, every attempt to extract information about their message will result in a disturbance that they can in principle detect.

7.8 Teleportation

Another major consequence of purification is teleportation. The task of teleportation is essentially the task of a fax: to transfer data from a sender to a receiver using only a classical transmission line and some pre-established correlations between sender and receiver. However, faxing data carried by a system that is entangled with another one doesn't seem a trivial task.

Suppose that we want to transfer the data carried by system A and that the receiver has a copy of system A in an entangled state with another system, B, in the sender's lab. The teleportation protocol has the following general structure:

1. The sender measures the input system A together with system B, obtaining an outcome x.
2. The outcome x is communicated to the receiver through the classical transmission line.
3. Depending on the outcome x, the receiver performs an operation on the output.

The protocol works if, after these three steps, the state of the input system is transferred to the output system. Let us see how to construct a working teleportation protocol starting just from first principles.

For the entangled state shared between sender and receiver, let us choose a purification of the invariant state χ of system A – call it $\Phi \in \mathsf{PurSt}(AB)$. Let us also choose an ensemble of reversible transformations $\{p_x \mathcal{U}_x\}$ that averages to the twirling channel, i.e. $\sum_x p_x \mathcal{U}_x = \mathcal{T}$. Then, assuming local discriminability, by Eq. (7.8) we have

$$\sum_{x \in \mathsf{X}} p_x \left(\mathcal{U}_x \otimes \mathcal{I}_B \right) \Phi = \left(\mathcal{T} \otimes \mathcal{I}_B \right) \Phi$$

$$= \chi \otimes \beta, \tag{7.14}$$

where β is the marginal of Φ on system B. Now, note that, by definition

that is, $\Phi \otimes \Phi$ is a purification of $\chi \otimes \beta$. Using Eq. (7.14) and the steering property of Proposition 7.5 (with $\Psi \equiv \Phi \otimes \Phi$ in Eq. (7.12)), we have that there exists a measurement $\{B_x\}$ such that

$$
\left(\Phi \begin{array}{c} A \\ B \end{array} \; \Phi \begin{array}{c} A \\ B \end{array} B_x \right) = p_x \left(\Phi \begin{array}{c} A \quad \boxed{U_x} \quad A \\ B \end{array} \right) \qquad \forall x \in \mathsf{X}.
$$

Since the correspondence $A \mapsto (A \otimes I_B)\Phi$ is injective assuming also local discriminability (see Section 7.6), we conclude that

$$
\left(\Phi \begin{array}{c} A \\ B \\ A \end{array} B_x \right) = p_x \quad A \boxed{U_x} A \qquad \forall x \in \mathsf{X}. \tag{7.15}
$$

We are done: the above equation says that, if a sender performs the measurement $\{B_x\}$ on the input system and on half of the entangled state Φ, then the state of the input system will be transferred on the receiver's side and will undergo a reversible transformation depending on the outcome. Using the classical transmission line, the sender can communicate the outcome to the receiver, who can undo the reversible transformation by applying its inverse U_x^{-1}. As a result of this procedure, the state of system A has been transferred from the sender's to the receiver's end.

7.9 A Reversible Picture of an Irreversible World

In a world satisfying purification, irreversible processes can be simulated by reversible ones, pretty much in the same way in which the preparation of mixed states can be simulated by the preparation of pure states. Suppose that you observe a deterministic process C acting on system A. We will see now that, thanks to purification, the process can be simulated as

$$
A \boxed{C} A \quad = \quad \begin{array}{c} \boxed{\eta} \; E \qquad E \; \boxed{e} \\ A \quad \boxed{U} \quad A \end{array}, \tag{7.16}
$$

where E is a suitable system (the "environment"), η is a pure state, and U is a reversible transformation. We refer to this simulation as a *pure and reversible simulation* of the process C.

The pure and reversible simulation is the analog of the purification of mixed states: when a process is irreversible, we can always claim that, in fact, the system is interacting with an environment in a reversible way, and the irreversibility results only from the fact that the environment has been discarded.

Let us see how Eq. (7.16) follows from purification. The basic idea is to use the tomographic correspondence between states and transformations: pick a state $\Psi \in \mathsf{PurSt}(AB)$ that can be used for process tomography, apply the process C on system A, and take a purification of the state $(C \otimes I_B)\Psi$ – call this purification $\Psi' \in \mathsf{PurSt}(EAB)$. By construction, Ψ' and Ψ are two purifications of the same state of system B; indeed, we have

where in the last equality we used the fact that $\mathcal{C}$ is deterministic. Now, if we pick a pure state of system E, call it η, the uniqueness of purification implies that

for some reversible transformation $\mathcal{U}$. Discarding E on both sides, we obtain

Since the map $\mathcal{A} \mapsto (\mathcal{A} \otimes \mathcal{I}_B)\Psi$ is injective, this proves Eq. (7.16).   □

We know that the purification of mixed states is unique up to reversible transformations. A similar result can be obtained for the pure and reversible simulation of processes. This result is easy to obtain once you learned the trick in the proof of Eq. (7.16) and it is left to you as an exercise:

Exercise 7.4   (Uniqueness of the pure and reversible simulation) Show that if

then there exists a reversible transformation $\mathcal{V}$, acting on system E, such that

Finally, with a little bit of extra effort you can prove a pure and reversible simulation for arbitrary processes, where the output can be different from the input:

Exercise 7.5   Show that every deterministic process $\mathcal{C}$ transforming A into A′ has a pure and reversible simulation of the form

where η is a pure state of E and $\mathcal{U}$ is a reversible transformation that converts system AE into system A′E′.

7.10 Displacing the Von Neumann's Cut

In the previous section we considered deterministic processes. We now turn our attention to tests that can have random outcomes, proving the analog of steering for general tests.

Suppose that an experimenter performs a test on system A, say $\{A_x \mid x \in X\}$. When the outcome x occurs, the experimenter will know it, e.g. because the device prints it on a display. A natural question is: "The display is a physical system too, why is it not described like any other system in the theory?" The obvious answer is that, if we want, we can always include it in the description: we can describe the device that provides the information about outcome x through the deterministic process

$$C := \sum_{x \in X} A_x \otimes \beta_x, \qquad (7.17)$$

where $\{\beta_x\}$ is a set of perfectly discriminable states of a suitable system B that carries information about the outcome. Informally, we will refer here to system B as the "display." Now, "looking at the display" means performing the test that discriminates among the states $\{\beta_x\}$. When the experimenter performs such a test – call it $\{b_x\}$ – she will obtain back the original test:

$$\forall x \in X.$$

There is nothing mysterious in this equation, it is just an equivalent way of representing the test $\{A_x\}$ as the result of the measurement done by the experimenter on the display. What is interesting, however, is that we can take a pure and reversible simulation of the process C, and regard our test as the result of a reversible interaction between the tested system A, the display B, and, possibly, an environment E. In formula,

$$\forall x \in X, \qquad (7.18)$$

where E is a suitable system, η is a pure state, and U is a reversible transformation. The proof of this fact is left to you as an exercise:

Exercise 7.6 Prove Eq. (7.18) and generalize it to tests with different input and output systems. [Hint: use the result of Exercise 7.5.]

The cut between the physical systems included in the description and those that are omitted is known as *von Neumann's cut*. In general, the cut can be done in different ways: we can imagine that there are photons going from the display to the eye of the experimenter, and, again, we can include them in the description, adding one more system in the interaction U that gives rise to the test. Of course, this game can go on forever: we can include into the description the experimenter herself, and we can even include an infinite chain of experimenters, each of them making tests on the previous one. Thanks to Eq. (7.18), we can always displace the cut between the systems that evolve reversibly and the system that undergoes the final measurement. Due to purification, each experimenter can claim that she is doing a measurement, while all the other systems evolve deterministically according to some fundamentally reversible dynamics.[5]

7.11 The State-transformation Isomorphism

In a theory satisfying purification there is a special correspondence between states and transformations, essentially based on the idea of process tomography. The steps to set up the correspondence are the following – for a given system A:

1. take a set of pure states $\{\alpha_x\}$ that spans the whole state space;
2. take a mixed state $\rho = \sum_x p_x \alpha_x$, where all probabilities $\{p_x\}$ are positive;
3. take a purification of ρ, say $\Psi \in \mathsf{PurSt}(AB)$ for some purifying system B.

Once these choices are made, we can define the *state-transformation isomorphism* as the linear map

$$\mathcal{A} \longmapsto \Psi_{\mathcal{A}} := (\mathcal{A} \otimes \mathcal{I}_B)\Psi , \tag{7.19}$$

which sends transformations in $\mathsf{Transf}(A \to A')$ into (unnormalized) states in $\mathsf{St}(A'B)$. Since the state Ψ can be used for process tomography, the map $\mathcal{A} \to \Psi_{\mathcal{A}}$ is injective. In addition to being injective, the correspondence of Eq. (7.19) has much more peculiar features, illustrated in the following paragraphs.

7.11.1 Deterministic Transformations

Here we characterize the deterministic transformations of our theory in terms of the corresponding states. By definition, a deterministic transformation $\mathcal{A} \in \mathsf{Transf}(A \to A')$, satisfies the condition

$$e_{A'} \circ \mathcal{A} = e_A .$$

From this condition, it is immediate to see that the state $\Psi_{\mathcal{A}}$ satisfies the relation

$$(e_A \otimes \mathcal{I}_B) \Psi_{\mathcal{A}} = \beta , \tag{7.20}$$

where β is the marginal of Ψ on system B. In other words, $\Psi_{\mathcal{A}}$ is an extension of β. Much more interestingly, the converse statement is also true: all the bipartite states that are extensions of β are of the form $\Psi_{\mathcal{A}}$ for some deterministic transformation $\mathcal{A}$.

Proposition 7.6 *If a state* $\Sigma \in \mathsf{St}(A'B)$ *is an extension of* β, *then there exists a deterministic transformation* $\mathcal{A} \in \mathsf{Transf}(A \to A')$ *such that* $\Sigma = (\mathcal{A} \otimes \mathcal{I}_B)\Psi$.

---

[5] Here we carefully avoid to make any statement on how things "really" are, which would lead to the so-called measurement problem.

Proof Take a purification of Σ, say $\Psi' \in \mathsf{PurSt}(\mathrm{CA'B})$. By the uniqueness of purification, there exists a deterministic transformation $\mathcal{C} \in \mathsf{Transf}(\mathrm{A} \to \mathrm{CA'})$ such that $\Psi' = (\mathcal{C} \otimes \mathcal{I}_B)\Psi$ (cf. Proposition 7.1). Hence, we obtained

having defined $\mathcal{A} := (e_C \otimes \mathcal{I}'_A)\mathcal{C}$. By definition, $\mathcal{A}$ is a deterministic transformation. □

7.11.2 General Tests

Consider now the case of a general test $\{\mathcal{A}_x\}_{x \in \mathsf{X}}$. By definition, if we coarse-grain over all outcomes, we obtain a deterministic transformation $\mathcal{A} := \sum_{x \in \mathsf{X}} \mathcal{A}_x$. Hence, the corresponding preparation test $\{\Psi_{\mathcal{A}_x}\}$ must satisfy the condition

$$\sum_{x \in \mathsf{X}} (e_A \otimes \mathcal{I}_B)\, \Psi_{\mathcal{A}_x} = \beta\,,$$

which follows directly from Eq. (7.20). Again, the converse result also holds. Since the case of general tests includes the case of deterministic transformations, this is the general statement of the isomorphism between states and transformations:

Theorem 7.7 (The State-transformation Isomorphism) *If a preparation test $\{\Sigma_x\}_{x \in \mathsf{X}}$ of system* $\mathrm{A'B}$ *satisfies the condition*

$$\sum_{x \in \mathsf{X}} (e_A \otimes \mathcal{I}_B)\, \Sigma_x = \beta \qquad \forall x \in \mathsf{X}\,,$$

then there exists a test $\{\mathcal{A}_x\}_{x \in \mathsf{X}}$, with input A *and output* $\mathrm{A'}$, *such that*

$$\Sigma_x = (\mathcal{A}_x \otimes \mathcal{I}_B)\, \Psi \qquad \forall x \in \mathsf{X}\,.$$

Proof Proving this fact requires a trick that we already used a couple of times in this chapter: encoding the outcome x into a state γ_x chosen from set of perfectly discriminable states of some system C. Defining the state

$$\Gamma := \sum_{x \in \mathsf{X}} \gamma_x \otimes \Sigma_x \in \mathsf{St}(\mathrm{CA'B})\,,$$

we have that, by construction, Γ is an extension of β, the marginal of Ψ on system B. Using the result of Proposition 7.6 we then obtain

$$\Gamma = (\mathcal{A} \otimes \mathcal{I}_B)\Psi\,,$$

where $\mathcal{A}$ is a deterministic transformation from A to CA'. Moreover, by definition of Γ we have

$$(c_x \otimes \mathcal{I}_{A'B})\Gamma = \Sigma_x \qquad \forall x \in X,$$

where $\{c_x\}_{x \in X}$ is the measurement that discriminates among the states $\{\gamma_x\}$. Combining the two equations above we finally obtain that, for every x,

$$\begin{aligned}
\Sigma_x &= (c_x \otimes \mathcal{I}_{A'B})\Gamma \\
&= (c_x \otimes \mathcal{I}_{A'B})(\mathcal{A} \otimes \mathcal{I}_B)\Psi \\
&= (\mathcal{A}_x \otimes \mathcal{I}_B)\Psi,
\end{aligned}$$

having defined $\mathcal{A}_x := (c_x \otimes \mathcal{I}_{A'})\mathcal{A}$.   □

In summary, the state-transformation isomorphism tells us that there is a one-to-one correspondence between tests and convex decompositions of the extensions of β, the marginal of the state used to set up the isomorphism.

7.12 Everything Not Forbidden is Allowed

The state-transformation isomorphism has a striking consequence: every hypothetical transformation that is compatible with the set of states in our theory *must* be a physical transformation allowed by the theory.

Suppose that, for some reason, you have only partial knowledge of a physical theory: you know the set of normalized states, but you have no idea of what are the transformations and the measurements. What can you say about them?

One thing that you can do is to guess the transformations and measurements using the constraints that come from your knowledge of the states. For example, you can try to guess the deterministic transformations from A to A' in the following way. First, we know that physical transformations act linearly on the state space. Hence, your candidate transformation $\mathcal{A}$ must be a linear map from $\mathsf{St}_{\mathbb{R}}(A)$ to $\mathsf{St}_{\mathbb{R}}(A')$.[6] Second, a deterministic transformation should map normalized states into normalized states. This condition must hold not only for states of system A, but also for states of arbitrary composite systems AB, because your candidate transformation should behave well also when applied to one part of a composite system. In other words, the candidate map $\mathcal{A}$ must satisfy

$$\rho \in \mathsf{St}_1(AB) \quad \Longrightarrow \quad (\mathcal{A} \otimes \mathcal{I}_B)\rho \in \mathsf{St}_1(A'B).$$

Linearity and the preservation of normalized states are two necessary requirements for a map $\mathcal{A}$ to be a deterministic transformation. But are they sufficient? Quite strikingly, the answer is "yes," thanks to the state-transformation isomorphism:

---

[6] We assume here the local discriminability principle, which guarantees that, for every system B the extension $\mathcal{A} \otimes \mathcal{I}_B$ is defined uniquely for every linear map $\mathcal{A}$ from $\mathsf{St}_{\mathbb{R}}(A)$ to $\mathsf{St}_{\mathbb{R}}(A')$.

Proposition 7.8 *If a linear map $\mathcal{A} : \mathsf{St}_{\mathbb{R}}(\mathrm{A}) \to \mathsf{St}_{\mathbb{R}}(\mathrm{A}')$ transforms normalized states into normalized states, then it is a deterministic transformation.*

Proof Since $\mathcal{A}$ transforms deterministic states into deterministic states, we must have $(e|\mathcal{A}\rho) = (e|\rho)$ for every state $\rho \in \mathsf{St}(\mathrm{A})$, or, equivalently

$$e_{\mathrm{A}'} \circ \mathcal{A} = e_{\mathrm{A}} .$$

Now, define the state $\Psi_{\mathcal{A}} := (\mathcal{A} \otimes \mathcal{I}_{\mathrm{B}})\Psi$. By construction, we have

$$(e_{\mathrm{A}'} \otimes \mathcal{I}_{\mathrm{B}})\Psi_{\mathcal{A}} = (e_{\mathrm{A}'} \otimes \mathcal{I}_{\mathrm{B}})(\mathcal{A} \otimes \mathcal{I}_{\mathrm{B}})\Psi$$
$$= [(e_{\mathrm{A}'} \circ \mathcal{A}) \otimes \mathcal{I}_{\mathrm{B}}] \Psi$$
$$= (e_{\mathrm{A}} \otimes \mathcal{I}_{\mathrm{B}})\Psi$$
$$\equiv \beta .$$

In other words, $\Psi_{\mathcal{A}}$ is an extension of the marginal of Ψ on system B. By Proposition 7.6, we have $\Psi_{\mathcal{A}} = \Psi_{\mathcal{A}'}$, where $\mathcal{A}'$ is a deterministic transformation allowed by the theory. But the correspondence $\mathcal{A} \mapsto \Psi_{\mathcal{A}}$ is injective, and, therefore, $\mathcal{A}$ coincides with $\mathcal{A}'$. In conclusion, $\mathcal{A}$ is a deterministic transformation allowed by the theory. □

Thanks to this result, we are finally in position to show that the group of reversible transformations acting on a given finite system is compact. Since by local discriminability the transformations acting on A form a finite-dimensional vector space, it is enough to show that the group $\mathbf{G}_{\mathrm{A}}$ is closed:

Proposition 7.9 *For every system A, the group of reversible transformations $\mathbf{G}_{\mathrm{A}}$ is closed.*

Proof Let $\{\mathcal{U}_n\}_{n \in \mathbb{N}} \subset \mathbf{G}_{\mathrm{A}}$ be a sequence converging to some linear map $\mathcal{M}$.[7] Convergence means that for every system B and for every state $\rho \in \mathsf{St}(\mathrm{AB})$ the sequence $\{(\mathcal{U}_n \otimes \mathcal{I}_{\mathrm{B}})\rho\}$ converges to $(\mathcal{M} \otimes \mathcal{I}_{\mathrm{B}})\rho$. Since the state space is closed, $(\mathcal{M} \otimes \mathcal{I}_{\mathrm{B}})\rho$ must be a state. Moreover, $(\mathcal{M} \otimes \mathcal{I}_{\mathrm{B}})\rho$ is normalized whenever ρ is normalized. Hence, Proposition 7.8 implies that $\mathcal{M}$ is a deterministic transformation. Now, consider the sequence $\{\mathcal{U}_n^{-1}\}_{n \in \mathbb{N}}$. Since physical transformations form a compact subset of a finite-dimensional vector space, the sequence must have a converging subsequence, say $\{\mathcal{U}_{n_k}^{-1}\}$. Calling $\mathcal{N}$ the limit of this subsequence, we can run again the argument made for $\mathcal{M}$, thus showing that $\mathcal{N}$ is a deterministic transformation. Moreover, it is clear that $\mathcal{M}\mathcal{N} = \mathcal{N}\mathcal{M} = \mathcal{I}_{\mathrm{A}}$. Hence, $\mathcal{M}$ is a reversible transformation and $\mathcal{N}$ is its inverse. □

The success that we had with deterministic transformations suggests that the approach of guessing physical transformations from the set of states may work for arbitrary transformations and arbitrary measurements. And indeed, this is the case.

Let $\{\mathcal{A}_x\}_{x \in \mathsf{X}} \subset \mathsf{Transf}(\mathrm{A} \to \mathrm{A}')$ be a candidate test. Like before, we know that physical transformations act linearly on the state space, and, therefore, each $\mathcal{A}_x$ must be a linear

---

[7] Convergence can be defined using the operational norm for maps defined in Section 3.7. Clearly, since the space $\mathsf{Transf}_{\mathbb{R}}(\mathrm{A})$ is finite dimensional, it is a Banach space. Moreover all metrics are equivalent.

map from $St_{\mathbb{R}}(A)$ to $St_{\mathbb{R}}(A')$. Moreover, each transformation $\mathcal{A}_x$ should map normalized states into (possibly subnormalized) states, namely

$$\rho \in St_1(AB) \quad \Longrightarrow \quad (\mathcal{A}_x \otimes \mathcal{I}_B)\rho \in St(A'B)\,, \tag{7.21}$$

and summing over all possible outcomes one should get a normalized state, namely

$$\rho \in St_1(AB) \quad \Longrightarrow \quad \sum_{x \in X}(\mathcal{A}_x \otimes \mathcal{I}_B)\rho \in St_1(A'B)\,. \tag{7.22}$$

Again, the conditions of Eqs. (7.21) and (7.22) are necessary requirements for a collection of maps to be called a test. Thanks to the state-transformation isomorphism, these conditions are also sufficient: every collection of linear maps satisfying Eqs. (7.21) and (7.22) is a test allowed by the theory.

Proposition 7.10 *Let $\{\mathcal{A}_x\}_{x \in X}$ be a collection of linear maps from $St_{\mathbb{R}}(A)$ to $St_{\mathbb{R}}(A')$. If $\{\mathcal{A}_x\}_{x \in X}$ satisfies Eqs. (7.21) and (7.22), then it is a test allowed by the theory.*

Proof Define the linear map $\mathcal{A} := \sum_{x \in X}\mathcal{A}_x$. By Eq. (7.22) we know that $\mathcal{A}$ transforms normalized states into normalized states, and, therefore $e_{A'} \circ \mathcal{A} = e_A$. Now, define the states $\Psi_{\mathcal{A}_x} := (\mathcal{A}_x \otimes \mathcal{I}_B)\Psi$. Clearly, they satisfy the condition

$$\sum_{x \in X}(e_{A'} \otimes \mathcal{I}_B)\Psi_{\mathcal{A}_x} = (e_{A'} \otimes \mathcal{I}_B)\Psi_{\mathcal{A}}$$

$$= (e_{A'} \otimes \mathcal{I}_B)(\mathcal{A} \otimes \mathcal{I}_B)\Psi$$

$$= [(e_{A'} \circ \mathcal{A}) \otimes \mathcal{I}_B]\Psi$$

$$= (e_A \otimes \mathcal{I}_B)\Psi$$

$$\equiv \beta\,.$$

Hence, by Theorem 7.7 there exists a test $\{\mathcal{A}'_x\}$ allowed by the theory such that $\Psi_{\mathcal{A}_x} = (\mathcal{A}'_x \otimes \mathcal{I}_B)\Psi$ for every x. Since the correspondence $\mathcal{A} \mapsto \Psi_{\mathcal{A}}$ is injective, we conclude that $\mathcal{A}_x = \mathcal{A}'_x$ for every x. In conclusion, $\{\mathcal{A}_x\}$ is a test allowed by the theory. $\square$

In summary, everything that is not forbidden *must* be allowed. This means that the set of normalized states determines in a unique way all the transformations, measurements, and general tests allowed by the theory: the whole theory is encoded in the set of normalized states. Our derivation of quantum theory will use this fact: first, we will show that the normalized states allowed by the theory are density matrices over Hilbert spaces. Then, we will use the fact that the theory is determined uniquely by the set of states. In other words, we will know for sure that there is only one theory that satisfies purification and has density matrices as the possible states: this theory is quantum theory.

7.13 Purification in a Nutshell

In the previous sections we have taken you through a tour of the most important consequences of the purification principle. Since the tour has been dense of ideas, we'll give here a bird eye view of the conceptual landscape that we have explored. We started from the statement of the purification principle: every mixed state can be prepared, in an essentially unique way, by discarding a part of a pure state of a composite system. The purification of mixed states led us first to entanglement and to the property of steering, which allows one to generate arbitrary ensemble decompositions of a mixed state by performing measurements on the purifying system. Steering is the feature that enables the tomography of processes; thanks to this feature, when we purify a completely mixed state we obtain a bipartite state that can be used to encode physical transformations faithfully. The key feature of this bipartite state is that it is *pure*: due to this fact, we have proven the property of "no information without disturbance," with fundamental implications for cryptography.

The consequences of purification do not involve only states, but also transformations. First of all, purification implies that every pure state can be transformed into any other pure state by a reversible transformation, a property known as transitivity. Transitivity implies that, for every finite-dimensional system, there is a unique maximally mixed state, i.e. a unique state that is invariant under the action of arbitrary reversible transformations. A rather spectacular application of this fact and of the steering property is teleportation. The reconstruction of teleportation is perhaps the best example of how reasoning from first principles allows us to derive classic results of quantum information without the need of the Hilbert-space formalism.

In addition to teleportation, purification implies (in fact, is equivalent to) the pure and reversible simulation of physical transformations. Every deterministic process can be simulated as a reversible transformation that affects jointly a system and its environment, which is originally in a pure state. In this picture, the only source of irreversibility is in the fact that the environment is discarded. This fact is the statement of the conservation of information, according to which at the fundamental level information cannot be destroyed, but only discarded. The pure and reversible simulation can be adapted to the case of general tests, including random outcomes. In this case, the result is that every test can be modeled as a reversible interaction between the input system and the testing apparatus, which is later measured to read out the outcome. In principle, the cut between the systems interacting reversibly and the system that is measured can be displaced indefinitely, as in von Neumann's treatment of the quantum measurement process.[8]

Finally, a striking consequence of purification is the fact that the theory is completely specified once we assign the sets of normalized states. In other words, there is no need to specify the measurements, the transformations, or, more generally, the tests: every hypothetical transformation that is compatible with the set of normalized states is necessarily included in the theory. This result guarantees that, if we want to derive quantum theory, it is enough to prove that the possible normalized states can be represented as density matrices over a complex Hilbert space. The fact that a theory satisfying purification is uniquely identified by the set of normalized states is a high-level consequence of the isomorphism between states and transformations that is set up by process tomography.

---

[8] von Neumann (1996).

7.14 Summary

In this chapter we presented the purification principle, showing that it leads to entanglement, steering, tomography of transformations, no information without disturbance, and teleportation. In a world satisfying purification, the pure states of a given system can be transformed into one another by reversible transformations. In addition, irreversibility can be attributed to the loss of information in an environment and the cut between measured systems and measurement apparatuses can be displaced arbitrarily, while still modeling the interaction between system and apparatus as a reversible transformation. Finally, we showed that purification establishes an isomorphism between transformations and bipartite states and we used this isomorphism to show that every linear map that preserves the structure of the state space must be a physical transformation allowed by the theory.

Notes

Purification and Von Neumann's Requirement   The requirement that a subjective element, like ignorance, admits an objective physical description reminds immediately of an observation by von Neumann, who wrote (von Neumann, 1996): "It is a fundamental requirement of the scientific viewpoint – the so-called principle of the psycho-physical parallelism – that it must be possible so to describe the extra-physical process of the subjective perception as if it were in reality in the physical world – i.e. to assign to its parts equivalent physical processes in the objective environment, in ordinary space." Here von Neumann was referring to the task of gathering information through a measurement, but it is quite natural to extend his observation to other subjective experiences, like the experience of ignorance. If we do that, then purification is nothing but von Neumann's requirement: the purifying system B is the "objective environment" that is used to explain the origin of our ignorance as the result of an "equivalent physical process" – the preparation of AB in a pure state.

Does God Play Dice?   Everyone knows Einstein's complaint about the randomness of quantum theory, expressed in the popular mantra "God does not play dice." However, the pure and reversible simulation provides an interesting rebuttal: God does not *need* to play dice. Instead, He can prepare the world in a pure state and let it evolve reversibly. If this is the case, the randomness and irreversibility that we observe in our experiments are just the result of our limited angle of observation on the world and of our limited degree of control over the interactions with the environment.

Purification and Causality   The formulation of the purification principle presented in this chapter requires the validity of causality. Indeed, Eq. (7.1) contains explicitly the unique deterministic effect. Nevertheless, it is possible to formulate purification in a way that works also in the absence of causality. This more general formulation is discussed in Chiribella *et al.* (2014).

Appendix 7.1 Carathéodory's Theorem

Theorem 7.11 (Carathéodory) *If a point $x \in \mathbb{R}^d$ belongs to the convex hull $\mathrm{Conv}(P)$ of a set $P \subseteq \mathbb{R}^d$, there is a subset $P' \subseteq P$ consisting of at most $d+1$ points such that $x \in \mathrm{Conv}(P')$ of P'.*[9]

Proof Let $x \in \mathrm{Conv}(P) \subseteq \mathbb{R}^d$. Then, x is a convex combination of a finite number of points in P:

$$\mathbf{x} = \sum_{j=1}^{k} \lambda_j \mathbf{x}_j, \qquad x_j \in P, \ \lambda_j \geq 0, \quad \forall j = 1, \ldots, k$$

For $k \leq d$ the statement holds trivially. Suppose $k > d + 1$. Then, the points $x_2 - x_1, \ldots, x_k - x_1$ are linearly dependent, hence there exist real scalars $\mu_2, \ldots, \mu_k$ not all zero, such that

$$\sum_{j=2}^{k} \mu_j(\mathbf{x}_j - \mathbf{x}_1) = \mathbf{0}.$$

Upon defining $\mu_1 := -\sum_{j=2}^{k} \mu_j$ one has

$$\sum_{j=1}^{k} \mu_j \mathbf{x}_j = \mathbf{0}, \quad \sum_{j=1}^{k} \mu_j = 0,$$

and not all of the μ_j are equal to zero. Therefore, at least one $\mu_j > 0$. Then,

$$\mathbf{x} = \sum_{j=1}^{k} \lambda_j \mathbf{x}_j - \alpha \sum_{j=1}^{k} \mu_j \mathbf{x}_j = \sum_{j=1}^{k} (\lambda_j - \alpha \mu_j) \mathbf{x}_j$$

for any real α. In particular, the equality will hold for α given by

$$\alpha := \min_{1 \leq j \leq k} \left\{ \frac{\lambda_j}{\mu_j} : \mu_j > 0 \right\} = \frac{\lambda_i}{\mu_i}.$$

Note that $\alpha > 0$, and for every $j = 1, \ldots, k$ one has

$$\lambda_j - \alpha \mu_j \geq 0.$$

In particular, $\lambda_i - \alpha \mu_i = 0$ by definition of α. Therefore,

$$\mathbf{x} = \sum_{j=1}^{k} (\lambda_j - \alpha \mu_j) \mathbf{x}_j,$$

where every $\lambda_j - \alpha \mu_j$ is non-negative, their sum is one, and furthermore, $\lambda_i - \alpha \mu_i = 0$. In other words, x is represented as a convex combination of at most $k - 1$ points of P. This

---

[9] Equivalently, x lies in an r-simplex with vertices in P, where $r \leq d$. The above proof is reported from Wikipedia, which also suggests an alternative proof using Helly's theorem.

process can be repeated until x is represented as a convex combination of at most $d + 1$ points in P. $\square$

Solutions to Selected Problems and Exercises

Exercise 7.1

Assuming causality, point 1 follows from the fact that all pure states are connected by a reversible transformation, hence upon decomposing convexly a state ρ into deterministic pure states as $\rho = \sum_x p_x \psi_x$ one has $\rho = \sum_x p_x \mathcal{U}_x \psi_0$, hence using invariance of the Haar measure, one has $\mathcal{T}\rho = p(\rho)\mathcal{T}\psi_0$. Point 2 is trivial. Equation (7.8) follows then immediately assuming local discriminability.

Exercise 7.2

Being χ_{AB} invariant under reversible transformations in $\mathsf{Transf}(A \to B)$, in particular is invariant under the parallel compositions $\mathcal{U} \otimes \mathcal{V}\chi_{AB} = \chi_{AB}$, hence

$$\int_{G_A} d\mathcal{U} \int_{G_B} d\mathcal{V}\, \mathcal{U} \otimes \mathcal{V}\chi_{AB} = \chi_{AB},$$

namely $\mathcal{T}_A \otimes \mathcal{T}_B \chi_{AB} = \chi_{AB}$. Using identity (7.8) one then has

$$\chi_{AB} = (|\chi)(e| \otimes |\chi)(e|)\chi_{AB} = \chi_A \otimes \chi_B.$$

Exercise 7.3

Point 1 follows easily from the fact that all pure states are connected to each other by a reversible transformation, whence any pure state is in the refinement set of the invariant state. Point 2 uses Carathéodory's theorem, along with the fact that any pure state belongs to the refinement of the invariant state.

PART III

QUANTUM INFORMATION
WITHOUT HILBERT SPACES

Encoding Information

8

In the previous chapters we presented a new set of principles for quantum theory, which we will assume from now on. Now starts the real fun: letting the principles interact with one another and watching what comes out. In this chapter, we start with the first interactions: putting together purification and local discriminability, we establish a fundamental equality between the processing of data and the processing of entanglement. We then use this equality to explore how information carried by a physical system can be encoded into another system. In particular, we study the example of ideal compression, where the system used for encoding is of the smallest possible size. Putting together ideal compression and purification, we show that every mixed state has a minimal purification, with the smallest purifying system. Finally, the ideas developed in the chapter are applied to the transmission of information to noisy channels and to the study of error correction.

8.1 Processing Data = Processing Entanglement

In information theory it is common to talk about information sources. An information source is nothing but a preparation device, which emits a given system A in some state ρ_x, chosen with probability $p_x > 0$ from set of possible states $\{\rho_x\}_{x \in X}$. Here, each of the states $\{\rho_x\}_{x \in X}$ represents a different piece of data, and the goal of the information theorist is to achieve some desired processing of the data – e.g. to store it on a hard disk, or to transmit it to a distant location. Let us say that the desired processing is described by the physical transformation $\mathcal{A}$. Then, every other transformation $\tilde{\mathcal{A}}$ that acts like $\mathcal{A}$ on the states $\{\rho_x\}_{x \in X}$ will be equally good. In the following we analyze the conditions under which two transformations act in the same way on a given source, showing a deep relationship between processing of data and processing of entanglement.

If the source is perfectly known, the condition for $\mathcal{A}$ and $\tilde{\mathcal{A}}$ to act in the same way is trivially

$$\mathcal{A}\rho_x = \tilde{\mathcal{A}}\rho_x \qquad \forall x \in X. \tag{8.1}$$

In this case, we say that $\mathcal{A}$ and $\tilde{\mathcal{A}}$ are *equal upon input of the source* $\{p_x\rho_x\}_{x \in X}$. Suppose instead that we do not know exactly the states in the source, but we know that, on average, the states produces the state $\rho = \sum_{x \in X} p_x \rho_x$. In order to guarantee that $\tilde{\mathcal{A}}$ acts in the same way as $\mathcal{A}$, we now need to check the validity of Eq. (8.1) *for every source* $\{p_x\rho_x\}_{x \in X}$ that has ρ as its average state. This condition can be rephrased in an elegant way in terms of the *face identified by* ρ, i.e. of the set F_ρ consisting of all states that appear in some convex decomposition of ρ:

$$F_\rho := \left\{ \sigma \in \mathsf{St}(A) \mid \exists \tau \in \mathsf{St}(A), \exists p > 0 : \rho = p\sigma + (1-p)\tau \right\}. \qquad (8.2)$$

Notice that the above definition corresponds to $F_\rho \equiv \mathsf{RefSet}_1(\rho)$ as explained in Chapter 2. The condition that $\mathcal{A}$ and $\widetilde{\mathcal{A}}$ act in the same way for every source averaging to ρ is equivalent to

$$\mathcal{A}\sigma = \widetilde{\mathcal{A}}\sigma \qquad \forall \sigma \in F_\rho. \qquad (8.3)$$

When this is the case, we say that $\mathcal{A}$ and $\widetilde{\mathcal{A}}$ are *equal upon input of* ρ, and denote it as $\mathcal{A} =_\rho \widetilde{\mathcal{A}}$. Note the obvious property (see Exercise 2.14):

Proposition 8.1 *If $\mathcal{A} =_\rho \widetilde{\mathcal{A}}$ and $\sigma \in F_\rho$, then $\mathcal{A} =_\sigma \widetilde{\mathcal{A}}$.*

According to the last proposition, we can define the notion of equality of two transformations upon a face $F \subseteq \mathsf{St}(A)$ as follows.

Definition 8.2 (Equality upon a Face) We say that two transformations $\mathcal{A}, \mathcal{B} \in \mathsf{Transf}(A \to B)$ are *equal upon the face* $F \subset \mathsf{St}_1(A)$, and write $\mathcal{A} =_F \mathcal{B}$, if $\mathcal{A} =_\rho \mathcal{B}$ for every $\rho \in F$.

We are now ready to establish the link between processing of data and processing of entanglement:

Theorem 8.3 *Let $\rho \in \mathsf{St}_1(A)$ be a state and $\Psi \in \mathsf{PurSt}_1(AB)$ be a purification of ρ. Then, two transformations $\mathcal{A}, \widetilde{\mathcal{A}} \in \mathsf{Transf}(A \to A')$ are equal upon F_ρ if and only if*

$$\qquad\qquad (8.4)$$

Proof Thanks to the steering property (Proposition 7.5), we know that for every state $\sigma \in F_\rho$ there exists a non-zero probability p and an effect $b \in \mathsf{Eff}(B)$ such that

The converse is also true: every effect $b \in \mathsf{Eff}(B)$ induces on system A a state that belongs to F_ρ. Hence, the condition $\mathcal{A} =_\rho \widetilde{\mathcal{A}}$ is equivalent to the condition

$\forall b \in \mathsf{Eff}(B),$

which in turn is equivalent to

$\forall a \in \mathsf{Eff}(A')$
                          $\forall b \in \mathsf{Eff}(B).$

By local discriminability, this is equivalent to the condition $(\mathcal{A} \otimes \mathcal{I}_B)\Psi = (\widetilde{\mathcal{A}} \otimes \mathcal{I}_B)\Psi$. □

In words, two transformations act in the same way on all the sources with average state ρ if and only if they act in the same way on a given purification of ρ: processing the data emitted by the source is the same as processing locally a suitable entangled state.

The equivalence between processing data and processing entanglement has many important consequences. The simplest is a strengthened version of the no information without disturbance property that we encountered in Section 7.7:

Proposition 8.4 *If a test* $\{A_x\}_{x\in X}$ *is non-disturbing upon* F, *i.e. if*

$$\sum_{x\in X} A_x =_F \mathcal{I}_A \,,$$

then it is non-informative upon F, *i.e. there exists a probability distribution* $\{p_x\}_{x\in X}$ *such that*

$$A_x =_F p_x \mathcal{I}_A \,.$$

Proof For a purification of ρ, say $\Psi \in \mathsf{PurSt}(AB)$, the no disturbance condition reads $\sum_{x\in X}(A_x \otimes \mathcal{I}_B)\Psi = \Psi$ (by Theorem 8.3). Since Ψ is pure, this means that there exists a probability distribution $\{p_x\}$ such that $(A_x \otimes \mathcal{I}_B)\Psi = p_x \Psi$. But this condition implies $A_x =_F p_x \mathcal{I}_A$ (by Theorem 8.3, again). $\qquad\square$

Exercise 8.1 Let $\rho \in \mathsf{St}_1(A)$, and $a \in \mathsf{Eff}(A)$. Prove that if $(a|\rho) = 1$ then $a =_\rho e$, and if $(a|\rho) = 0$ then $a =_\rho 0$.

8.2 Ideal Encodings

It is often useful to encode the information carried by a system A into another physical system A'. Think for example of data compression, where one wants to encode the information in a smaller physical support, or of the example of error correcting codes, where one wants to spread the information over many systems in order to fight unwanted errors. In all these cases, a good encoding is required to be lossless: the encoding $\mathcal{E}$ is *lossless upon input of* ρ iff there exists a decoding operation $\mathcal{D}$ such that $\mathcal{DE} =_\rho \mathcal{I}_A$.

Now, the encoding $\mathcal{E}$ transforms sources with average state ρ into sources with average state $\rho' := \mathcal{E}\rho$. Clearly, this means that every state in the face F_ρ is encoded into a state in the face $F_{\rho'}$, namely

$$\mathcal{E}F_\rho \subseteq F_{\rho'} \,, \tag{8.5}$$

We say that the encoding is *efficient* if one has the equality, i.e. if *every* state in $F_{\rho'}$ can be obtained by encoding some state in F_ρ, i.e. iff

$$\forall \tau \in F_{\rho'} \ \exists \sigma \in F_\rho : \quad \boxed{\tau} \!\!-\!\!\!\frac{A'}{}\!\!\!- \ = \ \boxed{\sigma} \!\!-\!\!\frac{A}{}\!\!\!-\!\!\boxed{\mathcal{E}}\!\!-\!\!\frac{A'}{}\!\!\!- \,. \tag{8.6}$$

This means that the encoding does not waste space: the face $F_{\rho'}$ in which the input states are encoded is as small as it could possibly be.

We call *ideal* an encoding that is both lossless and efficient. For example, the compression protocols postulated in the ideal compression principle are all examples of ideal encodings. In general, ideal encodings can be characterized as follows.

Proposition 8.5 *A transformation $\mathcal{E}$ is an ideal encoding for the state ρ iff there exists a decoding transformation $\mathcal{D}$ such that*

$$\mathcal{D}\mathcal{E} =_\rho \mathcal{I}_A \qquad \text{and} \qquad \mathcal{E}\mathcal{D} =_{\rho'} \mathcal{I}_{A'}, \tag{8.7}$$

with $\rho' = \mathcal{E}\rho$.

Proof Suppose that Eq. (8.7) holds. The first condition in Eq. (8.7) states that $\mathcal{E}$ is lossless, and, in particular $\mathcal{D}\rho' = \rho$. Clearly, this implies

$$\mathcal{D}F_{\rho'} \subseteq F_\rho. \tag{8.8}$$

The second condition in Eq. (8.7) guarantees that $F_{\rho'} = \mathcal{E}\mathcal{D}F_{\rho'} \subseteq \mathcal{E}F_\rho$. Hence, $\mathcal{E}$ is efficient. Summarizing, we have proven that $\mathcal{E}$ is ideal.

Conversely, suppose that $\mathcal{E}$ is ideal. In particular, $\mathcal{E}$ is lossless, i.e. satisfies the first condition in Eq. (8.7). Now, let τ be a generic state in $F_{\rho'}$. Since $\mathcal{E}$ is efficient, τ can be written as $\tau = \mathcal{E}\sigma$ for some state $\sigma \in F_\rho$. Since $\mathcal{E}$ is lossless, we have $\sigma = \mathcal{D}\mathcal{E}\sigma$. Combining these two facts, we get

$$\mathcal{E}\mathcal{D}\tau = \mathcal{E}\mathcal{D}(\mathcal{E}\sigma) = \mathcal{E}(\mathcal{D}\mathcal{E}\sigma) = \mathcal{E}\sigma = \tau.$$

Since the condition holds for every $\tau \in F_{\rho'}$, we proved that $\mathcal{E}\mathcal{D} =_{\rho'} \mathcal{I}_{A'}$. □

The above result tells us that ρ can be ideally encoded into ρ' if and only if ρ' can be ideally encoded into ρ. Moreover, combining it with Proposition 8.1 we obtain the following proposition.

Proposition 8.6 *If $\mathcal{E}$ is an ideal encoding for ρ, then $\mathcal{E}$ is an ideal encoding for every state $\sigma \in F_\rho$.*

It is then clear that the notion of ideal encoding applies to a face rather than to a single state, since by Proposition 8.6 every state σ in the face F_ρ has the same ideal encoding scheme $(C, \mathcal{E}, \mathcal{D})$ as ρ. We can then define the *ideal encoding scheme for a face F*, referring to the ideal compression scheme $(C, \mathcal{E}, \mathcal{D})$ for any state in F.

Ideal encodings can be characterized through an even simpler condition:

Theorem 8.7 *Let $\Psi \in \mathsf{PurSt}(AB)$ be a purification of ρ. Then, $\mathcal{E}$ is an ideal encoding for ρ if and only if the state $\Psi' := (\mathcal{E} \otimes \mathcal{I}_B)\Psi$ is pure.*

Proof Suppose that $(\mathcal{E} \otimes \mathcal{I}_B)\Psi$ is pure. Then, Ψ and Ψ' are two purifications of the same state on system B. Hence, the uniqueness of purification implies that there exists a deterministic transformation $\mathcal{D}$ such that $(\mathcal{D} \otimes \mathcal{I}_B)\Psi' = \Psi$ for some transformation $\mathcal{D}$. In other words, we have $(\mathcal{D}\mathcal{E} \otimes \mathcal{I}_B)\Psi = \Psi$, which is equivalent to $\mathcal{D}\mathcal{E} =_\rho \mathcal{I}_A$, thanks to Theorem 8.3. Moreover, we have

$$(\mathcal{E}\mathcal{D}\otimes\mathcal{I}_B)\Psi' \equiv (\mathcal{E}\mathcal{D}\mathcal{E}\otimes\mathcal{I}_B)\Psi$$
$$= (\mathcal{E}\otimes I_B)\Psi$$
$$\equiv \Psi',$$

which is equivalent to $\mathcal{E}\mathcal{D} =_{\rho'} \mathcal{I}_{A'}$. Hence, $\mathcal{E}$ is an ideal encoding.

Conversely, suppose that $\mathcal{E}$ is an ideal encoding for ρ, i.e. that there exists a decoding $\mathcal{D}$ such that $\mathcal{D}\mathcal{E} =_{\rho} \mathcal{I}_A$ and $\mathcal{E}\mathcal{D} =_{\rho'} \mathcal{I}_{A'}$ for $\rho' := \mathcal{E}\rho$. Now, suppose that $(\mathcal{E}\otimes\mathcal{U}_B)\Psi$ can be written as a convex combination of pure states, say

$$(\mathcal{E}\otimes\mathcal{U}_B)\Psi = \sum_x p_x \Gamma_x \tag{8.9}$$

for some probabilities $\{p_x\}$ and some pure states $\{\Gamma_x\}$. Applying $\mathcal{D}$ on both sides of the equality, we obtain

$$\sum_{i\in X} p_x (\mathcal{D}\otimes\mathcal{I}_B)\Gamma_x = (\mathcal{D}\mathcal{E}\otimes\mathcal{I}_B)\Psi \equiv \Psi,$$

having used the relation $\mathcal{D}\mathcal{E} =_{\rho} \mathcal{I}_A$. Since Ψ is pure, this relation implies

$$(\mathcal{D}\otimes\mathcal{I}_B)\Gamma_x = \Psi \qquad \forall x \in X. \tag{8.10}$$

Finally, Eq. (8.9) implies that the marginal of each pure state Γ_x – call it τ_x – belongs to the face identified by $\rho' := \mathcal{E}\rho$. Hence, the relation $\mathcal{E}\mathcal{D} =_{\rho'} \mathcal{I}_{A'}$ implies $\mathcal{E}\mathcal{D} =_{\tau_x} \mathcal{I}_{A'}$ (cf. Proposition 8.1) and, in turn, $(\mathcal{E}\mathcal{D}\otimes\mathcal{I}_B)\Gamma_x = \Gamma_x$. In conclusion, we obtained

$$\Gamma_x = (\mathcal{E}\mathcal{D}\otimes\mathcal{I}_B)\Gamma_x$$
$$= (\mathcal{E}\otimes\mathcal{I}_B)\Psi,$$

having used Eq. (8.10). Summarizing, $(\mathcal{E}\otimes\mathcal{I}_B)\Psi$ admits only trivial convex decompositions, i.e. it is pure. $\qquad\square$

Thanks to this result, ideal encodings are identified with the deterministic transformations that map pure states into pure states. An easy consequence is the following.

Corollary 8.8 *Let $\mathcal{E}$ be an ideal encoding for ρ and let α be a pure state in F_ρ. Then, $\mathcal{E}\alpha$ is pure.*

Proof Since α belongs to F_ρ, $\mathcal{E}$ is an ideal encoding for α (Proposition 8.6). Since α is pure, its purifications are only of the product form $\Psi = \alpha \otimes \beta$ for some $\beta \in \mathsf{PurSt}(B)$. Hence, Theorem 8.7 implies $(\mathcal{E}\otimes\mathcal{I}_A)(\alpha\otimes\beta) = \alpha'\otimes\beta'$, for some pure states $\alpha' \in \mathsf{PurSt}(A')$ and $\beta' \in \mathsf{PurSt}(B)$. In turn, this implies $\mathcal{E}\alpha = \alpha'$ and $\beta' = \beta$. $\qquad\square$

To practice the notions in this section, try the following exercises.

Exercise 8.2 Show that if $\mathcal{E} \in \mathsf{Transf}(A \to A')$ is an ideal encoding for ρ and $\mathcal{E}' \in \mathsf{Transf}(A'\to A'')$ is an ideal encoding for $\rho' := \mathcal{E}\rho$, then $\mathcal{E}'\mathcal{E}$ is an ideal encoding for ρ.

Exercise 8.3   Show that if $\mathcal{E} \in \mathsf{Transf}(A \to A')$ is an ideal encoding for ρ and $\mathcal{F} \in \mathsf{Transf}(B \to B')$ is an ideal encoding for σ, then $\mathcal{E} \otimes \mathcal{F}$ is an ideal encoding for $\rho \otimes \sigma$.

Exercise 8.4   Show that $\Psi \in \mathsf{PurSt}(AB)$ and $\Psi' \in \mathsf{PurSt}(AB')$ are two purifications of the same state $\rho \in \mathsf{St}(A)$ if and only if there exists an ideal encoding $\mathcal{E} \in \mathsf{Transf}(B \to B')$ such that $\Psi' = (\mathcal{I}_A \otimes \mathcal{E})\Psi$.

Exercise 8.5   Show that if $\mathcal{E} \in \mathsf{Transf}(A \to A')$ is an ideal encoding for ρ with decoding map $\mathcal{D} \in \mathsf{Transf}(A' \to A)$, then if $\beta \in \mathsf{PurSt}(A')$ also $\mathcal{D}\beta$ is pure.

8.3 Ideal Compression

Our ideal compression principle requires that every state ρ can be ideally encoded in a suitable physical system, here denoted by C. What makes ideal compression special compared to other ideal encodings is that the encoded state $\rho' = \mathcal{E}\rho$ is completely mixed – that is, that the face identified by ρ' is the whole state space of system C. Hence, the condition $\mathcal{E}\mathcal{D} =_{\rho'} \mathcal{I}_C$ becomes

$$\mathcal{E}\mathcal{D}\tau = \tau \qquad \forall \tau \in \mathsf{St}(C), \tag{8.11}$$

and, due to local discriminability,

$$\mathcal{E}\mathcal{D} = \mathcal{I}_C. \tag{8.12}$$

Another special feature of ideal compression is that the decoding operation is atomic:

Proposition 8.9   *In every ideal compression protocol* $(C, \mathcal{E}, \mathcal{D})$, *the decoding is an atomic transformation.*

Proof   Let $\rho \in \mathsf{St}(A)$ be the state that is compressed and let $\Psi \in \mathsf{PurSt}(AB)$ be a purification of ρ. By Theorem 8.7, the state $\Psi' := (\mathcal{E} \otimes \mathcal{I}_B)\Psi$ must be pure, and, by Eq. (8.12) we have $(\mathcal{D} \otimes \mathcal{I}_B)(\Psi') = \Psi$. Now, suppose that $\mathcal{D} = \sum_x \mathcal{D}_x$ for some set of transformations $\{\mathcal{D}_x\}$. Then, we must have $\sum_x (\mathcal{D}_x \otimes \mathcal{I}_B)\Psi' = \Psi$, and, since Ψ is pure,

$$(\mathcal{D}_x \otimes \mathcal{I}_B)\Psi' = \Psi = (\mathcal{D} \otimes \mathcal{I}_B)\Psi'$$

for some probabilities $\{p_x\}$. Hence, $\mathcal{D}_x =_{\rho'} p_x \mathcal{D}$. Since ρ' is completely mixed, by local discriminability this implies $\mathcal{D}_x = p_x \mathcal{D}$. Since $\mathcal{D}$ admits only trivial decompositions, it is atomic.   □

Intuitively, ideal compression allows us to *identify* the face F_ρ with the state space $\mathsf{St}_1(C)$. This identification is essentially unique:

Proposition 8.10   *If two systems* C *and* C' *allow for ideal compression of a state* $\rho \in \mathsf{St}_1(A)$, *then they are operationally equivalent.*

Proof   Suppose that $(C, \mathcal{E}, \mathcal{D})$ and $(C', \mathcal{E}', \mathcal{D}')$ are two ideal compression protocols for the state ρ. Then, define the transformations $\mathcal{U} := \mathcal{E}'\mathcal{D} \in \mathsf{Transf}(C \to C')$ and $\mathcal{V} = \mathcal{E}\mathcal{D}' \in \mathsf{Transf}(C' \to C)$. It is easy to see that $\mathcal{U}$ and $\mathcal{V}$ are reversible and $\mathcal{U}^{-1} = \mathcal{V}$. Indeed, for every state τ of system C one has

$$
\begin{aligned}
\mathcal{V}\mathcal{U}\tau &= \mathcal{E}(\mathcal{D}'\mathcal{E}')\mathcal{D}\tau \\
&= \mathcal{E}(\mathcal{D}'\mathcal{E}')\sigma \qquad \sigma := \mathcal{D}\tau \\
&= \mathcal{E}\sigma \\
&= \mathcal{E}\mathcal{D}\tau \\
&= \tau,
\end{aligned}
$$

where in the third and fifth equalities we used the fact that $\sigma := \mathcal{D}\tau$ belongs to F_ρ [cf. Eq. (8.8)], and, therefore, $\sigma = \mathcal{D}'\mathcal{E}'\sigma = \mathcal{D}\mathcal{E}\sigma$. Since $\mathcal{V}\mathcal{U}$ is equal to the identity on every input state, by local discriminability it is the identity on C. The same argument proves that $\mathcal{U}\mathcal{V}$ is the identity on C'. Since there exists a reversible transformation from $\mathcal{C}$ to $\mathcal{C}'$, the two systems are operationally equivalent.   □

In summary, every face of the state space of system A is identified with the full state space of one – and essentially only one – system C. Analogously to the case of ideal encodings, it is then clear that the notion of ideal compression applies to a face rather than to a single state. In the following, we will often use the notion of an *ideal compression scheme for a face*, referring to the ideal compression scheme $(C, \mathcal{E}, \mathcal{D})$ for any state in F, which is essentially unique thanks to Proposition 8.6 and Lemma 8.10.

Exercise 8.6   Let $(C, \mathcal{E}, \mathcal{D})$ be an ideal compression scheme for the face $F \subseteq \mathsf{St}_1(A)$. Prove that if $\rho \in \mathsf{St}_1(C)$ is maximally mixed, then the refinement set of its image under the decoding map is F, i.e. $F = \mathsf{RefSet}_1(\mathcal{D}\rho)$.

Exercise 8.7   Consider a causal theory with local discriminability and ideal compression. Show that for a state $\psi \otimes \sigma \in \mathsf{St}(AB)$ with $\psi \in \mathsf{PurSt}(A)$ and $\sigma \in \mathsf{St}(B)$ completely mixed, an ideal compression scheme is $(B, e_A \otimes \mathcal{I}_B, \psi' \otimes \mathcal{I}_B)$, with $\psi' := \psi/(e|\psi)$.

8.4 The Minimal Purification

We know that every fact about a source can be translated into a feature about its purification. What is the translation of ideal compression? Interestingly, ideal compression is equivalent to the existence of a *minimal purification*, namely a purification where the marginal state of the purifying system is completely mixed.

Let us see what this means. Purification tells us that we can always think of a mixed state as the marginal of some pure state, provided that we enlarge the description and include a purifying system. Still, there are many possible choices for the purifying system and, among all of them, it would be good to know which one is the smallest. Ideal compression provides the answer: for a given purification, we can compress the purifying system by applying the encoding operation $\mathcal{E}$. As a result, we obtain a new purification where the purifying system is as small as possible. Put in more formal terms, we have the following:

Theorem 8.11 (Minimal Purification)   *For every system* A *and every state* $\rho \in \mathsf{St}(A)$ *there exists a system* C *and a purification* $\Psi \in \mathsf{PurSt}(AC)$ *of* ρ *such that the marginal state on system* C *is completely mixed. The system* C *is unique up to operational equivalence.*

Proof   Choose a purification of ρ, say $\Psi' \in \mathsf{PurSt}(AB)$ for some purifying system B. Choose a compression protocol $(C, \mathcal{E}, \mathcal{D})$ for ρ_B, the marginal of Ψ on system B. Compress B into a system C, thus obtaining the state $\Psi := (\mathcal{I}_A \otimes \mathcal{E})\Psi'$. By Theorem 8.7, Ψ is pure, and, by construction, Ψ is a purification of ρ. The marginal of Ψ on system C is given by $\rho_C := \mathcal{E}\rho_B$ and is completely mixed by the definition of ideal compression. Hence, we constructed a minimal purification of ρ. It remains to show that the system C is fixed up to operational equivalence by the requirement that ρ_C is completely mixed. To this purpose, let $\Psi \in \mathsf{PurSt}(AC)$ be a minimal purification of ρ and $\Psi' \in \mathsf{St}(AB)$ be another purification of ρ. Since Ψ and Ψ' are two purifications of the same state, the uniqueness of purification requires that there must exist a channel $\mathcal{E} \in \mathsf{Transf}(B \to C)$ such that $\Psi = (\mathcal{I}_A \otimes \mathcal{E})\Psi'$. Now, by Theorem 8.7 we have that $\mathcal{E}$ is an ideal encoding of the state ρ_B into the state ρ_C. Since ρ_C is maximally mixed, $\mathcal{E}$ is an ideal compression. But we know from Theorem 8.10 that the system C in an ideal compression protocol is uniquely defined up to operational equivalence. $\square$

We have just seen that the ideal compression principle implies that every state has a minimal purification. If you are wondering about the converse, try the following exercise.

Exercise 8.8   Consider a theory that satisfies causality, local discriminability, and a stronger version of the purification principle, stating that every mixed state has a *minimal* *purification* (plus the fact that purifications are unique up to reversible transformations, of course). Show that such a theory satisfies ideal compression.

8.5 Sending Information Through a Noisy Channel

We conclude the chapter with an elementary discussion of noisy channels and error correction. The results presented in the following are not necessary for the derivation of the Hilbert-space formalism, but provide an excellent illustration of the notions developed in the chapter.

Suppose that you want to send the information carried by a system S through a noisy channel $\mathcal{N} \in \mathsf{Transf}(A \to A')$. Ideally, your goal is to find an encoding operation $\mathcal{E}$ and a decoding operation $\mathcal{D}$ such that in the end the information is left intact, namely

$$\mathcal{D}\mathcal{N}\mathcal{E} = \mathcal{I}_S . \tag{8.13}$$

We say that $\mathcal{E}$ is a *good encoding for* $\mathcal{N}$ if the above condition is satisfied for some decoding operation $\mathcal{D}$. Finding good encodings is a difficult task, so it is useful to have some clue on what these operations look like. The first clue is that, without loss of generality, a good encoding can be chosen to be ideal:

Proposition 8.12 *Every good encoding for $\mathcal{N}$ is a convex combination of ideal encodings, each of which is a good encoding for $\mathcal{N}$.*

Proof Let $\mathcal{E}$ be a good encoding for $\mathcal{N}$, let $\Psi \in \mathsf{St}(\mathrm{SB})$ be the purification of a completely mixed state of S, and $\Psi_{\mathcal{E}}$ be the state $\Psi_{\mathcal{E}} := (\mathcal{E} \otimes \mathcal{I}_{\mathrm{B}})\Psi$. If $\Psi_{\mathcal{E}}$ is pure, then $\mathcal{E}$ is an ideal encoding (by Theorem 8.7). If $\Psi_{\mathcal{E}}$ is mixed, we decompose it as a mixture of pure states, say $\Psi_{\mathcal{E}} = \sum_x p_x \Psi_x$. Then, by the state-transformation isomorphism of Theorem 7.7, there exists a test $\{\mathcal{E}_x\}$ such that $(\mathcal{E}_x \otimes \mathcal{I}_{\mathrm{B}})\Psi = p_x \Psi_x$. Hence, we have

$$\sum_x (\mathcal{D}\mathcal{N}\mathcal{E}_x \otimes \mathcal{I}_{\mathrm{B}})\Psi = (\mathcal{D}\mathcal{N}\mathcal{E} \otimes \mathcal{I}_{\mathrm{B}})\Psi$$

$$= \Psi,$$

having used Eq. (8.13) for the last equality. Since Ψ is pure, we must have $(\mathcal{D}\mathcal{N}\mathcal{E}_x \otimes \mathcal{I}_{\mathrm{B}})\Psi = p_x \Psi$. Recalling that Ψ is faithful, we then obtain $\mathcal{D}\mathcal{N}\mathcal{E}_x = p_x \mathcal{I}_{\mathrm{S}}$. This implies that *i)* $\mathcal{E}_x$ is proportional to a deterministic transformation $\widetilde{\mathcal{E}}_x$ and *ii)* $\widetilde{\mathcal{E}}_x$ is a good encoding. In addition, since the state $(\widetilde{\mathcal{E}}_x \otimes \mathcal{I}_{\mathrm{S}})\Psi$ is pure, the transformation $\widetilde{\mathcal{E}}_x$ is an ideal encoding by Theorem 8.7.   □

Thanks to the above result we can restrict our attention to ideal encodings, which reversibly encode the states of system S into a suitable face of the state space of system A – say, the face F_ρ identified by some state $\rho \in \mathsf{St}(\mathrm{A})$. Now, it is easy to see that, in order to have a good encoding, we must be able to undo the action of the noise on all states in the face F_ρ: indeed, defining the recovery operation $\mathcal{R} := \mathcal{E}\mathcal{D}$ we have

$$\mathcal{R}\mathcal{N}\mathcal{E}\sigma = \mathcal{E}(\mathcal{D}\mathcal{N}\mathcal{E})\sigma = \mathcal{E}\sigma \qquad \forall \sigma \in \mathsf{St}(\mathrm{S}),$$

having used Eq. (8.13). Since $\mathcal{E}\sigma$ is a generic state in F_ρ, this implies the relation

$$\mathcal{R}\mathcal{N} =_\rho \mathcal{I}_{\mathrm{A}}, \tag{8.14}$$

meaning that, upon input of ρ, the recovery operation removes the noise. Hence, from now on we will regard the noise $\mathcal{N}$ as an *error* and we will say that the error $\mathcal{N}$ is *correctable upon input of ρ* if Eq. (8.14) is satisfied for some recovery $\mathcal{R}$.

8.6 The Condition for Error Correction

Given the error $\mathcal{N}$, which conditions need to be met for the error to be correctable upon input of ρ? Let us now take advantage of the equivalence between processing data and processing entanglement. Using Theorem 8.3, the correctability condition becomes

$$\tag{8.15}$$

where $\Psi \in \mathsf{PurSt}(AB)$ is a fixed purification of ρ. What does this tell us about the error? The answer is clear if we think of $\mathcal{N}$ as the result of a reversible interaction between the input system and an environment, as

$$
\begin{array}{c}
\underline{\hspace{0.3cm}A\hspace{0.3cm}} \boxed{\mathcal{N}} \underline{\hspace{0.3cm}A'\hspace{0.3cm}} = \quad \overset{\eta}{\underset{A}{\boxed{}}} \underbrace{ E }_{} \boxed{\mathcal{U}} \overset{E' \, \boxed{e}}{\underset{A'}{}}
\end{array} \quad , \tag{8.16}
$$

for some systems E and E', some pure state $\eta \in \mathsf{PurSt}(E)$, and some reversible transformation $\mathcal{U}$. In terms of the reversible transformation $\mathcal{U}$, we obtain immediately the following criterion for error correction.

Proposition 8.13   *The error $\mathcal{N}$ is correctable upon input of ρ if and only if the reversible transformation $\mathcal{U}$ does not generate correlations between the environment and the purifying system, i.e. if and only if*

$$
\eta \underline{\hspace{0.2cm}E\hspace{0.2cm}} \boxed{\mathcal{U}} \underline{\hspace{0.2cm}E'\hspace{0.2cm}} \quad , \quad \Psi \underline{\hspace{0.2cm}A\hspace{0.2cm}} \, \boxed{\mathcal{U}} \, \underline{\hspace{0.2cm}A'\hspace{0.2cm}} \boxed{e} \quad \Psi \underline{\hspace{0.2cm}B\hspace{0.2cm}} \; = \; \overset{\eta'}{\boxed{}} \underline{\hspace{0.2cm}E'\hspace{0.2cm}} \quad \overset{\beta}{\boxed{}} \underline{\hspace{0.2cm}B\hspace{0.2cm}} \quad , \tag{8.17}
$$

where η' and β are two (generally mixed) states of E' and B'.

Proof   Suppose that $\mathcal{N}$ is correctable. Then, plugging Eq. (8.16) into the correctability condition of Eq. (8.15) we obtain

$$
\eta \underline{\hspace{0.2cm}E\hspace{0.2cm}} \boxed{\mathcal{U}} \underline{\hspace{0.2cm}E'\hspace{0.2cm}} \boxed{e} \; , \; \Psi \underline{\hspace{0.2cm}A\hspace{0.2cm}} \boxed{\mathcal{U}} \underline{\hspace{0.2cm}A'\hspace{0.2cm}} \boxed{\mathcal{R}} \underline{\hspace{0.2cm}A\hspace{0.2cm}} \; , \; \Psi \underline{\hspace{0.2cm}B\hspace{0.2cm}} \; = \; \Psi \underline{\hspace{0.2cm}A\hspace{0.2cm}} \, \underline{\hspace{0.2cm}B\hspace{0.2cm}} \quad .
$$

Since Ψ is a pure state, this implies

$$
\eta \underline{\hspace{0.2cm}E\hspace{0.2cm}} \boxed{\mathcal{U}} \underline{\hspace{0.2cm}E'\hspace{0.2cm}} \; , \; \Psi \underline{\hspace{0.2cm}A\hspace{0.2cm}} \boxed{\mathcal{U}} \underline{\hspace{0.2cm}A'\hspace{0.2cm}} \boxed{\mathcal{R}} \underline{\hspace{0.2cm}A\hspace{0.2cm}} \; , \; \Psi \underline{\hspace{0.2cm}B\hspace{0.2cm}} \; = \; \eta' \underline{\hspace{0.2cm}E'\hspace{0.2cm}} \; , \; \Psi \underline{\hspace{0.2cm}A\hspace{0.2cm}} \, \underline{\hspace{0.2cm}B\hspace{0.2cm}}
$$

for some state $\eta' \in \mathsf{St}(E')$. Discarding system A' and using the relation $e_A \mathcal{R} = e_A$ we obtain the desired condition:

$$
\eta \underline{\hspace{0.2cm}E\hspace{0.2cm}} \boxed{\mathcal{U}} \underline{\hspace{0.2cm}E'\hspace{0.2cm}} \; , \; \Psi \underline{\hspace{0.2cm}A\hspace{0.2cm}} \boxed{\mathcal{U}} \underline{\hspace{0.2cm}A'\hspace{0.2cm}} \boxed{\mathcal{R}} \underline{\hspace{0.2cm}A\hspace{0.2cm}} \boxed{e} \; , \; \Psi \underline{\hspace{0.2cm}B\hspace{0.2cm}} \; = \; \eta' \underline{\hspace{0.2cm}E'\hspace{0.2cm}} \; , \; \Psi \underline{\hspace{0.2cm}A\hspace{0.2cm}} \boxed{e} \; , \; \Psi \underline{\hspace{0.2cm}B\hspace{0.2cm}} \; = \; \eta' \underline{\hspace{0.2cm}E'\hspace{0.2cm}} \; , \; \beta \underline{\hspace{0.2cm}B\hspace{0.2cm}} \quad ,
$$

where β is the marginal of Ψ on system B.

Conversely, suppose that Eq. (8.17) holds. Then, the state β must be the marginal of Ψ on system B (this can be easily checked by discarding system E' on both sides of the equation). Hence, Eq. (8.17) can be rephrased as

where $\Gamma \in \mathsf{PurSt}(E'F)$ is an arbitrary purification of η'. By the uniqueness of purification (Proposition 7.1), we must have

for some deterministic transformation $\mathcal{C} \in \mathsf{Transf}_1(A' \to FA)$. Defining the recovery operation $\mathcal{R} := (e_F \otimes \mathcal{I}_A)\mathcal{C}$ we then obtain

Hence, the correctability condition of Eq. (8.15) is satisfied.   □

The condition that the environment and the purifying system remain uncorrelated can be expressed in an interesting, equivalent way, in terms of the *complementary channel* $\mathcal{N}_c$, defined as

$$(8.18)$$

Using this definition, the factorization between environment and purifying system can be rephrased as

$$(8.19)$$

This is equivalent to the equality $(\mathcal{N}_c \otimes \mathcal{I}_B)\Psi = (\mathcal{N}_0 \otimes \mathcal{I}_B)\Psi$, where $\mathcal{N}_0$ is the *erasure channel* defined by

$$\begin{array}{c} \text{A} \boxed{\mathcal{N}_0} \text{E}' \end{array} := \begin{array}{c} \text{A} \boxed{e} \end{array} \quad \begin{array}{c} \boxed{\eta'} \text{E}' \end{array}.$$

Thanks to the equality between processing data and processing entanglement [8.3], Eq. (8.19) can be restated as follows:

Proposition 8.14 *The error $\mathcal{N}$ is correctable upon input of ρ if and only if its complementary channel $\mathcal{N}_c$ acts as an erasure channel upon input of ρ.*

This correctability condition has an intuitive interpretation based on the idea of conservation of information – in this case, the information carried by the input system A: If no information goes to the environment, then all the information must be in the output system A', and therefore the error must be correctable.

Exercise 8.9 Suppose that $\mathcal{N}$ is a mixture of different errors, i.e. $\mathcal{N} = \sum p_x \mathcal{N}_x$ where $\{p_x\}_{x \in \mathsf{X}}$ is a probability distribution and each $\mathcal{N}_x$ is a deterministic transformation. Show that, if the error $\mathcal{N}$ is correctable upon input of ρ, then each error $\mathcal{N}_x$ is correctable upon input of ρ. In particular, show that the *same* recovery operation $\mathcal{R}$ works for all possible error $\mathcal{N}_x$, namely $\mathcal{R}\mathcal{N}_x =_\rho \mathcal{I}_A \forall x \in \mathsf{X}$.

8.7 Summary

This chapter provided the foundation for the encoding of information in physical systems. We started from the key result, namely the equivalence between processing data and processing entanglement. The result has been applied to the study of ideal encodings, and, in particular, of ideal compression. In particular, we observed that ideal compression guarantees that every state has a minimal purification, where the purifying system has the smallest possible size. Finally, we applied the ideas discussed in the chapter to the study of error correction, showing that a physical process is correctable if and only if it does not leak information into the environment.

Solutions to Selected Problems and Exercises

Exercise 8.1

Let $(a|\rho) = 1$, and take $\sigma \in \mathsf{RefSet}_1(\rho)$. Then there exists $0 < p < 1$ and $\tau \in \mathsf{St}_1(A)$ such that $\rho = p\sigma + (1 - p)\tau$. In this case, we have

$$p(a|\sigma) + (1 - p)(a|\tau) = 1, \tag{8.20}$$

and this implies that $(a|\sigma) = (a|\tau) = 1$. Then, $(a|\rho) = 1$ implies that $(a|\sigma) = 1 = (e|\sigma)$ for every $\sigma \in \mathsf{RefSet}_1(\rho)$, and finally $(a|\sigma) = (e|\sigma)$ for every $\sigma \in \mathsf{RefSet}(\rho)$. Thus, $a =_\rho e$. The same argument can be used for the case of $(a|\rho) = 0$.

Exercise 8.5

Let us consider an arbitrary convex decomposition of $\mathcal{D}\beta = \sum_i p_i \eta_i$. By Eq. 8.7, we have $\mathcal{E}\mathcal{D}\beta = \beta = \sum_i p_i \mathcal{E}\eta_i$, and since β is pure, we must have $\mathcal{E}\eta_i = \beta$ $\forall i$. Finally, again by Eq. 8.7, we must have $\eta_i = \mathcal{D}\mathcal{E}\eta_i = \mathcal{D}\beta$, which implies purity of $\mathcal{D}\beta$.

Exercise 8.6

Let $\sigma \in F$. Then, since $\rho \in \mathsf{St}_1(\mathrm{C})$ is completely mixed, $\mathcal{E}\sigma \in \mathsf{RefSet}_1(\rho)$. Explicitly, there exists a collection of states $\{v_i\}_{i=1}^k \subseteq \mathsf{RefSet}_1(\rho)$ such that

$$\mathcal{E}\sigma = \sum_{i=1}^k c_i v_i, \tag{8.21}$$

with $c_i \geq 0$ for all i. By definition of $\mathsf{RefSet}_1(\rho)$, for every v_i there exist $0 \leq p_i \leq 1$ and $\gamma_i \in \mathsf{St}_1(\mathrm{C})$ such that $\rho = p_i v_i + (1 - p_i)\gamma_i$. Applying the decoding map on both sides, we have $\mathcal{D}\rho = p_i \mathcal{D}v_i + (1 - p_i)\mathcal{D}\gamma_i$, namely $\{\mathcal{D}v_i\}_{i=1}^k \subseteq \mathsf{RefSet}_1(\mathcal{D}\rho)$. Now, applying the decoding map $\mathcal{D}$ on both sides of Eq. (8.21), and recalling that $\mathcal{D}\mathcal{E} =_F \mathcal{I}_\mathrm{A}$, we have

$$\sigma = \sum_{i=1}^k c_i \mathcal{D}v_i,$$

which finally implies $\sigma \in \mathsf{RefSet}_1(\mathcal{D}\rho)$. Since $\mathsf{RefSet}_1(\mathcal{D}\rho) \subseteq F$ by definition of ideal compression, and $F \subseteq \mathsf{RefSet}_1(\mathcal{D}\rho)$ by the above argument, we have $F = \mathsf{RefSet}_1(\mathcal{D}\rho)$.

Exercise 8.7

By Exercise 5.14, one has

$$\mathsf{Span}[\mathsf{RefSet}\,\rho \otimes \sigma] = \mathsf{Span}[\mathsf{RefSet}\,\rho \otimes \mathsf{RefSet}\,\sigma].$$

In the special case where $\rho = \psi \in \mathsf{PurSt}(\mathrm{A})$ and σ is completely mixed, one has $\mathsf{RefSet}(\rho) = \{\psi\}$, and then

$$\mathsf{Span}[\mathsf{RefSet}\,\psi \otimes \sigma] = \mathsf{Span}[\{k\psi\} \otimes \mathsf{St}(\mathrm{B})]$$
$$= \{\psi \otimes \tau | \tau \in \mathsf{St}_\mathbb{R}(\mathrm{B})\}.$$

This implies that $\mathsf{RefSet}[\psi \otimes \sigma] = \{\psi\} \otimes \mathsf{St}(\mathrm{B}) \simeq \mathsf{St}(\mathrm{B})$. Let us then verify that $\mathcal{E} := e_\mathrm{A} \otimes \mathcal{I}_\mathrm{B}$ and $\mathcal{D} := \psi \otimes \mathcal{I}_\mathrm{B}$ satisfy equation (8.7). Indeed, for any $\tau \in \mathsf{RefSet}[\psi \otimes \sigma]$ one has $\tau = \psi \otimes v$, with $v \in \mathsf{St}(\mathrm{B})$. Thus

$$\mathcal{D}\mathcal{E}\tau = \psi \otimes [(e|\psi)v]$$
$$= \psi \otimes v$$
$$= \tau.$$

On the other hand, $\mathcal{E}(\psi \otimes \sigma) = \sigma$ is completely mixed, and then $\mathcal{E}\mathcal{D} =_\sigma \mathcal{I}_\mathrm{B}$ iff $\mathcal{E}\mathcal{D} = \mathcal{I}_\mathrm{B}$. Let us then consider a general state $v \in \mathsf{St}(\mathrm{B})$. We have

$$\mathcal{E}\mathcal{D}v = (e_\mathrm{A} \otimes \mathcal{I}_\mathrm{B})(\psi \otimes v)$$
$$= v,$$

namely $\mathcal{E}\mathcal{D} = \mathcal{I}_\mathrm{B}$.

9

Three No-go Theorems

Quantum information theory is a wide area of investigation, which includes two main subjects: quantum computation and quantum communication. The success of quantum computation theory is due to a positive result: Shor's algorithm.[1] The algorithm provides factorization of integers in a much more efficient way with respect to what is achieved not only by present-day computers, but even by the most powerful conceivable computers based on classical physics. On the other hand, the success of quantum communication is due to the intrinsic security of ideal quantum cryptographic protocols, and thus relies on a negative result, that is the impossibility of reading quantum information without scrambling it in an unrecoverable way.

The no information without disturbance property spawns other no-go theorems, which represent some of the most famous and classic results in quantum information theory. Among these results one can certainly list the no-cloning theorem,[2] stating the impossibility of copying quantum information. This impossibility is clearly a consequence of the impossibility of extracting information without disturbing it. Indeed, if we could copy exactly the state of a system, reproducing it on another system of the same type, then we would also be able to iterate this process, and produce as many copies as we want. We could then extract all the information we are interested in from the copies, keeping the original intact.

This argument is so simple, but hides a very deep point about quantum information, and more generally about any type of information that can be purified. Suppose indeed that there exists a measurement allowing us to determine the state of a system, with absolute precision. We could then use such a measurement on the system, and once we know its exact state, we can in principle re-prepare the same state as many times as we want. If copying the state is forbidden within a theory, we must then conclude that, in that theory, measuring the state is forbidden, too! This is exactly the case in theories with purification, and in particular in quantum theory.[3]

In quantum theory, determining the state of a system is impossible. We can only have some information about what state our system is prepared in. The information may be accurate, but it will never be certain. This could be practically the case in classical theory, too. However, in our quantum world, we have something stronger than a practical impossibility. Suppose indeed we have some prior knowledge about our state, and we want to test it against a measurement. In the classical case we can confirm such an expectation,

---

[1] Shor (1997).

[2] Dieks (1982); Wootters and Zurek (1982); Yuen (1986).

[3] D'Ariano and Yuen (1996).

while in the quantum case we can only falsify our knowledge. In other words, even if an event occurs which is compatible with our prior knowledge, this will never be sufficient to confirm us in our belief about the state. What is certain are only negative answers.

A second relevant no-go that follows from the impossibility of extracting information without disturbing the state is the no-programming theorem,[4] which states the impossibility of building a quantum device that accepts two input states, a *data state* and a *program state*, and depending on the program performs on the data any unitary transformation exactly. This negative result shows us a further difference between classical and quantum information theory. The success of modern computation is indeed based – besides the astonishing progress in silicon technology – on the paradigm of von Neumann of "program as data," which implements the idea of the universal Turing machine: a classical computer can be programmed to do any computation, by supplying it with a suitable program encoded as if it was a piece of data. While processing this program along with the actual data through a universal program, the computer performs the desired processing on the input. This simple architecture is impossible in quantum theory. As a matter of fact, universal quantum computation can be achieved approximately, the approximation improving as the dimension of the program is increased.

Finally, we will discuss a no-go theorem which characterizes quantum cryptography: the no-bit-commitment theorem. This theorem is very popular in the quantum information community, and was even used as a principle in attempts at a reconstruction of quantum theory.[5] A bit-commitment protocol is meant to allow one party, Alice, to send a bit to a second party, Bob, in such a way that Bob cannot read the bit until Alice allows for its disclosure, while Alice cannot change the value of the bit after she encoded it.

The bit-commitment protocal is a very important primitive in cryptography, and perfectly secure protocols are known to be impossible in classical information theory. This is the case in quantum theory as well, but the proof requires rather sophisticated tools to take into account every possible sequence of transformations that each party can apply to the quantum systems involved in the protocol. The proof for general theories with purification involves a very important characterization theorem that provides a handy mathematical representation of strategies and is thus very instructive, both for its simplicity and for its insightful nature.

It is now time to see how all these astonishing features come out of the principles that we stated.

9.1 No Cloning

A *cloning channel* for a set of states $\{\rho_i\}_{i \in X}$ of a system A is a channel $\mathcal{C}$ from A to the composite system AA′, with A′ operationally equivalent to A, such that

$$\mathcal{C}|\rho_i)_A = |\rho_i)_A|\rho_i)_{A'}. \tag{9.1}$$

---

[4] Nielsen and Chuang (1997).
[5] Clifton *et al.* (2003).

If such a cloning channel exists, we say that the states $\{\rho_i\}_{i \in X}$ are *perfectly cloneable*. Notice that, if a cloning channel exists, we can apply it twice to the unknown state ρ_i, thus getting three identical copies of it, namely

$$(\mathcal{C} \otimes \mathcal{I}_{A'})\mathcal{C}|\rho_i) = |\rho_i)_A|\rho_i)_{A'}|\rho_i)_{A''}, \tag{9.2}$$

and clearly, iterating the procedure, we can produce as many copies as we desire.

We now show that in a theory with purification a spanning set of states (in particular, the set of pure states of a system A) cannot be perfectly cloned. The proof is based on the equivalence of perfect cloneability and perfect discriminability.[6] This result actually holds in any convex theory where all "measure-and-prepare" channels are allowed, without the need of causality and local discriminability.

Theorem 9.1 (Cloning-discrimination Equivalence)   *In a convex theory where all "measure-and-prepare" channels are allowed, the deterministic states* $\{\rho_i\}_{i \in X} \subset \mathsf{St}_1(A)$ *are perfectly cloneable if and only if they are perfectly discriminable.*

Proof   Suppose that the states $\{\rho_i\}_{i \in X}$ can be perfectly cloned, and consider a general binary discrimination strategy between two states $\rho_j, \rho_l, j \neq l$, corresponding to a binary observation test $\{a_j, a_l\}$. Define the worst-case error probability as

$$p_{wc} := \max\{p(l|j), p(j|l)\} \qquad p(k|i) := (a_k|\rho_i)_A, \tag{9.3}$$

and take the optimal discrimination, corresponding to the minimum over all binary tests

$$p_{wc}^{(opt)} := \min_{a_j, a_l} p_{wc}, \tag{9.4}$$

that is achieved by the optimal test $\{a_j^{(opt)}, a_l^{(opt)}\}$. Now, if a cloning channel exists we can apply it twice to the unknown state ρ_i, thus getting three identical copies as in Eq. (9.2). After performing three times the optimal test, we can use a majority voting discrimination strategy, corresponding to a new binary discrimination test $\{b_j, b_l\}$ defined as

$$(b_j|_A := [(A_{jjl}|_{AA'A''} + (A_{jlj}|_{AA'A''} + (A_{ljj}|_{AA'A''} + (A_{jjj}|_{AA'A''}](\mathcal{C} \otimes \mathcal{I}_{A'})\mathcal{C},$$
$$(b_l|_A := [(A_{jll}|_{AA'A''} + (A_{ljl}|_{AA'A''} + (A_{llj}|_{AA'A''} + (A_{lll}|_{AA'A''}](\mathcal{C} \otimes \mathcal{I}_{A'})\mathcal{C}, \tag{9.5}$$

where $(A_{xyz}|_{AA'A''} := (a_x^{(opt)}|_A (a_y^{(opt)}|_{A'} (a_z^{(opt)}|_{A''}$. The error probabilities for the observation test $\{b_j, b_l\}$ are given by

$$p'(i|j) = f(p^{(opt)}(i|j)) \qquad f(x) := x^2(3 - 2x), \tag{9.6}$$

where $p^{(opt)}(i|j) := (a_i^{(opt)}|\rho_j)_A$. Since f is a non-decreasing function for $x \in [0, 1]$, we also have $p'_{wc} = f\left(p_{wc}^{(opt)}\right)$. Then, since $p_{wc}^{(opt)}$ is the minimum error probability, by definition $p'_{wc} \geq p_{wc}^{(opt)}$. The only solutions of the inequality $f(x) \geq x$ are $x = 0$ and $x \in [1/2, 1]$, and, since $p_{wc}^{(opt)}$ must be in the interval $[0, 1/2)$ (see Problem 3.1), we obtain $p_{wc}^{(opt)} = 0$. This means that any pair of states from the set $\{\rho_i\}_{i \in X}$ can be perfectly discriminated. In turn, this implies that using $|X| - 1$ pairwise tests we can perfectly discriminate all the states $\{\rho_i\}_{i \in X}$

---

[6] The proof was originally provided in Chiribella *et al.* (2008b).

(see Exercise 9.1). This proves that perfect cloneability of a set of states implies its perfect discriminability. If the theory contains all possible "measure-and-prepare" channels, the converse is obviously true: if the states can be perfectly discriminated by an observation test $\{a_i\}_{i\in X}$, then the measure-and-prepare channel $\mathcal{C} := \sum_{i\in X}|\rho_i)_A|\rho_i)_{A'}(a_i|_A$ is a cloning channel.

□

Since measure-and-prepare channels can be obtained by conditioning the choice of a preparation test on the outcome of an observation test, any causal theory satisfies the hypotheses of the previous theorem, which then implies:

Corollary 9.2 (Cloning-discrimination Equivalence in Causal Theories) *In a causal theory the states $\{\rho_i\}_{i\in X} \subset \mathsf{St}_1(A)$ are perfectly cloneable if and only if they are perfectly discriminable.*

In the case of causal theories with purification, the results proved so far imply the following no-cloning statement.

Corollary 9.3 (No Cloning of Pure States) *In a causal theory with purification, a cloning channel for a spanning set of states cannot exist. In particular, pure states cannot be cloned.*

This is indeed an immediate consequence of Corollary 9.2 combined with Theorem 9.1 and Exercise 10.1, where we prove that in a causal theory with purification it is impossible to discriminate all pure states by a single observation test.

Exercise 9.1 Prove that if the set of states $\{\rho_i\}_{i\in X} \subseteq \mathsf{St}_1(A)$ is such that any pair ρ_j, ρ_l with $j \neq l$ is perfectly discriminable, then the set $\{\rho_i^{\otimes(|X|-1)}\}_{i\in X} \subseteq \mathsf{St}_1(A_1 A_2 \ldots A_{|X|-1})$ is perfectly discriminable.

Exercise 9.2 [Cloning and signaling] Prove that Nick Herbert made a mistake.[7]

9.2 No Programming

The second no-go theorem that we introduce is no programming. This theorem states that, differently from the classical case, in quantum theory there is no *universal processor*. A universal processor for a system A is a channel $\mathcal{C} \in \mathsf{Transf}_1(AP)$ that can be instructed to perform any processing $\mathcal{D}$ on system A by providing it with a suitable *program* state σ of the system P. In mathematical terms, for every target channel $\mathcal{D} \in \mathsf{Transf}_1(A)$ there must exist a program state $\sigma \in \mathsf{St}_1(P)$ such that the following equality holds:

$$
\sigma - P\boxed{\mathcal{C}}\,{}^{A}_{P} - e \;\; = \;\; {}^{A}\!-\!\boxed{\mathcal{D}}\!-\!{}^{A} . \tag{9.7}
$$

As a consequence of the quantum no-programming theorem, it is impossible to encode all algorithms involving a finite number of input target systems on the states of a finite

[7] Herbert (1982).

program system P. This result[8] can be extended to causal theories with purification and local discriminability where systems have infinitely many pure states – or equivalently, by transitivity, an infinite group of reversible transformations.

As a matter of fact, in causal theories with purification and local discriminability *probabilistic* programming is possible: the program state for a channel $\mathcal{D}$ is simply obtained by applying the channel to a faithful state, namely

$$\sigma \begin{array}{c} P_1 \\ P_2 \end{array} = \Phi \begin{array}{c} P_1 \quad P_1 \\ P_2 \boxed{\mathcal{D}} \end{array} ,$$

and then performing the probabilistic teleportation protocol. Indeed, this scheme provides a probabilistic universal processor with the following structure:

$$\begin{array}{c} A \quad\quad A \\ P_1 \boxed{\mathcal{C}} P_1 \\ P_2 \quad\quad P_2 \end{array} = \begin{array}{c} A \boxed{B_x} \Psi \quad S \; P_1 \\ P_1 \quad\quad\quad\quad\quad\quad\quad \\ P_2 \quad\quad\quad S \quad\quad P_2 \end{array} ,$$

where $P_1 \sim P_2 \sim A$, $\mathcal{S}$ denotes the swap gate $\mathcal{S}|\psi\rangle|\varphi\rangle = |\varphi\rangle|\psi\rangle$, Ψ is an arbitrary state, and $\{B_x\}_{x\in X}$ is an observation test such that

$$\Phi \begin{array}{c} A \\ P_1 \boxed{B_0} \\ A \end{array} = p_0 \; A \boxed{\mathcal{I}} A ,$$

as in Eq. (7.15). However, the following theorem tells us that the above-mentioned programming procedure cannot in general be extended to a deterministic one by any means. Indeed, the theorem proves that programming a set of reversible channels requires perfectly discriminable program states. It is then clear that for theories where the group of reversible transformations of a system is infinite, perfect programming is impossible.

Theorem 9.4 (Perfect Discriminability of Program States)   *Let $\{\mathcal{U}_i\}_{i\in X}$ be a set of reversible channels on A, and $\{\eta_i\}_{i\in X}$ be a set of pure states of P. If there exists a channel $\mathcal{R} \in$ Transf(AP) such that*

$$\eta_i \begin{array}{c} A \quad A \\ P \boxed{\mathcal{C}} P \\ e \end{array} = A \boxed{\mathcal{U}_i} A \tag{9.8}$$

then the states $\{\eta_i\}_{i\in X}$ are perfectly discriminable.

Proof   Take a unitary dilation of $\mathcal{C}$, with pure state $\varphi_0 \in St_1(C)$ and reversible channel $\mathcal{U} \in$ Transf(APC). Upon defining the pure states $\varphi_i := \eta_i \otimes \varphi_0$ we have

$$\varphi_i \begin{array}{c} A \quad A \\ PC \boxed{\mathcal{U}} PC \\ e \end{array} = A \boxed{\mathcal{U}_i} A . \tag{9.9}$$

---

[8] The theorem was originally proved by Nielsen and Chuang (1997).

Since the left-hand side member of Eq. (9.8) is a dilation of the reversible transformation $\mathcal{U}_i$, by the uniqueness of the reversible dilation (see Exercise 7.4) there must be a pure state $\psi_i \in \mathsf{St}_1(\mathrm{PC})$ such that

$$
\begin{array}{c}
\underset{\varphi_i}{\overset{A}{\boxed{\mathcal{U}}}} \;=\; \underset{\psi_i}{\overset{A\;\boxed{\mathcal{U}_i}\;A}{}}
\end{array}
\tag{9.10}
$$

By applying $\mathcal{U}_i^{-1}$ to the left and $\mathcal{U}^{-1}$ to the right on both sides of Eq. (9.10), one has

$$
\begin{array}{c}
\overset{A\;\boxed{\mathcal{U}_i^{-1}}\;A}{\underset{\varphi_i}{}} \;=\; \underset{\psi_i}{\overset{A\;\boxed{\mathcal{U}^{-1}}\;A}{}}
\end{array}
\tag{9.11}
$$

Composing Eqs. (9.10) and (9.11) we then obtain

$$
\begin{array}{c}
\underset{\varphi_i}{\overset{A\;\boxed{\mathcal{U}}\;A}{}}\;\boxed{\mathcal{U}^{-1}} \;=\; \underset{\varphi_i}{\overset{A\;\boxed{\mathcal{U}_i}\;A}{}}\;\boxed{\mathcal{U}_i^{-1}}
\end{array}
\tag{9.12}
$$

This means that by iterating the application of $\mathcal{U}$ and $\mathcal{U}^{-1}$ we can obtain an unbounded number of copies of $\mathcal{U}_i$ and $\mathcal{U}_i^{-1}$. Now, if $\mathcal{U}_i$ and $\mathcal{U}_j$ are different, the probability of error in discriminating among them using N copies decreases to zero as N runs to infinity (this can be seen by repeating N times the optimal test and using majority voting, as in the proof of Theorem 9.1). Since programming the transformations $\{(\mathcal{U}_i \otimes \mathcal{U}_i^{-1})^{\otimes N}\}$ and discriminating among them is a particular strategy for discrimination among the program states φ_i, the latter must be perfectly discriminable. Finally, since the states $\varphi_i = \eta_i \otimes \varphi_0$ are perfectly discriminable, the program states $\{\eta_i\}$ must also be.   □

Corollary 9.5 (No-programming Theorem)   *In a causal theory with purification and local discriminability, exact programming is forbidden for systems having infinitely many pure states.*

Exercise 9.3   Prove that using mixed program states $\{\rho_i\}$ cannot help in reducing the number of perfectly discriminable states needed in the program system P.

9.3 No Bit Commitment

Bit commitment is an important primitive cryptographic task, with various applications in communication and computation theory. The task involves two parties, Alice and Bob. Alice is supposed to store one bit for Bob to access later, after Alice's *disclosure*. A bit-commitment protocol is *binding* if Alice cannot change the value of the stored bit before disclosing it to Bob, while it is *concealing* if Bob cannot access the bit until Alice discloses it. Bit commitment is possible within a given theory if there exists a protocol that is both binding and concealing.

In the classical case, bit-commitment protocols exist, whose binding and concealing properties rely on the practical impossibility of efficiently solving some given problems, like calculating discrete logarithms or inverting functions in a given class. In the quantum case a bit-commitment protocol that is both concealing and binding is not possible (see the notes at the end of the chapter for a brief historical account).

We will now show that the impossibility proof can be carried out in the framework of operational probabilistic theories under the assumptions of causality, local discriminability, and purification.[9] The proof will be given for the impossibility of a protocol that is *exactly* binding and concealing. Extending the analysis to *arbitrarily* binding and concealing protocols, as was done in the quantum cryptography literature, is then just a technical matter, and will not be explored here.

9.3.1 Quantum Strategies: Channels with Memory

A quantum protocol for a two-party multi-round processing task is a sequence of operations performed by the two parties, who exchange a system at each round, while keeping some system with themselves for use in subsequent rounds. This implies that the full protocol can be depicted as in the following diagram:

$$(9.13)$$

If we now split the above circuit into two parts, the top one representing Alice's sequence of operations and the bottom one representing Bob's, we end up with two diagrams representing Alice's and Bob's *strategies*, respectively. As an example, Alice's strategy is represented by the following diagram:

$$(9.14)$$

[9] The proof follows the lines of that provided in Chiribella *et al.* (2010a).

The above scheme can be applied also to protocols involving the exchange of classical information. Indeed, classical messages can be modeled by perfectly discriminable states, while classical channels can be modeled by measure-and-prepare channels where the observation test is discriminating, and the prepared states are perfectly discriminable. The fact that some systems can only be prepared in perfectly discriminable states is usually referred to as the "communication interface" of the protocol.[10]

The structure of diagram (9.14) is the general structure of a *channel with memory*, because it exploits memory systems – the systems that in diagram (9.14) are labeled F_i with $i \geq 2$ – that store some information output by C_i at a given step i to be used at the following steps.[11]

In a causal theory with purification and local discriminability, a channel with memory $C^{(N)}$ as the one in diagram (9.14) can always be dilated to a reversible one by dilating the channels $C_1, \ldots, C_N$, including the output dilating systems D_{j+1} in the memory systems $F'_{j+1} := F_{j+1}D_{j+1}$ and discarding all of them after the final step N, as in the following diagram:

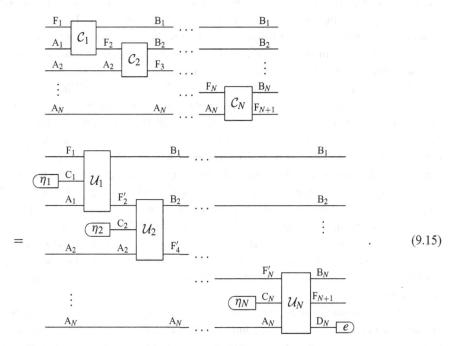

$$\text{(9.15)}$$

Since the realization provided in diagram (9.15) is the pure and reversible dilation of $C^{(n)}$, this realization is essentially unique, thanks to the uniqueness of the pure and reversible dilation of Exercise 7.4. This observation is crucial for the no-bit-commitment theorem provided in the next section.

---

[10] See D'Ariano *et al.* (2007) and Chiribella *et al.* (2013b).

[11] Channels with memory were studied in the quantum case from different points of view in Kretschmann and Werner (2005); Johnson and Feige (2007); Chiribella *et al.* (2009).

9.3.2 The No-bit-commitment Theorem

Sequences of channels with memory can be used to describe sequences of moves of a given party in a cryptographic protocol or in a multiparty game.[12] In this scenario, the memory systems are the private systems available to a party, while the other input–output systems are the systems exchanged in the communication with other parties. In this context, the uniqueness of the realization of a causal channel directly implies the impossibility of tasks like perfectly secure bit commitment.[13] A proof in the general case is given in the following.

Theorem 9.6 (No Bit Commitment)   *In a theory with purification, if an N-round protocol is perfectly concealing, then there is perfect cheating.*

Proof   We first prove the thesis for protocols that do not involve the exchange of classical information. Let $\mathcal{A}_0, \mathcal{A}_1 \in \mathsf{Transf}(A_1 \ldots \to A_N, B_1 \ldots B_{N-1}B_N F_{N+1})$ be two N-partite channels with memory (here the last output system of both channels is the bipartite system $B_N F_{N+1}$), representing Alice's moves to encode the bit value $b = 0, 1$, respectively. The system F_{N+1} is the system sent from Alice to Bob at the disclosure phase of the protocol (called the *opening*) in order to unveil the value of the bit. Now, take two pure and reversible dilations for $\mathcal{A}_0$ and $\mathcal{A}_1$, given by $(C, D, H_0, \mathcal{V}_0)$ and $(C, D, H_1, \mathcal{V}_1)$, respectively, as in diagram (9.15) where $C := C_1 C_2 \ldots C_N$, $D := D_N$, and $H_i := \eta_1^{(i)} \otimes \eta_2^{(i)} \cdots \otimes \eta_N^{(i)} \in \mathsf{PurSt}_1(C)$, while $\mathcal{V}_i$ correspond to the reversible channels with memory in the realization of diagram (9.15), obtained by the concatenation of the reversible channels $\mathcal{U}_j^{(i)}$ for $1 \leq j \leq N$. If the protocol is perfectly concealing, then the reduced channels $(e|_{F_{N+1}} \mathcal{A}_i$ before the opening phase must be indistinguishable, namely $(e|_{F_{N+1}} \mathcal{A}_0 = (e|_{F_{N+1}} \mathcal{A}_1 =: \mathcal{C}$. Since $\mathcal{V}_0$ and $\mathcal{V}_1$ are also two dilations of the channel $\mathcal{C}$, by the uniqueness of the reversible dilation of Exercise 7.4 there is a channel $\mathcal{R} \in \mathsf{RevTransf}(F_N C_{\mathrm{out}})$ such that $\mathcal{V}_1 = \mathcal{R}\mathcal{V}_0$. Applying the channel $\mathcal{R}$ to her private systems, Alice can switch from $\mathcal{V}_0$ to $\mathcal{V}_1$ just before the opening, and conversely applying $\mathcal{R}^{-1}$ she can switch from $\mathcal{V}_1$ to $\mathcal{V}_0$. Alice then just discards the auxiliary system D, yielding the channel $\mathcal{A}_1$ in the first case and $\mathcal{A}_0$ in the second case, respectively. The cheating is perfect, since Alice can play the strategy $\mathcal{V}_0$ until the end of the commitment, and decide the bit value before the opening without being detected by Bob.

The above reasoning can be extended to N-round protocols involving the exchange of classical information. Indeed, if the communication interface of the protocol involves non-trivial exchange of classical information, we can proceed as follows. We first take the reversible dilations $\mathcal{V}_0, \mathcal{V}_1$ and the channel $\mathcal{R}$ such that $\mathcal{V}_1 = \mathcal{R}\mathcal{V}_0$. In order to comply with the communication interface of the protocol, one then composes $\mathcal{V}_0$ and $\mathcal{V}_1$ with classical channels on all systems that carry classical information, thus obtaining two channels $\mathcal{E}_0$ and $\mathcal{E}_1$ that are no longer reversible, as represented in the following diagram:

---

[12] See Johnson and Feige (2007) for the case of quantum games.

[13] For the exact definition of the problem see D'Ariano *et al.* (2007); Chiribella *et al.* (2013b) and references therein.

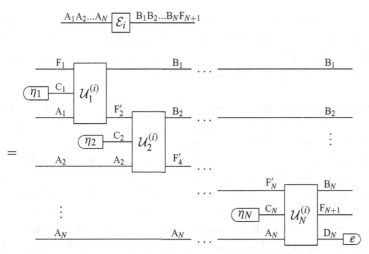

Since $\mathcal{V}_1 = \mathcal{R}\mathcal{V}_0$, also the channels $\mathcal{E}_0, \mathcal{E}_1$ still satisfy $\mathcal{E}_1 = \mathcal{R}\mathcal{E}_0$. Discarding the auxiliary system C_{out} and applying a measure-and-prepare channel $\mathcal{F}$ on the system F_N if this is required by the communication interface, Alice can then obtain the channels $\mathcal{E}_0$ or $\mathcal{E}_1$. Again, the existence of a reversible channel $\mathcal{R}$ such that $\mathcal{E}_1 = \mathcal{R}\mathcal{E}_0$ allows Alice to decide the value of the bit right before the opening, without being detected. $\qquad\square$

9.4 Summary

In this chapter we proved three important no-go theorems of quantum information: no-cloning, no-programming and no-bit-commitment. The theorems are proved in the framework of OPTs, using only our principles, thus highlighting their conceptual power.

Notes

No Cloning in Non-causal Theories As discussed in Section 9.1, a sufficient condition for the no-cloning theorem to hold in a theory with purification is causality. However, this result is just a corollary of Theorem 9.1 which has the weaker hypotheses of convexity and free measure-and-prepare tests. There exists a relevant class of non-causal theories that satisfy these hypotheses, that are *second-order theories*. From every causal theory Θ where the states-transformations isomorphism holds, one can construct its second-order theory Θ' by regarding the set of transformations $\mathsf{Transf}(A \to B)$ from A to B as the set of "states" $\mathsf{St}'(A \to B)$ of the second-order system "$A \to B$," with deterministic states corresponding to channels. An example is provided by the theory introduced in Section 5.7, where states are quantum operations. One can prove that performing an observation test on a state $\mathcal{C} \in \mathsf{St}'(A \to B)$ in this case can be interpreted in the underlying causal theory Θ as applying the transformation $\mathcal{C} \in \mathsf{Transf}(A \to B)$ to an input state

$\sigma \in St_1(AC)$, and subsequently performing an observation test $\{b_i\}_{i \in X}$ on the output state $(\mathcal{C} \otimes \mathcal{I}_C)|\sigma)_{AC}$ (see e.g. Chiribella *et al.*, 2015). Of course, since the theory Θ is causal, one can use conditioning and perform a channel $\mathcal{C}_i$ that depends on the outcome i. This provides the realization of an arbitrary measure-and-prepare channel in the non-causal theory Θ'.

No Programming and Universal Quantum Computation   The no-programming theorem apparently clashes with the existence of a universal quantum Turing machine, which is by definition a programmable quantum processor. However, it was proved in Bernstein and Vazirani (1993) that a universal quantum Turing machine exists, and universal quantum computation is also possible in the quantum circuit model, which was proved to be equivalent to the quantum Turing machine by Yao (1993). The reason for this apparent clash of results is that the whole quantum computation is based on the notion of approximate programming: in other words, a universal quantum Turing machine is required to approximate any quantum channel not exactly, but with arbitrarily small error, with a cost in term of resources that grows polynomially with respect to the inverse of the size of the error. For the case of programmable POVMs, the optimal universal quantum processor was derived by D'Ariano and Perinotti (2005). The universality theorems of quantum computation theory prove that such a universal processor exists, thus reducing the practical relevance of the no-programming theorem, while preserving its foundational importance.

Unconditionally Secure Quantum Bit Commitment   Analogously to what happens in the case of the universal processor, the fact that a perfect bit-commitment protocol does not exist in causal theories with purification and local discriminability does not mean that, if we allow for small probabilities of a successful cheating, we cannot find an effective protocol. Indeed, the main technical difficulties in proving the no bit commitment in the quantum case arise from relaxing the requirement that the protocol is perfectly secure. A family of increasingly secure protocols allowing for arbitrary security is called *unconditionally secure* bit commitment. In the quantum case, unconditionally secure bit commitment can be proved impossible, and this is the actual quantum no-go theorem. However, extending the impossibility proof to unconditionally secure protocols in the case of causal theories with purification is not only a matter of technicalities. The missing ingredient to this end is a theorem that ensures that, given two states that are close in the operational norm distance, one can find a pair of purifications of the same states that have the same distance. This theorem holding in the quantum case was proved by Uhlmann (1977).

Historically, the analysis of the quantum protocol started in the 1990s, with the results of Mayers (1997) and Lo and Chau (1997) showing the impossibility of quantum protocols that are both concealing and binding. Criticisms of these security proofs have been raised by H. P. Yuen (see Yuen, 2012 and references therein). A complete impossibility proof required a thorough classification of all protocols that can be performed in the quantum scenario, and a complete account of all the possibilities was finally considered in the proof of D'Ariano *et al.* (2007). A different proof, with a significantly shorter analysis, was provided by Chiribella *et al.* (2013b), thanks to the use of the theory of *quantum combs* (Chiribella *et al.*, 2008c, 2009), which is particularly suited to the description of multi-round quantum communication protocols.

Solutions to Selected Problems and Exercises

Exercise 9.1

Suppose that the program state ρ_i for the unitary $\mathcal{U}_i$ as in diagram (9.9) corresponds to the following mixture $\rho_i = \sum_j p_j^{(i)} \psi_j^{(i)}$. Since reversible transformations are atomic, this means that each pure state $\psi_j^{(i)}$ must work as a program for $\mathcal{U}_i$. But the above theorem implies that, whichever choice we make, the pure states $\{\varphi_{j_i}^{(i)}\}_{i \in X}$ must be perfectly discriminable.

10

Perfectly Discriminable States

Discriminating states is a fundamental task in both physics and information theory. In this chapter we will analyze the most favorable instance of this task, involving states that can be discriminated perfectly, i.e. without error. We will start by introducing the notion of perfect discriminability of states and by linking it with the perfect discriminability of faces of the state space. Then, we will use our principles to prove a key property, known as no disturbance without information. This property will lead us to some pivotal features of perfect discriminability. For example, it will allow us to show that every set of perfectly discriminable states can be refined to a set of perfectly discriminable pure states. In addition, the no disturbance without information property will lead to a mathematical characterization of perfect discriminability as orthogonality. Finally, we will introduce the notion of maximal set of perfectly discriminable states and will prove a necessary and sufficient condition for maximality.

10.1 Perfect Discriminability of States

Suppose that Alice wants to communicate to Bob one out of N possible messages. To this purpose, she encodes the message into the state of a physical system, say A, and sends it to Bob. Upon receiving system A, Bob will measure it, trying to find out which message was encoded in it. The natural question is: can Bob decode Alice's message without errors?

Let us describe this situation in the language of operational theories. Alice's encoding is just an assignment of a deterministic state $\alpha_i \in \mathsf{St}_1(A)$ to every possible message $i \in \{1, \ldots, N\}$. Bob's decoding consists of a measurement with N outcomes, say $\{b_i\}_{i=1}^N$, with the outcome j corresponding to the j-th message. If Alice encoded the i-th message, then the probability that Bob reads out the j-th message is given by

$$p(j|i) = (b_j|\alpha_i). \tag{10.1}$$

Clearly, the message is decoded without errors if and only if one has

$$(b_j|\alpha_i) = \delta_{ij} \qquad i,j \in \{1, \ldots, N\}. \tag{10.2}$$

When this is the case we say that the states $\{\alpha_i\}_{i=1}^N$ are *perfectly discriminable*. Note that, since the states are deterministic, the perfect discriminability condition of Eq. (10.2) is equivalent to the condition

$$(b_i|\alpha_i) = 1 \qquad \forall i \in \{1, \ldots, N\}. \tag{10.3}$$

This condition is trivially necessary for Eq. (10.2). It is also easy to see that it is sufficient: since each state α_i is deterministic, one has

$$\sum_{j=1}^{N}(b_j|\alpha_i) = 1 \qquad \forall i \in \{1,\ldots,N\},$$

which, combined with Eq. (10.3), implies

$$\sum_{j \neq i}(b_j|\alpha_i) = \sum_{j=1}^{N}(b_j|\alpha_i) - (b_i|\alpha_i)$$

$$= 1 - 1$$

$$= 0 \qquad \forall i \in \{1,\ldots,N\}.$$

In turn, this relation implies that $(b_j|\alpha_i) = 0$ whenever i and j are distinct, hence Eq. (10.2).

At the deepest level, perfect discriminability is a feature of faces, rather than states. Let us clarify this point: we call the faces $\{F_i\}_{i=1}^{N}$ *perfectly discriminable* iff for every choice of states $\{\alpha_i\}_{i=1}^{N}$ with $\alpha_i \in F_i, \forall i$, the states $\{\alpha_i, i = 1,\ldots,N\}$ are perfectly discriminable. Then, we have the following:

Proposition 10.1 *Let $\{\alpha_i\}_{i=1}^{N}$ be a set of states and, for every i, let F_i be the face identified by α_i. Then, the following statements are equivalent:*

1. the states $\{\alpha_i\}_{i=1}^{N}$ are perfectly discriminable;
2. the faces $\{F_i\}_{i=1}^{N}$ are perfectly discriminable.

Proof Suppose that the states $\{\alpha_i\}_{i=1}^{N}$ are perfectly discriminable and let $\{b_i\}_{i=1}^{N}$ be a measurement that discriminates perfectly among them. By the perfect discriminability condition, one must have $(b_i|\alpha_i) = 1$ for every $i \in \{1,\ldots,N\}$. By Exercise 8.1 this condition implies $(b_i|\alpha_i') = 1$ for every $\alpha_i' \in F_i$. By Eq. (10.3), one concludes that the states $\{\alpha_i'\}_{i=1}^{N}$ are perfectly discriminable. Since each state α_i' is an arbitrary element of F_i, this means that the faces $\{F_i\}_{i=1}^{N}$ are perfectly discriminable. The converse is trivial: if the faces $\{F_i\}_{i=1}^{N}$ are perfectly discriminable, then also the states $\{\alpha_i\}_{i=1}^{N}$ are perfectly discriminable. $\qquad\square$

10.2 No Disturbance Without Information

The features of state discrimination discussed in the previous paragraph were generic to every operational theory. We now turn our attention to features that depend critically on the validity of purification, local discriminability and the atomicity of composition.

Suppose that you are given a system in an unknown state ρ_x, which is promised to be one of the three states ρ_1, ρ_2, and ρ_3. The states have the property that (i) ρ_1 is perfectly discriminable from ρ_2, and (ii) ρ_3 is perfectly discriminable from every mixture of ρ_1 and ρ_2. Can you identify the unknown state?

One strategy would be to perform first a test that tells ρ_3 apart from ρ_1 and ρ_2, and later, if the outcome indicates that the state was either ρ_1 or ρ_2, perform a second test that discriminates between these two states. But can we find a test that tells ρ_3 apart from ρ_1 and ρ_2 without compromising the discriminability of these two states? In a generic theory, this is not the case.[1] However, it *is* the case in a theory satisfying our principles. Thanks to purification, local discriminability and atomicity of composition, every measurement that does not extract information about the states in a face of the state space can be implemented without disturbing those states. In particular, a measurement that tells ρ_3 apart from ρ_1 and ρ_2 but does not provide any information on whether the state was ρ_1 or ρ_2 can be implemented without disturbing these two states.

Before establishing the result, let us make precise the notions that come into play. Consider a random source of states, with average state ρ. Relative to this source, a measurement $\{a_x\}_{x\in X}$ extracts no information if the probabilities of its outcomes do not allow one to discriminate among the possible states, that is, if there exists a probability distribution $\{p_x\}_{x\in X}$ such that

$$(a_x|\sigma) = p_x \qquad \forall \sigma \in F_\rho , \forall x \in X .$$

Equivalently, the condition can be restated as

$$a_x =_\rho p_x \, e_A . \tag{10.4}$$

This means that the measurement outcomes have the same probability for every state in the face F_ρ. When this is the case, we say that the measurement *extracts no information upon input of ρ*.

In order to talk about disturbance, we have to specify what is the state after the measurement, i.e. we have to specify a test. We say that a test $\{A_x\}_{x\in X}$ implements the measurement $\{a_x\}_{x\in X}$ iff the probability of the events in the test are equal to the probabilities of the outcomes in the measurement, that is, iff

$$\underline{\quad A\quad}\boxed{a_x} = \underline{\quad A\quad}\boxed{A_x}\underline{\quad A'\quad}\boxed{e} \qquad \forall x \in X . \tag{10.5}$$

In general, the same measurement can be implemented by different tests. The question is whether, among all possible implementations, there is one that does not disturb the states in the source. Precisely, we say that the test $\{A_x\}_{x\in X}$ *has no disturbance upon input of ρ* iff $A' \equiv A$ in Eq. (10.5) and

$$\sum_{x\in X} A_x =_\rho \mathcal{I}_A . \tag{10.6}$$

This means that, if we ignore the outcome of the test, every state in the face F_ρ is left untouched.

---

[1] Think for example of a hypothetical system where normalized states form a square and the set of measurements consists of (a) measurements that discriminate states on one side of the square from states on the opposite side, and (b) measurements obtained by choosing at random among the measurement of type (a) and by coarse-graining over some of the outcomes. If ρ_1, ρ_2, and ρ_3 are three pure states on the vertices of the square, there is no way to discriminate them perfectly. Still, ρ_1 is perfectly discriminable from ρ_2 and ρ_3 is perfectly discriminable from every mixture of ρ_1 and ρ_3.

We are now ready to present the main result of the section:

Theorem 10.2 (No Disturbance Without Information) *If a measurement extracts no information upon input of ρ then it can be implemented through a test that has no disturbance upon input of ρ.*

As the first step towards the proof, we derive a lemma that is interesting in its own right:

Lemma 10.3 *Every measurement can be implemented through an atomic test, i.e. a test where each transformation is pure.*

Proof Pick a minimal purification of χ_A, say $\Psi \in \mathrm{PurSt}(AB)$ and define $\beta_x := (a_x \otimes \mathcal{I}_B)\Psi$. Take a purification of β_x, say Ψ_x for some[2] purifying system A'. By the state-transformation isomorphism (Theorem 7.7), there exists a test $\{\mathcal{A}_x\}_{x \in X}$ such that

$$
\begin{array}{c}
\Psi \!-\!\!\begin{array}{c} A \\ B \end{array}\!\!\!-\boxed{\mathcal{A}_x}\!-\! A' \\
\end{array}
\;=\;
\begin{array}{c}
\Psi_x \!\begin{array}{c} A' \\ B \end{array}
\end{array}
$$

Since each Ψ_x is pure and the state Ψ is faithful, each transformation $\mathcal{A}_x$ must be atomic. Moreover, applying the deterministic effect on system A', we obtain

$$
\begin{array}{c}
\Psi \!-\!\!\begin{array}{c} A \\ B \end{array}\!\!\!-\boxed{\mathcal{A}_x}\!-\! A' \!-\!\boxed{e}
\end{array}
\;=\;
\begin{array}{c}
\Psi_x \!-\! A' \!-\!\boxed{e} \\ B
\end{array}
\;\equiv\;
\begin{array}{c}
\Psi \!-\!\!\begin{array}{c} A \\ B \end{array}\!\!\!-\boxed{a_x}
\end{array}
$$

Since Ψ is faithful, this implies $e_{A'} \mathcal{A}_x = a_x$, meaning that the measurement $\{a_x\}$ is implemented through the test $\{\mathcal{A}_x\}$, as in Eq. (10.5). $\square$

Note that Lemma 10.3 expresses a highly non-classical feature: in classical theory, it is clearly impossible to implement a coarse-grained measurement through a pure test.

We are now ready to prove the no disturbance without information result

Proof of Theorem 10.2 Suppose that the measurement $\{a_x\}_{x \in X}$ extracts no information upon input of ρ, i.e. $a_x =_\rho p_x\, e_A$ for every outcome x. Due to the steering property (Proposition 7.5), the condition $a_x =_\rho p_x\, e_A$ is equivalent to

$$
\begin{array}{c}
\Psi \!-\!\!\begin{array}{c} A \!-\!\boxed{a_x} \\ B \end{array}
\end{array}
\;= p_x
\begin{array}{c}
\Psi \!-\!\!\begin{array}{c} A \!-\!\boxed{e} \\ B \end{array}
\end{array}
, \tag{10.7}
$$

where Ψ is a purification of ρ. Now, let $\{\mathcal{A}_x\}$ be the pure test that implements the measurement $\{a_x\}$ according to Lemma 10.3. For each x, define the state

$$
\begin{array}{c}
\Psi_x \!\begin{array}{c} A' \\ B \end{array}
\end{array}
:= \frac{1}{p_x}
\begin{array}{c}
\Psi \!-\!\!\begin{array}{c} A \\ B \end{array}\!\!\!-\boxed{\mathcal{A}_x}\!-\! A'
\end{array}
.
$$

---

[2] A priori, the purifying system could depend on x. However, by appending dummy pure states one can always choose a purifying system A' that is independent of x.

Thanks to the Atomicity of Composition, the state Ψ_x is pure. Moreover, its marginal on system B is equal to the marginal of the state Ψ; indeed, we have

$$
\left(\Psi_x\right)_{B}^{A'}\!-\!\boxed{e} \;=\; \frac{1}{p_x}\,\left(\Psi\right)_{B}^{A}\!-\!\boxed{\mathcal{A}_x}\!-\!^{A'}\!\boxed{e}
$$

$$
=\; \frac{1}{p_x}\,\left(\Psi\right)_{B}^{A}\!-\!\boxed{a_x}
$$

$$
=\; \left(\Psi\right)_{B}^{A}\!-\!\boxed{e}\;,
$$

having used Eq. (10.5) in the second equality and Eq. (10.7) in the third. Hence, Ψ_x and Ψ are two purifications of the same mixed state on system B. By the uniqueness of purification (Proposition 7.1), there exists a deterministic transformation $\mathcal{C}_x$ such that $(\mathcal{C}_x \otimes \mathcal{I}_B)\Psi_x = \Psi$. In other words, we have

$$
\frac{1}{p_x}\,\left(\Psi\right)_{B}^{A}\!-\!\boxed{\mathcal{A}_x}\!-\!^{A'}\!\boxed{\mathcal{C}_x}\!-\!^{A}\;=\;\left(\Psi\right)_{B}^{A}\;.
$$

which implies $\mathcal{C}_x\mathcal{A}_x =_\rho p_x \mathcal{I}_A$. This relation tells us that the test $\{\mathcal{A}'_x\}_{x\in X}$ defined by $\mathcal{A}'_x :=$ $\mathcal{C}_x\mathcal{A}_x$ is non-disturbing upon input of ρ. This test provides the desired realization of the measurement $\{a_x\}_{x\in X}$; indeed, we have

$$
^{A}\!-\!\boxed{a_x} \;=\; ^{A}\!-\!\boxed{\mathcal{A}_x}\!-\!^{A'}\!\boxed{e}
$$

$$
=\; ^{A}\!-\!\boxed{\mathcal{A}_x}\!-\!^{A'}\!\boxed{\mathcal{C}_x}\!-\!^{A}\boxed{e}
$$

$$
\equiv\; ^{A}\!-\!\boxed{\mathcal{A}'_x}\!-\!^{A}\boxed{e} \qquad \forall x \in X,
$$

having used the fact that $\mathcal{C}_x$ is deterministic in the second equality.   □

10.3 Perfect Discriminability Implies No Disturbance

The no disturbance without information property has a strong impact on the discriminability of states. First of all, it implies that the measurement that discriminates perfectly among a set of states can be implemented without disturbing all the faces corresponding to such states:

Proposition 10.4 (Perfect Discriminability Implies No Disturbance) *If the states $\{\rho_i\}_{i=1}^{N}$ are perfectly discriminable, then the measurement that discriminates between them can be implemented through a test that makes no disturbance upon input of ρ_i for every $i \in \{1,\ldots,N\}$.*

Proof Let $\{a_i\}_{i=1}^N$ be the measurement that discriminates perfectly among the states $\{\rho_i\}_{i=1}^N$, namely

$$(a_i|\rho_j) = \delta_{ij} \qquad \forall i,j \in \{1,\ldots,N\}. \tag{10.8}$$

Let F_j be the face identified by ρ_j and let ρ_j' be an arbitrary state in F_j. For every fixed j, it is immediate to see that Eq. (10.8) implies

$$(a_i \mid \rho_j') = \delta_{ij} \qquad \forall j \in \{1,\ldots,N\}.$$

Since ρ_j' is an arbitrary state in F_j, the above relation is equivalent to

$$a_i =_{\rho_j} \delta_{ij}\, e_{\mathrm{A}}. \tag{10.9}$$

In other words, the measurement extracts no information upon input of ρ_j. By the no disturbance without information property (Theorem 10.2), such a measurement can be implemented through a test $\left\{\mathcal{A}_i^{(j)}\right\}$ that has no disturbance upon input of ρ_j. Explicitly, this means that the test $\left\{\mathcal{A}_i^{(j)}\right\}$ satisfies the relations

$$e_{\mathrm{A}}\, \mathcal{A}_i^{(j)} = a_i \qquad \forall i \in \{1,\ldots,N\}. \tag{10.10}$$

and

$$\sum_{i=1}^N \mathcal{A}_i^{(j)} =_{\rho_j} \mathcal{I}_{\mathrm{A}}. \tag{10.11}$$

Combining Eqs. (10.9) and (10.10) we obtain $e_{\mathrm{A}}\mathcal{A}_i^{(j)} =_{\rho_j} \delta_{ij}\, e_{\mathrm{A}}$. In turn, combining this relation with Eq. (10.11) we obtain

$$\mathcal{A}_i^{(j)} =_{\rho_j} \delta_{ij}\mathcal{I}_{\mathrm{A}} \qquad \forall i,j \in \{1,\ldots,N\}. \tag{10.12}$$

Now, we constructed a test for each $j \in \{1,\ldots,N\}$, with the property that the j-th test does not disturb the states in F_j. However, what we want is a single test that has no disturbance on *every* face F_j. This is obtained by defining $\mathcal{A}_i := \mathcal{A}_i^{(i)}$. Such a definition implies the relation

$$\underset{\mathrm{A}}{\quad}\boxed{\mathcal{A}_i}\underset{\mathrm{A}}{\quad}\!\!\bigcirc\!\!\!e = \underset{\mathrm{A}}{\quad}\!\!\bigcirc\!\!\!a_i \qquad \forall i \in \{1,\ldots,N\},$$

and, therefore

$$\sum_{i=1}^N \underset{\mathrm{A}}{\quad}\boxed{\mathcal{A}_i}\underset{\mathrm{A}}{\quad}\!\!\bigcirc\!\!\!e = \underset{\mathrm{A}}{\quad}\!\!\bigcirc\!\!\!e\ .$$

Hence, the "everything not forbidden must be allowed" result (Proposition 7.10) ensures that the collection $\{\mathcal{A}_i\}_{i=1}^N$ is a test. By construction, the test $\{\mathcal{A}_i\}_{i=1}^N$ has no disturbance upon input of ρ_j for every $j \in \{1,\ldots,N\}$: indeed, Eq. (10.12) gives the relation

$$\mathcal{A}_j =_{\rho_j} \mathcal{I}_{\mathrm{A}} \qquad \forall j \in \{1,\ldots,N\},$$

which implies $\mathcal{A}_i =_{\rho_j} 0$ for $i \neq j$, hence the no disturbance condition $\sum_i \mathcal{A}_i =_{\rho_j} \mathcal{I}_{\mathrm{A}}$. $\quad\square$

Exercise 10.1   **[No joint discrimination of pure states]** Prove that it is impossible to discriminate all pure states of a system by a single observation test.

Equipped with the above proposition, we can prove a rather strong result:

Proposition 10.5 (Refined Discriminability Property)   *Let $\{\rho_i\}_{i=1}^N$ be a set of perfectly discriminable states and, for every i, let S_i be a set of perfectly discriminable states in F_i, the face identified by ρ_i. Then, the states in the set $S := \cup_{i=1}^N S_i$ are perfectly discriminable.*

Proof   It is enough to perform the test that discriminates among the states $\{\rho_i\}$ without disturbance and, depending on the outcome, perform the measurement that discriminates perfectly among the states in the set S_i.   □

Interestingly, the above property fails when the state space of a system is a square:

Exercise 10.2   Consider a hypothetical system for which the deterministic states form a square and the set of the allowed measurements includes all the measurements that discriminate perfectly between two opposite sides of the square. Show that this choice of state space and measurements is incompatible with the refined discriminability property of Proposition 10.5.

The refined discriminability property also allows one to show that the three states ρ_1, ρ_2, and ρ_3 in the example at the beginning of Section 10.2 are perfectly discriminable:

Exercise 10.3   Let ρ_1, ρ_2, and ρ_3 be three states of system A, with the properties that (i) ρ_1 is perfectly discriminable from ρ_2 and (ii) ρ_3 is perfectly discriminable from $\rho_{12} = \frac{1}{2}(\rho_1 + \rho_2)$. Show that ρ_1, ρ_2, and ρ_3 are perfectly discriminable.

Finally, the refined discriminability property implies that every set of perfectly discriminable mixed states can be refined to a set of perfectly discriminable pure states:

Proposition 10.6   *Let $\{\rho_i\}_{i=1}^M$ be a set of perfectly discriminable states. If one of the states in the set is not pure, then there exists another set of perfectly discriminable states with $N > M$ elements.*

Proof   Let ρ_1 be mixed, and let $(C, \mathcal{E}, \mathcal{D})$ be an ideal compression protocol for ρ_1. Note that, since ρ_1 is mixed, the system C contains at least two discriminable states, thanks to Proposition 4.1. Let $\{\gamma_k\}$ be a set of (more than two) perfectly discriminable states in C. Applying the decoding operation $\mathcal{D}$, one obtains the set $\{\mathcal{D}\gamma_k\}$, which consists of perfectly discriminable states in the face identified by ρ_1. By the refined discriminability property of Proposition 10.5, we obtain that $\{\mathcal{D}\gamma_k\} \cup \{\rho_i\}_{i=2}^M$ is a set of perfectly discriminable states, of cardinality larger than M.   □

Clearly, since a finite-dimensional system has a finite number of perfectly discriminable states, the process of refinement described in the previous proposition can be iterated only for a finite number of times. When no further refinement is possible one is left with a set of perfectly discriminable pure states. This argument proves the following.

Corollary 10.7 *Every set of perfectly discriminable states can be refined to a set of perfectly discriminable pure states.*

10.4 Orthogonality

Thanks to the no disturbance without information property, the operational notion of perfect discriminability can be proven to be equivalent to the mathematical notion of orthogonality.

Let us make this point explicit. We say that the states $\{\rho_i\}_{i=1}^{N}$ are *orthogonal* iff there exists a set of effects $\{a_i\}_{i=1}^{N}$ (not necessarily forming a measurement) such that

$$(a_i \mid \rho_j) = \delta_{ij} \qquad \forall i, j \in \{1, \dots, N\}. \tag{10.13}$$

Clearly, states that are perfectly discriminable are orthogonal. However, the converse may not be true, because the effects $\{a_i\}_{i=1}^{N}$ may not form a measurement (for example, one could have $\sum_{i=1}^{N} a_i \neq e_A$).

The notion of orthogonality has no clear operational meaning. Nevertheless, the refined discriminability property stated by Proposition 10.5 allows us to identify it with the operational notion of perfect discriminability:

Proposition 10.8 (Perfect Discriminability Equals Orthogonality) *Let $\{\rho_i\}_{i=1}^{N}$ be a set of states of a given system. The following are equivalent:*

1. the states $\{\rho_i\}_{i=1}^{N}$ are orthogonal;
2. the states $\{\rho_i\}_{i=1}^{N}$ are perfectly discriminable.

Proof Let $\{\rho_i\}_{i=1}^{N}$ be a set of orthogonal states and let $\{a_i\}_{i=1}^{N}$ be the corresponding set of effects. For each $i \in \{1, \dots, N\}$ consider the binary measurement $\{a_i, e - a_i\}$. Since by hypothesis $(a_i \mid \rho_j) = \delta_{ij}$, the test $\{a_i, e - a_i\}$ can perfectly discriminate ρ_i from any mixture of the states $\{\rho_j\}_{j \neq i}$. In particular, the state ρ_{M+1} can be perfectly discriminated from the mixture $\omega_M = \sum_{j=1}^{M} \rho_j / M$, for every $M < N$. Note that, by definition, the states $\{\rho_i\}_{i=1}^{M}$ belong to the face F_{ω_M}. We now prove by induction on M that the states $\{\rho_i\}_{i=1}^{M}$ are perfectly discriminable. This is true for $M = 1$ by hypothesis. Now, suppose that the states $\{\rho_i\}_{i=1}^{M}$ are perfectly discriminable. Since the state ρ_{M+1} is perfectly discriminable from ω_M, Proposition 10.5 guarantees that the states $\{\rho_i\}_{i=1}^{M+1}$ are perfectly discriminable. Taking $M = N - 1$ the thesis follows. □

The notion of orthogonality can be easily extended from states to faces: we say that the faces $\{F_i\}_{i=1}^{N}$ are *orthogonal* iff for every possible choice of states $\rho_i' \in F_i$, the states $\{\rho_i'\}_{i=1}^{N}$ are orthogonal. Collecting the results obtained in Propositions 10.1 and 10.8, we have the following:

Corollary 10.9 *Let $\{F_i\}_{i=1}^{N}$ be a set of faces of a given physical system. Then, the following are equivalent:*

1. *the faces $\{F_i\}_{i=1}^N$ are orthogonal;*
2. *the faces $\{F_i\}_{i=1}^N$ are perfectly discriminable.*

In other words, the notions of orthogonality and discriminability are equivalent even at the most fundamental level, involving faces of the state space.

Exercise 10.4   Let F be a face of $\mathsf{St}_1(F)$, and (C, B, A) an ideal encoding scheme for F. Then prove that $\{\alpha_i\}_{i=1}^k \subset F$ is a maximal set of perfectly discriminable states in F if and only if $\{A\alpha_i\}_{i=1}^k$ is a maximal set of perfectly discriminable states in AF.

10.5   Maximal Sets of Perfectly Discriminable States

A simple way to increase the number of perfectly discriminable states in a set $\{\rho_i\}_{i=1}^N$ is to append another state ρ_{N+1} that is orthogonal to all the others. When such a state exists, the original set can be extended to a larger set of perfectly discriminable states. Of course, such an extension is not always possible: since perfectly discriminable states are linearly independent, for a finite-dimensional system there can only be a finite number of perfectly discriminable states.

If there is no state ρ_{N+1} such that the states $\{\rho_i\}_{i=1}^{N+1}$ are perfectly discriminable, then we say that the set $\{\rho_i\}_{i=1}^N$ is *maximal*. By definition, a set of discriminable states is either maximal, or can be extended to a maximal set. For set of pure states, one can choose an extension that consists of pure states:

Proposition 10.10   *Every set of perfectly discriminable pure states can be extended to a maximal set of pure states. In particular, every pure state belongs to a maximal set of pure states.*

Proof   Let $\{\varphi_i\}_{i=1}^N$ be a non-maximal set of perfectly discriminable pure states. By definition, there exists a state σ such that $\{\varphi_i\}_{i=1}^N \cup \{\sigma\}$ is perfectly discriminable. Let φ_{N+1} be a pure state in F_σ. By Proposition 10.1, the states $\{\varphi_i\}_{i=1}^{N+1}$ will be perfectly discriminable. Since the dimension of $\mathsf{St}_\mathbb{R}(A)$ is finite and discriminable states are linearly independent, iterating this procedure for a finite number of times one must finally obtain a maximal set of pure states.   □

Intuitively, maximal sets are sets of states that occupy all the room available in the state space of the system. This intuition is made precise by the following necessary and sufficient condition:

Theorem 10.11   *A set of perfectly discriminable states $\{\rho_i\}_{i=1}^N$ is maximal if and only if the states $\omega = \sum_{i=1}^N p_i\rho_i$ with $p_i > 0$ for all i are completely mixed.*

Proof   We first prove that if ω is internal, then the set $\{\rho_i\}_{i=1}^N$ must be maximal. Indeed, if there existed a state ρ_{N+1} such that $\{\rho_i\}_{i=1}^{N+1}$ are perfectly discriminable, then clearly

ρ_{N+1} would be discriminable from ω. This is absurd because no state can be perfectly discriminated from an internal state.[3] Conversely, if $\{\rho_i\}_{i=1}^{N}$ is maximal, then ω is internal. If it were not, by the perfect discriminability principle, ω would be perfectly discriminable from some state ρ_{N+1}. By the refined discriminability property (Proposition 10.5), this would imply that the states $\{\rho_i\}_{i=1}^{N+1}$ are perfectly discriminable, in contradiction with the hypothesis that the set $\{\rho_i\}_{i=1}^{N}$ is maximal. $\qquad\square$

Maximal sets of pure states play a key role in the structure of quantum theory. In the Hilbert-space framework, they are identified with orthonormal bases. Operationally, we will see that they have a number of striking properties, such as:

1. if a measurement discriminates among a maximal set of pure states, then each effect in that measurement must be pure;
2. for every maximal set of pure states of system A, say $\{\alpha_i\}_{i=1}^{N}$, the invariant state χ_A can be written as

$$\chi_A = \frac{1}{N} \sum_{i=1}^{N} \alpha_i;$$

3. all the maximal sets of pure states for a given system have the same cardinality;
4. every maximal set of pure states for a given system can be converted into every other maximal set of pure states for the same system via a reversible transformation.

These properties, which are familiar from the Hilbert-space framework, are quite non-trivial in the general framework of OPTs. To discover how they can be proved from first principles, you will need to read further into the next few chapters. But before that, let us summarize what we have learnt in this one.

10.6 Summary

We first discussed the notion of perfectly discriminable states, which can be identified with the more fundamental notion of perfectly discriminable faces of the state space. The identification holds in arbitrary theories. We then proved the no disturbance without information property and derived a number of consequences from it. Precisely, we showed that (i) if a set of states are perfectly discriminable, then one can discriminate among them using a test that leaves them unaltered, and (ii) every set of perfectly discriminable mixed states can be refined to a set of perfectly discriminable pure states with larger cardinality. Using these results, we provided a characterization of perfect discriminability as orthogonality. Finally, we discussed the notion of maximal sets, i.e. sets of perfectly discriminable states that cannot be further extended. Using this notion, we showed that every set of perfectly discriminable pure states can be extended to a maximal set consisting exclusively of pure states.

---

[3] By definition of internal state, for every given state ρ, there exists a probability $p > 0$ and a state σ such that $\omega = p\rho + (1 - p)\sigma$. Clearly, we cannot discriminate perfectly between ρ and ω, because this would require us to discriminate perfectly between ρ and ρ itself.

Solutions to Selected Problems and Exercises

Exercise 10.1

First of all, notice that the pure states of a system A are a complete set in $St_\mathbb{R}(A)$. Since the dimension D_A is finite, we can find a finite set of pure states $\{\psi_i\}_{i=1}^{D_A}$ that is complete. Suppose now that all pure states can be jointly discriminated in a single test: this implies that by coarse-graining one can construct a single observation test $\{a_i\} \subseteq \text{Eff}(A)$ such that

$$(a_i|\psi_j) = \delta_{i,j}. \tag{10.14}$$

By Proposition 10.4 this implies that there exists a perfectly discriminating test that makes no disturbance on the complete set of states $\{\psi_i\}_{i=1}^{D_A}$, and thus makes no disturbance on $St(A)$. However, by the no information without disturbance property of theories with purification of Section 7.7, this implies that the test has the form $\mathcal{A}_i = p_i \mathcal{I}$, in contradiction with Eq. (10.14).

Exercise 10.4

We recall that $\mathcal{A} \in \text{Transf}(F \to C)$ and $\mathcal{B} \in \text{Transf}(C \to F)$ satisfy the condition (8.7), namely $\mathcal{B}\mathcal{A} =_F \mathcal{I}_F$ and $\mathcal{A}\mathcal{B} =_{\mathcal{A}F} \mathcal{I}_C$. Moreover, suppose that $\{\alpha_i\}_{i=1}^k$ is a perfectly discriminable set of states in $St_1(F)$ with discriminating test $\{a_i\}_{i=1}^k$. Then, by equation 8.7 one has $(a_i|\mathcal{A}\mathcal{B}|\alpha_j) = \delta_{i,j}$, and by proposition 10.8 the set $\{\mathcal{B}\alpha_j\}_{j=1}^k \subseteq St_1(C)$ is perfectly discriminable. Moreover, if the set is $\{\alpha_i\}_{i=1}^k$ maximal in F_ρ, then $\{\mathcal{B}\alpha_j\}_{j=1}^k$ is maximal in $F_{\mathcal{A}\rho}$. Indeed, otherwise there would exist $\beta \in F_{\mathcal{A}\rho}$ such that $\{\mathcal{B}\alpha_j\}_{j=1}^k \cup \{\beta\}$ is perfectly discriminable, and applying the previous argument also $\{\mathcal{A}\mathcal{B}\alpha_i\}_{i=1}^k \cup \{\mathcal{A}\beta\}$ is perfectly discriminable. Recalling Eq. (8.7), this would mean that $\{\alpha_i\}_{i=1}^k \cup \{\mathcal{A}\beta\}$ is perfectly discriminable, contradicting the maximality hypothesis for $\{\alpha_i\}_{i=1}^k$. Vice versa, using the same argument exchanging $\mathcal{A}$ with $\mathcal{B}$ and $\{\alpha_i\}_{i=1}^k$ with $\{\mathcal{B}\alpha_j\}_{j=1}^k$, we obtain the proof of the converse.

11

Identifying Pure States

How can we test the hypothesis that a system is in a given pure state? In this chapter we analyze this question, introducing the task of *state identification*. We show that, thanks to some of our principles, a pure state can be identified by an atomic effect, meaning that, when a binary test is performed, the atomic effect will take place with certainty only on the given pure state. In addition, we will show that the correspondence between pure states and atomic effects is one-to-one and will refer to this correspondence as the *dagger*. In order to establish the correspondence, we will analyze another one-to-one correspondence between states and effects, induced by the steering property. Such a correspondence will be the starting point for the results of the next chapters.

11.1 The State Identification Task

Suppose that Alice buys a new preparation device, which is claimed to prepare system A in the pure state α. Alice, however, does not trust the vendor, since she has reasons to believe that the device does not really prepare the state α. In order to check that the prepared state is really α the most obvious strategy is to prepare a large number of systems and to perform a tomography, i.e. to perform a complete set of measurements, whose outcome probabilities are in one-to-one correspondence with the state. Using tomography, Alice can find out what state is prepared by the device, up to a small error due to the finite statistics. However, if her goal is just to check whether the state is α or not, a full tomography reconstruction is not needed: it would be enough to have a binary measurement $\{a_{\text{yes}}, a_{\text{no}}\}$ which tests the hypothesis "the system is prepared in the state α." Reasonable requirements for such a test are that:

1. if the state is α, then the test should answer positively, i.e.

$$(a_{\text{yes}} \mid \alpha) = 1; \tag{11.1}$$

2. if the state is not α, then the test should have some chance of detecting this fact and answer negatively, i.e. $(a_{\text{no}} \mid \rho) > 0$ for every $\rho \neq \alpha$.[1]

The second requirement is equivalent to the condition

$$(a_{\text{yes}} \mid \rho) < 1 \qquad \forall \rho \in \text{St}(A), \, \rho \neq \alpha. \tag{11.2}$$

---

[1] Asking $(a_{\text{no}} \mid \rho) = 1$ for every $\rho \neq \alpha$ would be unreasonable, since the refinement set $\text{RefSet}_1(\rho)$ generally contains α (i.e. $\rho = (1 - \epsilon)\alpha + \epsilon\sigma$), making perfect discriminability impossible.

When conditions (11.1) and (11.2) are satisfied, we say that the effect a_{yes} *identifies* the pure state α. In a general theory, there is no guarantee that a given pure state can be identified by some effect, e.g. the theory may contain only "noisy measurements" for which no outcome takes place with probability 1. Nevertheless, the identification of pure states is possible in a large number of theories, including quantum theory, classical theory, and the theory of no-signaling boxes.[2] To familiarize yourself with this idea, try the following exercises:

Exercise 11.1   Show that in quantum and classical theory every pure state is identified by at least one effect. Moreover, show that if the effect is required to be atomic, then it is unique.

Exercise 11.2   Consider an hypothetical system whose states form a square and whose set of allowed measurements consists of (i) all measurements that discriminate one side of the square from the opposite and (ii) all mixtures and coarse-grainings of the measurements at point (i). Show that every pure state is identified by an effect. Note that, in fact, there are two distinct atomic effects that identify the same pure state.

In the following sections we will use our principles to show that every pure state is identified by one and only one atomic effect. Like many other consequences of our principles, this feature rules out systems with square state space.

11.2 Only One Pure State for Each Atomic Effect

Let us start by showing a property of atomic effects, namely that an atomic effect reaches its maximum probability on a specific pure state.

Theorem 11.1   *For every atomic effect a in $\mathsf{Eff}(A)$, there exists a unique pure state α in $\mathsf{St}(A)$ such that $(a|\alpha) = \|a\|$.*

The proof of the theorem makes use of the following:

Lemma 11.2   *If a is atomic and ρ is such that $(a|\rho) = \|a\|$, then ρ must be pure.*

Proof   Clearly, the condition $(a|\rho) = \|a\|$ implies $(a|\sigma) = \|a\|$ for every state σ in the face identified by ρ. In other words, one has $a =_\rho \|a\| \, e_A$. For every purification of ρ, say $\Psi \in \mathsf{PurSt}(AB)$, the condition $a =_\rho \|a\| \, e_A$ implies

$$
\left(\!\! \Psi \!\!\begin{array}{c} \text{—} A\,\boxed{a} \\ \text{—} B \text{—} \end{array} \right) \;=\; \|a\| \left(\!\! \Psi \!\!\begin{array}{c} \text{—} A\,\boxed{e} \\ \text{—} B \text{—} \end{array} \right) \tag{11.3}
$$

due to the equality on purifications (Theorem 8.3). Since a is atomic, the atomicity of composition implies that the state

[2] Barrett (2007).

$$\beta \!-\!\!\boxed{\text{B}} := \frac{1}{\|a\|} \; \left(\!\!\boxed{\Psi}\begin{matrix}\text{A}\!-\!\boxed{a}\\ \text{B}\end{matrix}\right)$$

is pure. On the other hand, Eq. (11.3) implies

$$\beta \!-\!\!\boxed{\text{B}} = \left(\!\!\boxed{\Psi}\begin{matrix}\text{A}\!-\!\boxed{e}\\ \text{B}\end{matrix}\right) ,$$

namely, β is the marginal of Ψ on system B. Since its marginal is pure, Ψ must be factorized, i.e. of the form $\Psi = \alpha \otimes \beta$ (this follows by reading Proposition 7.3 in the contrapositive) for some state α. Clearly, α must coincide with ρ and must be pure, otherwise $\Psi = \alpha \otimes \beta$ would be mixed. In conclusion, ρ must be pure. $\qquad\square$

Using the above lemma, it is immediate to prove Theorem 11.1.

Proof of Theorem 11.1   Let ρ be a state such that $(a|\rho) = \|a\|$. By Lemma 11.2, ρ must be pure. Moreover, this pure state must be unique: suppose that α and α' are pure states such that $(a|\alpha) = (a|\alpha') = \|a\|$. Then for $\omega = \frac{1}{2}(\alpha + \alpha')$ one has $(a|\omega) = \|a\|$. Since ω must be pure, one has $\alpha = \alpha'$. $\qquad\square$

As a consequence, *if* an atomic effect gives probability 1 on some pure state, then it identifies that state, in the sense of Eqs. (11.1) and (11.2). To show that every pure state can be identified, we have to show that the set of atomic effects that give probability 1 on a given pure state is non-empty. This will be done in the next section.

11.3 Every Pure State can be Identified

We know that every pure state can be part of a maximal set of perfectly discriminable states. Given a measurement that discriminates among these states, we now prove that the effect associated to the pure state must be atomic.

Lemma 11.3   *Let $\{\alpha_i\}_{i=1}^{N} \subset \mathsf{St}(A)$ be a maximal set of perfectly discriminable states and let $\{a_i\}_{i=1}^{N}$ be the measurement that discriminates among them. If the state α_{i_0} is pure, then the effect a_{i_0} is atomic.*

Proof   Consider the state $\omega = \sum_{i=1}^{N} \alpha_i/N$. By Theorem 10.11, ω is completely mixed. Let $\Psi \in \mathsf{PurSt}(AB)$ be a minimal purification of ω, as defined in Section 8.4. By the steering property, there exists a measurement $\{b_i\}_{i=1}^{N}$ on system B such that

$$\left(\!\!\boxed{\Psi}\begin{matrix}\text{A}\\ \text{B}\!-\!\boxed{b_i}\end{matrix}\right) = \frac{1}{N}\;\boxed{\alpha_i}\!-\!\text{A} \qquad i \in \{1,\dots,N\}. \tag{11.4}$$

Since Ψ is faithful on system B and the state α_{i_0} is pure, the effect b_{i_0} must be atomic.

Now, define the normalized states $\{\beta_i\}_{i=1}^N$ via the relation

$$\boxed{\beta_i}\!-\!\!\raisebox{0.5ex}{B}\;\; := N \;\; \Psi\!\!\begin{array}{l}\!-\!\raisebox{0.5ex}{A}\;\boxed{a_i}\\[2pt]\!-\!\raisebox{0.5ex}{B}\end{array} \qquad\qquad (11.5)$$

The normalization of the states is clear: if we apply the deterministic effect on both sides, we obtain

$$\boxed{\beta_i}\!-\!\raisebox{0.5ex}{B}\!-\!\boxed{e} = N \;\; \Psi\!\!\begin{array}{l}\!-\!\raisebox{0.5ex}{A}\;\boxed{a_i}\\[2pt]\!-\!\raisebox{0.5ex}{B}\!-\!\boxed{e}\end{array}$$

$$= N \;\; \boxed{\omega}\!-\!\raisebox{0.5ex}{A}\!-\!\boxed{a_i}$$

$$= \sum_{j=1}^N \boxed{\alpha_j}\!-\!\raisebox{0.5ex}{A}\!-\!\boxed{a_i}$$

$$= 1 \qquad\qquad \forall i \in \{1, \ldots, N\}.$$

Applying the effect b_j on both sides of Eq. (11.5) and using Eq. (11.4), we then obtain

$$\boxed{\beta_i}\!-\!\raisebox{0.5ex}{B}\!-\!\boxed{b_j} = N \;\; \Psi\!\!\begin{array}{l}\!-\!\raisebox{0.5ex}{A}\;\boxed{a_i}\\[2pt]\!-\!\raisebox{0.5ex}{B}\!-\!\boxed{b_j}\end{array}$$

$$= \boxed{\alpha_j}\!-\!\raisebox{0.5ex}{A}\!-\!\boxed{a_i}$$

$$= \delta_{ij} \qquad\qquad \forall i,j \in \{1, \ldots, N\},$$

meaning that the states $\{\beta_i\}$ are perfectly discriminable. In particular, we have $1 \geq \|b_{i_0}\| \geq (b_{i_0}|\beta_{i_0}) = 1$, and thus $\|b_{i_0}\| = (b_{i_0}|\beta_{i_0})$. Since the effect b_{i_0} is atomic, Lemma 11.2 forces the state β_{i_0} to be pure. Recalling the definition of β_{i_0} in Eq. (11.5), it is clear that the purity of β_{i_0} implies the purity of a_{i_0}. $\qquad\qquad\square$

Summing up, we have that:

1. every atomic effect reaches its maximum probability on a unique pure state (Theorem 11.1);
2. every pure state belongs to some maximal set (Proposition 10.10);
3. the measurement that discriminates the states in a maximal set must associate atomic effects to pure states in the set (Lemma 11.3).

Combining the last two facts, we obtain the desired result:

Theorem 11.4   *For every system* A, *every pure state can be identified by an atomic effect.*

11.4   For a Pure State, Only One Atomic Effect

We have seen that every pure state can be identified by an atomic effect. But is this effect unique? Or can there be two different atomic effects that identify the same pure state? The question is answered by the following theorem.

Theorem 11.5   *If two atomic effects identify the same pure state, then they coincide.*

The basic idea of the proof is to swap the role of states and effects, reducing the thesis of the theorem to the statement "If two pure states are identified by the same atomic effect, then they coincide" – a fact that had been already proven in Theorem 11.1.

The swap of roles between states and effects is itself a very fundamental result, achieved by the following:

Theorem 11.6 (Transposition Theorem)   *Let B be the purifying system in a minimal purification of the invariant state χ_A. Then, there exist two bijective maps*

$$\sharp : \mathsf{St}_\mathbb{R}(A) \to \mathsf{Eff}_\mathbb{R}(B), \quad \alpha \mapsto \alpha^\sharp, \tag{11.6}$$

and

$$\flat : \mathsf{Eff}_\mathbb{R}(A) \to \mathsf{St}_\mathbb{R}(B), \quad a \mapsto a^\flat, \tag{11.7}$$

such that

$$\left(\underline{\alpha} \!-\!\!\overset{A}{}\!\!-\! \underline{a} \right) \;=\; \left(\underline{a^\flat} \!-\!\!\overset{B}{}\!\!-\! \underline{\alpha^\sharp} \right), \tag{11.8}$$

for every $\alpha \in \mathsf{St}_\mathbb{R}(A)$ and every $a \in \mathsf{Eff}_\mathbb{R}(A)$. In particular, the map $\sharp$ transforms normalized pure states into normalized atomic effects, while the map $\flat$ transforms normalized atomic effects into normalized pure states.

The proof of the theorem and the construction of the linear maps $\sharp$ and $\flat$ are important in their own right and require some technical work. For this reason, we will postpone them to the concluding sections of the chapter. For the moment, we limit ourselves to observe that the transposition theorem (Theorem 11.6) gives a straightforward proof of Theorem 11.5.

Proof of Theorem 11.5   Let A be an arbitrary system, $\alpha \in \mathsf{PurSt}(A)$ be an arbitrary pure state, and let a and a' be two atomic effects such that $(a|\alpha) = (a'|\alpha) = 1$. Then, Eq. 11.8 gives the equalities

$$1 = \left(\underline{a^\flat} \!-\!\!\overset{B}{}\!\!-\! \underline{\alpha^\sharp} \right) \quad \text{and} \quad 1 = \left(\underline{a'^\flat} \!-\!\!\overset{B}{}\!\!-\! \underline{\alpha^\sharp} \right) .$$

Since an atomic effect can reach its maximum probability only on one pure state (Theorem 11.1) we conclude the equality $a^\flat = a'^\flat$. The injectivity of the map $\flat$ then implies $a = a'$.   □

11.5 The Dagger

Before proceeding further, it is worth stressing what we have achieved so far. By studying the task of identifying pure states we established a one-to-one correspondence between the set of normalized pure states and the set of normalized atomic effects of a given system A. More precisely we can define the *dagger map* $\dagger$ from deterministic pure states of A to normalized atomic effects of A via the relation $\alpha^\dagger := a$ normalized atomic effect of A such that $(a|\alpha) = 1$. With this definition, the one-to-one correspondence between pure states and atomic effects gives the following:

Proposition 11.7   *The map † is a bijection from the set of deterministc pure states of* A *to the set of normalized atomic effects of* A.

Since † is a bijection, we can also define its inverse, which maps normalized atomic effects into normalized pure states. With a little abuse of notation, we use the symbol † also for the inverse: for a normalized atomic effect a of A we define $a^\dagger := \alpha$ pure deterministc state of A such that $(a|\alpha) = 1$.

An important property of the dagger is that it sends maximal sets of perfectly discriminable pure states into maximally discriminating measurements:

Proposition 11.8   *If* $S = \{\alpha_i\}_{i=1}^N$ *is a maximal set of perfectly discriminable pure states, then* $\left\{\alpha_i^\dagger\right\}_{i=1}^N$ *is a measurement – precisely, it is the measurement that discriminates perfectly among the states in* S.

Proof   Let $\{a_i\}_{i=1}^N$ be the measurement that discriminates perfectly among the states in S. Since all the states in S are pure, each effect a_i must be atomic (Lemma 11.3). The condition $(a_i|\alpha_i) = 1$ then implies $a_i = \alpha_i^\dagger$.   □

More generally, we have the following:

Corollary 11.9   *If* $\{\alpha_i\}_{i=1}^r$ *are perfectly discriminable pure states, then the atomic effects* $\left\{\alpha_i^\dagger\right\}_{i=1}^r$ *coexist in a measurement.*

Proof   Just extend $\{\alpha_i\}_{i=1}^r$ to a pure maximal set and use Proposition 11.8.   □

We will see in the next chapter that the one-to-one correspondence between pure states and atomic effects has far reaching consequences. A first consequence is that the group G_A of reversible transformations is transitive on normalized atomic effects, as it is on pure states:

Lemma 11.10   *If* $a, a' \in \text{Eff}(A)$ *are two atomic effects with* $\|a\| = \|a'\| = 1$*, then there is a reversible channel* $\mathcal{U} \in G_A$ *such that* $(a'|_A = (a|_A\mathcal{U}$.

Proof   From Theorem 11.1 we know that there are two unique pure states φ and φ' such that $(a|\varphi) = 1$ and $(a'|\varphi') = 1$, respectively. Now, by Proposition 7.4 there is a reversible channel $\mathcal{U} \in G_A$ such that $|\varphi)_A = \mathcal{U}|\varphi')_A$. Hence, $(a'|\varphi') = (a|\varphi)_A = (a|\mathcal{U}|\varphi')$. Finally, by Theorem 11.5, one has $(a'|_A = (a|_A\mathcal{U}$.   □

To practice what you have learnt so far, you can try the following:

Exercise 11.3   Given a pure state α and a generic mixed state ρ, show that the following are equivalent:

1. α and ρ are perfectly discriminable;
2. $(\alpha^\dagger|\rho) = 0$.

11.6 Transposing States

In this section and in the following two sections we will prove the transposition theorem (Theorem 11.6), showing how to swap pure states with atomic effects. The key idea is to take a minimal purification of the invariant state and to use this purification to "transpose" states and effects.

The construction works as follows: for a system A and a *normalized* pure state $\alpha \in$ $\mathsf{PurSt}_1(A)$, we define the probability p_α as

$$p_\alpha := \max \left\{ p : \exists \rho \in \mathsf{St}_1(A), \chi_A = p\alpha + (1-p)\rho \right\} . \tag{11.9}$$

By definition, p_α is the maximum probability of the pure state α in a convex decomposition of the invariant state χ_A. Note that p_α is non-zero, because the invariant state χ_A is completely mixed, i.e. every pure state can stay in its convex decomposition with some non-zero probability. Moreover, p_α is independent of α:

Lemma 11.11 *For every pair of normalized pure states α and α', one has $p_\alpha = p_{\alpha'}$.*

Proof For every pure state α' there exists a reversible transformation $\mathcal{U}$ such that $\alpha' = \mathcal{U}\alpha$ (Proposition 7.4). Hence, we have $\chi = p\alpha + (1-p)\rho$ if and only if $\chi = p\alpha' + (1-p)\mathcal{U}\rho$. Maximizing over p we obtain $p_{\alpha'} = p_\alpha$. □

Since p_α is independent of α, we will denote it by $p_{\max}^A$.

We are now ready to construct the map $\sharp$, which turns states of A into effects of B. Let us pick a minimal purification of the invariant state χ_A, call it $\Phi \in \mathsf{PurSt}(AB)$. Then, the steering property of Proposition 7.5 ensures that for every normalized pure state $\alpha \in$ $\mathsf{PurSt}_1(A)$ there exists an effect $\alpha^\sharp$ such that

$$\begin{array}{c}\Phi \underset{B}{\overset{A}{\longrightarrow}}\!\!\boxed{\alpha^\sharp} := p_{\max}^A \boxed{\alpha}\!\!-\!\!A . \end{array} \tag{11.10}$$

Note that $\alpha^\sharp$ is uniquely defined by Eq. (11.10): since Φ is a minimal purification, one has

$$\begin{array}{c}\Phi \underset{B}{\overset{A}{\longrightarrow}}\!\!\boxed{b'} = \Phi \underset{B}{\overset{A}{\longrightarrow}}\!\!\boxed{b}\end{array} \implies b' = b .$$

Moreover, the effect $\alpha^\sharp$ must be atomic; indeed, $b_1 + b_2 = \alpha^\sharp$ implies

$$\begin{array}{c}\Phi \underset{B}{\overset{A}{\longrightarrow}}\!\!\boxed{b_1} + \Phi \underset{B}{\overset{A}{\longrightarrow}}\!\!\boxed{b_2} = \Phi \underset{B}{\overset{A}{\longrightarrow}}\!\!\boxed{\alpha^\sharp} = p_\alpha \boxed{\alpha}\!\!-\!\!A ,\end{array}$$

and, since α is pure,

$$\begin{array}{c}\Phi \underset{B}{\overset{A}{\longrightarrow}}\!\!\boxed{b_1} \propto \Phi \underset{B}{\overset{A}{\longrightarrow}}\!\!\boxed{b_2} \propto \Phi \underset{B}{\overset{A}{\longrightarrow}}\!\!\boxed{\alpha^\sharp} .\end{array}$$

Again, the fact that Φ is a minimal purification yields $b_1 \propto b_2 \propto \alpha^\sharp$, meaning that $\alpha^\sharp$ is atomic.

The action of the map $\sharp : \alpha \mapsto \alpha^\sharp$ can be extended by linearity to arbitrary vectors in $\mathsf{St}_\mathbb{R}(A)$: for every linear combination of normalized pure states, say $X = \sum_i c_i \alpha_i$, we set

$$X^\sharp := \sum_i c_i \alpha_i^\sharp . \tag{11.11}$$

It is easy to check that this definition is well posed: if $X = \sum_j c_j \alpha_j'$ is an alternative decomposition of X, one must have

$$\sum_j c_j' \left(\begin{array}{c} \Phi \end{array} \quad \begin{array}{c} A \\ B \quad \alpha_j'^\sharp \end{array} \right) = \left(X \right) - A = \sum_i c_i \left(\begin{array}{c} \Phi \end{array} \quad \begin{array}{c} A \\ B \quad \alpha_i^\sharp \end{array} \right) .$$

Since Φ is a minimal purification, this implies the relation

$$\sum_j c_j' \alpha_j'^\sharp = \sum_i c_i \alpha_i^\sharp \equiv b_X ,$$

meaning that the definition of $X^\sharp$ does not depend on the choice of decomposition used for X.

The key properties of the map $\sharp$ are the following:

Proposition 11.12 *The map $\sharp$ is injective. Moreover, every (normalized) atomic effect of B is the image of a (normalized) pure state of A.*

Proof Injectivity follows trivially from Eq. (11.10), since an effect $b \in \mathsf{Eff}(B)$ cannot steer two different states of system A when applied to Φ. Now, let b' be a generic atomic effect of B. We have to prove that b' is the image of a pure state of A. To this purpose, define the unnormalized state

$$\left(\alpha' \right) - A := \left(\begin{array}{c} \Phi \end{array} \quad \begin{array}{c} A \\ B \quad b' \end{array} \right) . \tag{11.12}$$

By the atomicity of composition, the state α' must be pure. Now, define the probability $p' := (e_A | \alpha')$ and note that, by definition $p' \leq p_{\max}^A$. Defining the normalized pure state $\tilde{\alpha} := \alpha'/p'$, Eq. (11.12) yields

$$\left(\begin{array}{c} \Phi \end{array} \quad \begin{array}{c} A \\ B \quad b' \end{array} \right) = \left(\alpha' \right) - A = p' \left(\tilde{\alpha} \right) - A = \frac{p'}{p_{\max}^A} \left(\begin{array}{c} \Phi \end{array} \quad \begin{array}{c} A \\ B \quad \tilde{\alpha}^\sharp \end{array} \right) ,$$

having used Eq. (11.10) in the last equality. Since Φ is a minimal purification, we conclude

$$b' = \frac{p'}{p_{\max}^A} \tilde{\alpha}^\sharp , \tag{11.13}$$

or, equivalently,

$$b' = \alpha_*^\sharp , \qquad\qquad \alpha_* := \frac{p'}{p_{\max}^A} \tilde{\alpha} . \tag{11.14}$$

Note that α_* is a (sub-normalized) state, since $p' \le p^A_{max}$. Equation (11.14) proves that every atomic effect $b' \in \mathsf{Eff}(B)$ is the image of a pure state $\alpha_* \in \mathsf{PurSt}(A)$.

In addition, if b' is normalized, then also α_* must be normalized: indeed, taking the norm of Eq. (11.13) one obtains

$$1 = \|b'\| = \frac{p'}{p^A_{max}} \|\tilde{\alpha}^\sharp\| \le \|\tilde{\alpha}^\sharp\| ,$$

which implies $\|\tilde{\alpha}^\sharp\| = 1$ and $p'/p^A_{max} = 1$. In conclusion we obtain $\alpha_* = \tilde{\alpha}$ in Eq. (11.14), meaning that α_* is normalized. Hence, b' is the image of a normalized pure state. $\square$

As a consequence of proposition 11.12, we have the following result.

Corollary 11.13 *Every atomic effect of* B *is proportional to a normalized atomic effect.*

Proof Since the map $\sharp$ is linear, and the set of atomic effects of B is contained in the image of $\mathsf{PurSt}(A)$, it is clear that for an atomic effect $b \in \mathsf{Eff}(B)$ there is $\alpha \in \mathsf{PurSt}(A)$ such that $b = \alpha^\sharp$. Now, since $\alpha = \lambda\tilde{\alpha}$ with $\tilde{\alpha}$ normalized and pure, one also has $b = \alpha^\sharp = \lambda\tilde{\alpha}^\sharp$, with $\|\tilde{\alpha}^\sharp\| = 1$ by Proposition 11.12. $\square$

11.7 Transposing Effects

We now define the map $\flat : \mathsf{Eff}_{\mathbb{R}}(A) \to \mathsf{St}_{\mathbb{R}}(B)$, which transposes the effects of A. In this case, we simply set

$$\begin{array}{c}\underbrace{a^\flat}\!-\!\!\text{B} \end{array} := \frac{1}{p^A_{max}} \left(\Phi \begin{array}{c}\text{A}\!-\!\underbrace{a}\\ \text{B}\end{array} \right) \qquad \forall a \in \mathsf{Eff}(A), \qquad (11.15)$$

which, by linearity, defines $\flat$ on every element of $\mathsf{Eff}_{\mathbb{R}}(A)$.

The key properties of the map $\flat$ are summarized in the following:

Proposition 11.14 *The map $\flat$ is injective. Moreover, it sends normalized atomic effects into normalized pure states.*

Proof Injectivity is obvious from the fact that Φ is a purification of the completely mixed state χ_A. Now, let a be a normalized atomic effect. By definition (11.15), $p^A_{max} a^\flat$ is an unnormalized state of system B. The state is pure due to the atomicity of composition. We now prove that $a^\flat$ is a *normalized* pure state. Let us define the probability

$$q_a := \left(\Phi \begin{array}{c}\text{A}\!-\!\underbrace{a}\\ \text{B}\!-\!\underbrace{e}\end{array} \right) \equiv \underbrace{\chi}\!-\!\text{A}\!-\!\underbrace{a} \qquad (11.16)$$

and the normalized pure state

$$\underbrace{\tilde{a}^\flat}\!-\!\text{B} := \frac{1}{q_a}\left(\Phi \begin{array}{c}\text{A}\!-\!\underbrace{a}\\ \text{B}\end{array} \right) \equiv \frac{p^A_{max}}{q_a}\,\underbrace{a^\flat}\!-\!\text{B} \,. \qquad (11.17)$$

Now, our goal is to prove that $q_a = p_{\text{max}}^A$, which implies that a^b is a normalized state. To this purpose, let us denote by α_* the pure state such that $(a|\alpha_*) = 1$ and decompose the invariant state as $\chi_A = p_{\text{max}}^A \alpha_* + (1 - p_{\text{max}}^A)\rho$, for some state ρ. From Eq. (11.16) we have

$$
\begin{aligned}
q_a &= (a|\chi_A) \\
&= p_{\text{max}}^A (a|\alpha_*) + (1 - p_{\text{max}}^A)(a|\rho) \\
&\geq p_{\text{max}}^A,
\end{aligned}
\tag{11.18}
$$

having used the conditions $(a|\alpha_*) = 1$ and $(a|\rho) \geq 0$. We now prove the reverse inequality. Let b' be an atomic effect such that

$$
\boxed{\tilde{a}^b} \!-\!\!\!\overset{\text{B}}{}\!\!\!-\! \boxed{b'} = 1
\tag{11.19}
$$

(such an effect exists because of Theorem 11.4). Define the probability p' by

$$
p' := \left(\Phi \begin{array}{l} \overset{\text{A}}{} \boxed{e} \\ \underset{\text{B}}{} \boxed{b'} \end{array} \right)
\tag{11.20}
$$

and the normalized pure state α' by

$$
\boxed{\alpha'} \!-\!\!\overset{\text{A}}{}\!\!-\! := \frac{1}{p'} \left(\Phi \begin{array}{l} \overset{\text{A}}{} \\ \underset{\text{B}}{} \boxed{b'} \end{array} \right).
\tag{11.21}
$$

Clearly, the above equations imply that the invariant state can be decomposed as $\chi_A = p'\alpha' + (1 - p')\rho'$, for some suitable state ρ'. Hence, we must have

$$
p' \leq p_{\text{max}}^A.
\tag{11.22}
$$

Moreover, applying the effect a on both sides of Eq. (11.21) we obtain

$$
\begin{aligned}
\boxed{\alpha'} \!-\!\!\overset{\text{A}}{}\!\!-\! \boxed{a} &= \frac{1}{p'} \left(\Phi \begin{array}{l} \overset{\text{A}}{} \boxed{a} \\ \underset{\text{B}}{} \boxed{b'} \end{array} \right) \\
&= \frac{q_a}{p'} \boxed{\tilde{a}^b} \!-\!\!\!\overset{\text{B}}{}\!\!\!-\! \boxed{b'} \\
&= \frac{q_a}{p'},
\end{aligned}
$$

having used Eq. (11.17) in the second equality and Eq. (11.19) in the last one. Since $(a|\alpha') \leq 1$, we obtained the inequality

$$
\frac{q_a}{p'} \leq 1,
$$

which combined with Eq. (11.22) yields

$$
q_a \leq p' \leq p_{\text{max}}^A.
$$

This inequality, combined with the reverse inequality of Eq. (11.18), yields $q_a = p_{\text{max}}^A$. Inserting this relation in Eq. (11.17) we obtain $\tilde{a}^b = a^b$, meaning that a^b is normalized. $\quad\square$

11.8 Playing with Transposition

In the previous two sections we worked hard to construct the maps $\sharp$ and $\flat$. Now it is time to enjoy the benefits of this construction.

First of all, it is easy to see that transposition does not change the probabilities:

Lemma 11.15 *For every state $\alpha \in \mathsf{St}(A)$ and for every effect $a \in \mathsf{Eff}(A)$, one has $(a|\alpha) = (\alpha^{\sharp}|a^{\flat})$.*

Proof If α is a normalized pure state, the equality follows trivially from the definitions (11.10) and (11.15). By linearity, the equality extends to arbitrary states. □

In addition, the normalized pure states of A are in one-to-one correspondence with the normalized atomic effects of B:

Lemma 11.16 *The map $\sharp$ establishes a bijection between pure states of A and atomic effects of B.*

Proof Let us start from normalized pure states. If α is a normalized pure state and a is the atomic effect such that $(a|\alpha) = 1$, then by Lemma 11.15, we have $(\alpha^{\sharp}|a^{\flat}) = 1$. Hence, $\alpha^{\sharp}$ is a normalized atomic effect. In short, the image of the set of normalized pure states of A is contained in the set of normalized atomic effects of B. On the other hand, Proposition 11.12 already proved the reverse inclusion. We conclude that $\sharp$ is a bijection between the set of normalized pure states of A and the set of normalized atomic effects of B. Using Corollary 11.13, stating that all atomic effects of B are proportional to some normalized atomic effect, the bijection is extended to arbitrary pure states and atomic effects by linearity. □

The bijection between pure states and atomic effects has important consequences, among which the most striking one is highlighted by the following lemma.

Lemma 11.17 *Let $\Phi \in \mathsf{St}(AB)$ be a minimal purification of the invariant state of system A. Then, the marginal of Φ on system B is the invariant state of system B.*

Proof Let b be a normalized atomic effect of B. By Proposition 11.12 there is a normalized pure state α of system A such that $b = \alpha^{\sharp}$. Applying Lemma 11.15 to the relation $(e_A|\alpha) = 1$ we obtain $(b|e_A^{\flat}) = (\alpha^{\sharp}|e_A^{\flat}) = 1$. Let now $(b'| = (b|\mathcal{V}$. Since the reversible transformation $\mathcal{V}$ preserves the norm of b, i.e. $\|b'\| = \|b\| = 1$, by Lemma 11.16 we have $b' = \alpha'^{\sharp}$ for some normalized pure state α' of A. Then $(b|\mathcal{V}|e_A^{\flat}) = (\alpha'^{\sharp}|e_A^{\flat}) = 1$ for every $\mathcal{V} \in \mathbf{G}_B$, and averaging on $\mathcal{V}$ we have

$$(b|e_A^{\flat}) = (b|\mathcal{T}|e_A^{\flat}) = \lambda(b|\chi_B),$$

where $\mathcal{T}$ is the twirling channel. Since the normalized atomic effect b is arbitrary, and every atomic effect of B is proportional to a normalized one, this implies that $e_A^{\flat} = \lambda \chi_B$ with $\lambda = (e|e_A^{\flat}) \geq 0$. Recalling the definition of the map $\flat$ [Eq. (11.15)], we have

$$\lambda \;\boxed{\chi}\!\!-\!\!^{B}\;=\;\boxed{e^{\flat}_{A}}\!\!-\!\!^{B}\;=\;\frac{1}{p^{A}_{\max}}\;\boxed{\Phi}\begin{smallmatrix}A\;\boxed{e}\\B\end{smallmatrix}\;.$$

Applying the deterministic effect on both sides we then obtain $\lambda = 1/p^{A}_{\max}$, which implies

$$\boxed{\chi}\!\!-\!\!^{B}\;=\;\boxed{\Phi}\begin{smallmatrix}A\;\boxed{e}\\B\end{smallmatrix}\;.$$

$\square$

Since Φ is a purification of the invariant state on B, we can exchange the roles of systems A and B in our construction. In this way, we can define the transposition maps from system B to system A. With a little abuse of notation, we still use the symbols $\sharp$ and $\flat$ for the maps $\sharp : \mathsf{St}_{\mathbb{R}}(B) \to \mathsf{Eff}_{\mathbb{R}}(A)$ and $\flat : \mathsf{Eff}_{\mathbb{R}}(B) \to \mathsf{St}_{\mathbb{R}}(A)$ defined by the relations

$$\boxed{\Phi}\begin{smallmatrix}A\;\boxed{\beta^{\sharp}}\\B\end{smallmatrix}\;:=\;p^{B}_{\max}\;\boxed{\beta}\!\!-\!\!^{B}\qquad\forall\beta\in\mathsf{St}(B)\qquad\qquad(11.23)$$

and

$$\boxed{b^{\flat}}\!\!-\!\!^{A}\;:=\;\frac{1}{p^{B}_{\max}}\;\boxed{\Phi}\begin{smallmatrix}A\\B\;\boxed{b}\end{smallmatrix}\qquad\forall b\in\mathsf{Eff}(B)\,.\qquad\qquad(11.24)$$

Comparing the definitions in equations (11.10) and (11.15) with those in equations (11.23) and (11.24), we obtain the following lemma.

Lemma 11.18   *The maximum steering probabilities $p^{A}_{\max}$ and $p^{B}_{\max}$ coincide.*

Proof   For two normalized atomic effects $a \in \mathsf{Eff}(A)$ and $b \in \mathsf{Eff}(B)$ we have

$$\boxed{b^{\flat}}\!\!-\!\!^{A}\!\!-\!\!\boxed{a}\;=\;\frac{1}{p^{B}_{\max}}\;\boxed{\Phi}\begin{smallmatrix}A\;\boxed{a}\\B\;\boxed{b}\end{smallmatrix}\;=\;\frac{p^{A}_{\max}}{p^{B}_{\max}}\;\boxed{a^{\flat}}\!\!-\!\!^{B}\!\!-\!\!\boxed{b}\;.$$

Choosing a to be an effect that identifies $b^{\flat}$, we then obtain

$$1 = (a|b^{\flat}) = \frac{p^{A}_{\max}}{p^{B}_{\max}}\,(b|a^{\flat}) \le \frac{p^{A}_{\max}}{p^{B}_{\max}}\,,$$

which implies the inequality $p^{B}_{\max} \le p^{A}_{\max}$. On the other hand, choosing b to be an effect that identifies $a^{\flat}$ we obtain the reverse inequality $p^{B}_{\max} \ge p^{A}_{\max}$. $\square$

We are now ready to prove that the maps $\flat$ and $\sharp$ are bijective. To this purpose, we prove the following:

Proposition 11.19   *The maps $\sharp : \mathsf{St}(A) \to \mathsf{Eff}(B)$ and $\flat : \mathsf{Eff}(B) \to \mathsf{St}(A)$ are inverse of each other. Similarly, the maps $\sharp : \mathsf{St}(B) \to \mathsf{Eff}(A)$ and $\flat : \mathsf{Eff}(A) \to \mathsf{St}(B)$ are inverse of each other.*

Proof Let α be a normalized pure state of A. Then, by definition we have

$$\left(\alpha^\sharp\right)^\flat \!\!-\!\! A = \frac{1}{p_{max}^B}\; \Phi\begin{array}{c} A \\ B\!\!-\!\!\alpha^\sharp \end{array} = \frac{p_{max}^A}{p_{max}^B}\; \alpha\!\!-\!\! A = \alpha\!\!-\!\! A\,,$$

having used the fact that p_{max}^A and p_{max}^B coincide (Lemma 11.18). Hence, $\flat \circ \sharp$ is the identity on $\mathsf{PurSt}_1(\mathrm{A})$. By linearity, it is immediate to see that $\flat \circ \sharp$ is the identity on $\mathsf{St}_{\mathbb{R}}(\mathrm{A})$.

On the other hand, let b be a normalized atomic effect on B. Then, one has

$$\Phi\begin{array}{c} A \\ B\!\!-\!\!(b^\flat)^\sharp \end{array} = p_{max}^A \quad b^\flat\!\!-\!\! A = \frac{p_{max}^A}{p_{max}^B} \quad \Phi\begin{array}{c} A \\ B\; b \end{array} = \Phi\begin{array}{c} A \\ B\; b \end{array}\,,$$

again, having used the fact that p_{max}^A and p_{max}^B coincide. Comparing the first and last term we obtain $\left(b^\flat\right)^\sharp = b$ for every b, meaning that $\sharp \circ \flat$ is the identity on the set of normalized atomic effects of B. By linearity, using Corollary (11.13), we obtain that $\sharp \circ \flat$ is also the identity on $\mathsf{Eff}_{\mathbb{R}}(\mathrm{B})$. In summary, $\sharp : \mathsf{St}(\mathrm{A}) \to \mathsf{Eff}(\mathrm{B})$ and $\flat : \mathsf{Eff}(\mathrm{B}) \to \mathsf{St}(\mathrm{A})$ are inverse of each other.

Exchanging the roles of A and B in the previous arguments we obtain the desired results for the maps $\sharp : \mathsf{St}(\mathrm{A}) \to \mathsf{Eff}(\mathrm{B})$ and $\flat : \mathsf{Eff}(\mathrm{B}) \to \mathsf{St}(\mathrm{A})$. □

We have now reached the conclusion of our construction, which aimed at proving the Transposition Theorem 11.6. The proof follows by collecting the results of Propositions 11.12, 11.14, and 11.19.

11.9 Summary

In this chapter we investigated the operational task of identifying pure states. We found out that in a theory satisfying our principles, one has that:

1. every normalized atomic effect identifies some normalized pure state;
2. every normalized pure state can be identified by one and only one normalized atomic effect.

These two results established a one-to-one correspondence between normalized pure states of a system and its normalized atomic effects, which we called the *dagger*. In order to construct the dagger, we took advantage of another one-to-one correspondence, called *transposition* and based on the notion of steering. As a biproduct of the construction, we also discovered that every minimal purification of the invariant state of A with purifying system B must be a purification of the invariant state of system B.

12 Diagonalization

Thanks to the spectral theorem, every quantum state can be decomposed into a random mixture of perfectly discriminable states. This is a remarkable property, which plays a key role in the definition of the von Neumann entropy and of other entropic quantities. In this chapter we show that this property can be reconstructed directly from first principles. Our strategy will be to associate every physical system A with a *conjugate system* $\overline{A}$, representing a mirror image that is maximally correlated with A. We will show that states and measurements on the conjugate system are in one-to-one correspondence with states and measurement on the original system. This correspondence will allow us to show two important facts: (i) all pure maximal sets of a system have the same number of elements, and (ii) every pure maximal set has the invariant state as its barycenter. Using these two facts we will prove that *every* state can be decomposed into a mixture of perfectly discriminable pure states, and obtain operational versions of the spectral theorem and of the Schmidt decomposition.

12.1 Conjugate Systems and Conjugate States

Definition 12.1 We say that system $\overline{A}$ is *conjugate* to A if the composite system $A\overline{A}$ allows for a minimal purification of the invariant state χ_A.

Clearly, every system A has a conjugate, unique up to operational equivalence: indeed, we know that every state has a minimal purification, and that minimal purifications are unique up to reversible transformations, by Theorem 8.11. In addition, it is easy to see that conjugation is an involution:

Proposition 12.2 *For every system* A, *one has* $\overline{\overline{A}} = A$.

Proof Let $\Phi \in \mathsf{PurSt}(A\overline{A})$ be a minimal purification of the invariant state χ_A. By Lemma 11.17, the marginal of Φ on system $\overline{A}$ is the invariant state $\chi_{\overline{A}}$. Hence, Φ is a purification of $\chi_{\overline{A}}$. Moreover, the purification is minimal because the marginal on the purifying system is completely mixed. $\qquad\square$

In addition to the conjugation of systems, it is convenient to introduce a conjugation of states. This can be done using the maps $\dagger$, $\sharp$, and $\flat$ defined in the previous chapter:

Definition 12.3 The conjugate of the state $\alpha \in \mathrm{St}(A)$ is the state $\overline{\alpha} \in \mathrm{St}(\overline{A})$ defined by

$$\overline{\alpha} := \left(\alpha^{\dagger}\right)^{\flat}.$$

It is easy to see that the correspondence $\alpha \to \overline{\alpha}$ is bijective: indeed, the map $\dagger$ is a bijection between the normalized pure states of A and the normalized atomic effects of A, while the map $\flat$ is a bijection between the normalized atomic effects of A and the normalized pure states of $\overline{A}$. Being the composition of two bijective maps, the conjugation is a bijective map between the normalized pure states of A and the normalized pure states of $\overline{A}$. The bijection is extended to general states by linearity.

An equivalent definition for the conjugation of states is provided by the following lemma:

Lemma 12.4 *For every state α of system A, one has $\overline{\alpha} = \left(\alpha^{\sharp}\right)^{\dagger}$.*

Proof By definition, we have

$$1 = \boxed{\alpha}\!\!-\!\!\mathrm{A}\!\!-\!\!\boxed{\alpha^{\dagger}} = \boxed{\left(\alpha^{\dagger}\right)^{\flat}}\!\!-\!\!\mathrm{A}\!\!-\!\!\boxed{\alpha^{\sharp}} \equiv \boxed{\overline{\alpha}}\!\!-\!\!\mathrm{A}\!\!-\!\!\boxed{\alpha^{\sharp}}.$$

Since the atomic effect $\alpha^{\sharp}$ identifies the pure state $\overline{\alpha}$, we have $\overline{\alpha} = \left(\alpha^{\sharp}\right)^{\dagger}$. ☐

The two alternative definitions of the conjugate of a state are illustrated by the commutative diagram

$$
\begin{array}{ccc}
\alpha & \xrightarrow{\;\dagger\;} & \alpha^{\dagger} \\
{\scriptstyle \sharp}\downarrow & & \downarrow{\scriptstyle \flat} \\
\alpha^{\sharp} & \xrightarrow[\;\dagger\;]{} & \overline{\alpha}
\end{array}
$$

Using the above result it is easy to show that the conjugation of states is an involution:

Lemma 12.5 *For every state α of system A, one has $\overline{\overline{\alpha}} = \alpha$.*

Proof One has

$$\overline{\overline{\alpha}} = \left\{\left[\left(\alpha^{\dagger}\right)^{\flat}\right]^{\sharp}\right\}^{\dagger} = \left(\alpha^{\dagger}\right)^{\dagger} = \alpha,$$

the first equality following from Lemma 12.4, the second from the fact that $\sharp$ is the inverse of $\flat$ and the third from the fact that the dagger of effects is the inverse of the dagger of states. ☐

The most important fact about the conjugation of states is that it sets up a one-to-one correspondence between the maximal sets of perfectly discriminable pure states of A and those of $\overline{A}$:

Lemma 12.6 *If $S = \{\alpha_i\}_{i=1}^{N}$ is a maximal set of perfectly discriminable pure states of A, then $\overline{S} = \{\overline{\alpha}_i\}_{i=1}^{N}$ is a maximal set of perfectly discriminable pure states of $\overline{A}$.*

Proof   By definition, we have

$$
\delta_{ij} = \left(\alpha_i \!-\!\boxed{A}\!-\! \alpha_j^\dagger \right)
$$

$$
= \left(\left(\alpha_j^\dagger\right)^b \!-\!\boxed{\overline{A}}\!-\! \alpha_i^\sharp \right)
$$

$$
\equiv \left(\overline{\alpha}_j \!-\!\boxed{\overline{A}}\!-\! \alpha_i^\sharp \right).
$$

Hence, the states in the set $\overline{S} = \left\{ a_i^b \right\}_{i=1}^N$ are orthogonal. But in our theory orthogonality is equivalent to perfect discriminability (Lemma 10.8), so $\overline{S}$ is a set of perfectly discriminable pure states. Finally, $\overline{S}$ must be maximal. Indeed, if $\overline{S}$ were contained in a larger set $\overline{T}$, then the original set $S = \overline{\overline{S}}$ would be contained in the larger set $T = \overline{\overline{T}}$, in contradiction with the fact that S is maximal.   □

The correspondence between pure maximal sets has an easy consequence, which will be useful in the next section:

Corollary 12.7   *For every maximal set of perfectly discriminable pure states $\{\alpha_i\}_{i=1}^N$, the effects $\left\{\alpha_i^\sharp\right\}_{i=1}^N$ form a measurement.*

Proof   We know that $\{\overline{\alpha}_i\}_{i=1}^N$ is a maximal set of perfectly discriminable pure states. The measurement that discriminates among them is $\left\{\overline{\alpha}_i^\dagger\right\}_{i=1}^N$ and satisfies

$$
\overline{\alpha}_i^\dagger = \left[\left(\alpha_i^\sharp\right)^\dagger\right]^\dagger = \alpha_i^\sharp \qquad \forall i \in \{1,\dots,N\}, \tag{12.1}
$$

the first equality following from Lemma 12.4.   □

12.2   A Most Fundamental Result

The properties of the conjugation of states lead us to a spectacular result, which provides the key to many of the developments in the next chapters. Let us state the result first and then see what is special about it:

Theorem 12.8   *For every system A, the following properties hold:*

1. All maximal sets of perfectly discriminable pure states have the same cardinality, given by

$$
d_A = \frac{1}{p_{max}^A}, \tag{12.2}
$$

where p_{max}^A is the maximum steering probability defined in Chapter 11.

2. *Every maximal set of perfectly discriminable pure states* $\{\alpha_i\}_{i=1}^{d_A}$ *has the invariant state as its barycenter, namely*

$$\chi_A = \frac{1}{d_A} \sum_{i=1}^{d_A} \alpha_i .$$ (12.3)

Proof Let $\{\alpha_i\}_{i=1}^N$ be a pure maximal set of A. By definition of the map $\sharp$, we have

(12.4)

Multiplying both members by $1/N$ and summing over i we obtain

(12.5)

having used the fact that the effects $\left\{\alpha_i^\sharp\right\}_{i=1}^N$ form a measurement (Corollary 12.7). Now, applying the deterministic effect on both sides we obtain

$$N p_{max}^A = 1 .$$

Since the pure maximal set $\{\alpha_i\}_{i=1}^N$ is generic, we have proven that all pure maximal sets have the same cardinality $N = 1/p_{max}^A =: d_A$. Inserting this equality in Eq. (12.5), we immediately get the decomposition of Eq. (12.3), valid for a generic pure maximal set $\{\alpha_i\}_{i=1}^N$. □

In a single shot, Theorem 12.8 proves three non-trivial facts:

1. it proves that all maximal sets of perfectly discriminable pure states have the same cardinality – a property that a priori may not have been satisfied;
2. it establishes a quantitative link between two different operational tasks – state discrimination and the remote steering of states;
3. it proves that the invariant state can be decomposed as a mixture of perfectly discriminable pure states.

All these facts will have major implications for our reconstruction of quantum theory: In the next chapter, the relation

$$p_{max}^A = \frac{1}{d_A}$$ (12.6)

will be used to find the maximum probability of *conclusive teleportation* and to show that states can be represented as Hermitian matrices on a Hilbert space. Finally, the decomposition of the invariant state will be the key to prove that every state can be represented as a mixture of perfectly discriminable pure states – thus obtaining the operational analog of the spectral theorem in quantum theory. In turn, this result will lead us to a notion of *orthogonal projection* and to an operational reconstruction of the superposition principle of quantum theory. In the next sections we work toward the derivation of these results, by proving the first consequences of Theorem 12.8.

Exercise 12.1 Prove that the distance between the invariant state χ_A and an arbitrary pure state $\varphi \in St_1(A)$ is

$$\|\chi_A - \varphi\| = \frac{2(d_A - 1)}{d_A}.$$

12.3 The Informational Dimension

We call the cardinality d_A the *informational dimension* of system A. Operationally, the informational dimension quantifies the maximum number of classical messages that can be encoded into system A without incurring errors in the decoding. Once the informational dimension is known, it can be used to establish whether a set of perfectly discriminable states is maximal and pure:

Corollary 12.9 *If a set of perfectly discriminable states has cardinality d_A, then the set is maximal and consists only of pure states.*

Proof If the set contained a mixed state, then it could be refined to a set of pure states (Proposition 10.5). The refined set would have more than d_A elements, in contradiction with Theorem 12.8. Thus the set must consist of pure states. If the set were not maximal, then it could be extended to a maximal set of cardinality $k > d_A$, contradicting Theorem 12.8. □

The informational dimension has a number of important properties. First of all, a system and its conjugate have the same informational dimension:

Corollary 12.10 *For every system A, one has $d_A = d_{\overline{A}}$.*

Proof Immediate from the correspondence between pure maximal sets of A and pure maximal sets of $\overline{A}$ (12.6). □

More interestingly, the informational dimension of a composite system is equal to the product of the informational dimensions of its components:

Corollary 12.11 *For every pair of systems A and B, one has $d_{AB} = d_A d_B$.*

Proof Let $\{\alpha_i\}_{i=1}^{d_A}$ and $\{\beta_j\}_{j=1}^{d_B}$ be pure maximal sets of A and B, respectively. Then, the product set

$$P = \{\alpha_i \otimes \beta_j \mid i = 1, \ldots, d_A, j = 1, \ldots, d_B\}$$

is a set of perfectly discriminable pure states of AB. In addition, it is maximal. The proof is by *reductio ad absurdum*: suppose that P is not maximal and let $P' = P \cup \{\Psi_k\}_{k=1}^K$ be a pure maximal set that extends P. Then, one has

$$(\alpha_i^\dagger \otimes \beta_j^\dagger | \Psi_k) = 0 \qquad \forall k \in \{1, \ldots, K\},$$

and, summing over i and j,

$$(e_A \otimes e_B | \Psi_k) = 0 \qquad \forall k \in \{1, \ldots, K\}.$$

This is absurd, because the states in the set P' are supposed to be normalized. Hence, the product set P must be a pure maximal set for AB. Since P has cardinality $d_A d_B$ and all pure maximal sets have the same cardinality d_{AB}, we conclude the equality $d_{AB} = d_A d_B$. □

Notice that Corollary 12.11 along with Theorem 12.8 implies the relation

$$\chi_{AB} = \frac{1}{d_A d_B} \sum_{i=1}^{d_A} \sum_{j=1}^{d_B} \alpha_i \otimes \beta_j = \chi_A \otimes \chi_B,$$

thus confirming the result of Exercise 7.2.

The above results are very important for our derivation of quantum theory. Eventually, in Chapter 13 they will be used to prove that the dimension of the state space is equal to the square of the informational dimension, thus allowing us to represent the states of A as $d_A \times d_A$ matrices.

12.4 The Informational Dimension of a Face

The notion of informational dimension can be applied not only to physical systems, but also to faces of the state space of a given system. The informational dimension of a face F, denoted by d_F, is simply the maximum number of perfectly discriminable states contained in F. The dimension can be evaluated by finding a *maximal set of perfectly discriminable pure states in F* (shortly, *pure maximal set for F*), that is a set $S \subset F$ such that:

1. all states in S are pure;
2. the states in S are perfectly discriminable;
3. for every state $\varphi \in F$, the states $S \cup \{\varphi\}$ are not perfectly discriminable.

In terms of this notion we have the following:

Theorem 12.12 *Let F be a face of the convex set $\mathrm{St}_1(A)$. All maximal sets of perfectly discriminable pure states in F have the same cardinality, equal to d_F.*

The proof follows easily from the compression axiom. The idea is to compress the face F into the state space of a suitable system C and to exploit the following correspondence between maximal sets in F and maximal sets in C:

Lemma 12.13   *Let* $(C, \mathcal{E}, \mathcal{D})$ *be an ideal compression scheme for the face F. Then,*

1. *a set of states* $\{\varphi_i\}_{i=1}^k$ *is a maximal in F if and only if the encoded set* $\{\mathcal{E}\varphi_i\}_{i=1}^k$ *is maximal in* $\mathsf{St}_1(C)$;
2. *a set of states* $\{\gamma_i\}_{i=1}^l$ *is maximal in* $\mathsf{St}_1(C)$ *if and only if the decoded set* $\{\mathcal{D}\gamma_i\}_{i=1}^l$ *is maximal in F;*
3. *the informational dimension of the face F is equal to the informational dimension of system C.*

The proof follows easily from Exercise 10.4.

Proof of Theorem 12.12   Let $\{\varphi_i\}_{i=1}^k$ be a pure maximal set for F. Then, the encoded set $\{\mathcal{E}\varphi_i\}_{i=1}^k$ is maximal in $\mathsf{St}_1(C)$ (Lemma 12.13) and consists of pure states (Corollary 8.8). Hence, its cardinality must be equal to the informational dimension of C (Theorem 12.8), namely $k = d_C$. Since the pure maximal set $\{\varphi_i\}_{i=1}^k$ was arbitrary, we conclude that all pure maximal sets in F have the same cardinality. $\square$

The notion of informational dimension of a face will be very useful in the next section. In preparation, you can try the following exercise.

Exercise 12.2   Prove that a face $F \subseteq \mathsf{St}_1(A)$ contains d_A perfectly discriminable pure states if and only if it contains a completely mixed state.

12.5   Diagonalizing States

We have seen that the invariant state can be decomposed into a mixture of perfectly discriminable pure states. Can we find a similar decomposition for arbitrary states?

Proving such a result is important for two reasons: first of all, in a reconstruction of quantum theory it is important to find out the operational meaning of the spectral theorem, which in the case of states is nothing but the decomposition of mixed states into mixtures of perfectly discriminable pure states. Furthermore, decomposing mixed states as mixtures of perfectly discriminable states is important per se. Conceptually, it allows us to interpret every mixed state as *ignorance about a classical random variable* encoded into the preparation of the system. Indeed, we can imagine that a source emits a classical message i with probability p_i and that the message is encoded into a state α_i from the pure maximal set $\{\alpha_i\}_{i=1}^{d_A}$. In this way, the average state of the source will be

$$\rho = \sum_{i=1}^{d_A} p_i \alpha_i, \tag{12.7}$$

that is, it will be a mixture of perfectly discriminable states. In this scenario, our initial question can be rephrased as: Can we interpret every mixed state as the average state of a classical information source, whose messages are encoded into perfectly discriminable pure states?

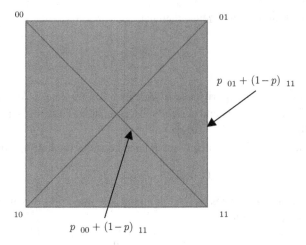

The set of deterministic states of a two-dimensional system in the *"boxworld"* theory. The highlighted edges and diagonals contain the only states that are mixtures of a maximal set of perfectly discriminable pure states.

Fig. 12.1

When a mixed state ρ can be decomposed as in Eq. (12.7), we say that it can be *diagonalized*. In a generic theory, not every state can be diagonalized. For example, consider the "boxworld" theory, where the two-dimensional systems have a square state space: for these systems, the pure states are the vertices of the square and the only states that can be decomposed as mixtures of two pure states are those on the edges and on the diagonals of the square (see Fig. 12.1). All the remaining states cannot be diagonalized.

In contrast to the boxworld example, we now show that our principles imply that every state can be diagonalized. Using this result, we will prove the operational version of the spectral theorem: every element in the linear span of the state space can be decomposed as a linear combination of perfectly discriminable pure states.

Theorem 12.14 (State Diagonalization) *For every system A, every normalized state can be decomposed into a convex mixture of perfectly discriminable pure states.*

Proof The proof is by induction on the informational dimension of the system. If $d_A = 1$, the thesis trivially holds. Now suppose that the thesis holds for any system B of dimension $d_B \leq N$, and take a mixed state ρ of a system A of dimension $d_A = N + 1$. Now, there are two possibilities: either (1) ρ is not completely mixed or (2) ρ is completely mixed.

Suppose first that (1) ρ is not completely mixed and denote by F the face identified by ρ. Since ρ is not completely mixed, F can contain at most N perfectly discriminable states (cf. Exercise 12.2) – hence, $d_F \leq N$. If we apply ideal compression to encode ρ into a smaller system C, then system C must satisfy the relation

$$d_C = d_F \leq N.$$

By the induction hypothesis, the encoded state $\mathcal{E}\rho$ can be diagonalized as

$$\mathcal{E}\rho = \sum_{i=1}^{d_C} p_i \gamma_i.$$

Applying the decoding operation $\mathcal{D}$ we then get

$$\rho = \mathcal{D}\mathcal{E}\rho = \sum_{i=1}^{d_C} p_i \mathcal{D}\gamma_i .$$

Since decoding preserves the purity and discriminability of states (Corollary 12.13), the decoded states $\{\mathcal{D}\gamma_i\}_{i=1}^{d_C}$ are pure and perfectly discriminable. Thus, we obtained the desired decomposition for ρ.

Suppose now that (2) ρ is completely mixed and consider the half-line defined by

$$\sigma_t = (1+t)\rho - t\chi_A \qquad t \geq 0 .$$

Since the set of normalized states $\mathsf{St}_1(A)$ is mixed, the line will cross its border at some point t_0. Therefore, one will have

$$\rho = \frac{1}{1+t_0}\sigma_{t_0} + \frac{t_0}{1+t_0}\chi_A$$

for some state σ_{t_0} on the border of $\mathsf{St}_1(A)$. By definition, states on the border are not completely mixed. Now, we know from point (1) that the state σ_{t_0} can be diagonalized as

$$\sigma_{t_0} = \sum_{i=1}^{d_C} p_i \alpha_i , \tag{12.8}$$

C being the system in the ideal encoding of σ_{t_0}. By Proposition 10.10, the set $\{\alpha_i\}_{i=1}^{d_C}$ can be extended to a maximal set of perfectly discriminable pure states $\{\alpha_i\}_{i=1}^{d_A}$. On the other hand, the invariant state can be decomposed as a uniform mixture of these states, namely

$$\chi_A = \frac{1}{d_A}\sum_{i=1}^{d_A}\alpha_i \tag{12.9}$$

having used Theorem 12.8. Combining Eqs. (12.8) and (12.9), we then obtain the desired decomposition

$$\rho = \sum_{i=1}^{d_A}\left(\frac{p_i}{1+t_0} + \frac{t_0}{d_A(1+t_0)}\right)\alpha_i,$$

having defined $p_i = 0$ for $i > d_C$.   □

In analogy with quantum theory, we call the probabilities $\{p_i \mid i = 1,\ldots,d_A\}$ the *spectrum* of ρ. Since some of the probabilities may be zero, the diagonalization of ρ can be written more concisely as

$$\rho = \sum_{i=1}^{r} p_i \alpha_i , \tag{12.10}$$

where $\{\alpha_i\}_{i=1}^{r}$ are perfectly discriminable pure states (not necessarily forming a maximal set) and $p_i > i$ for every $i \in \{1,\ldots,r\}$. In this case, we call r the *rank* of the state ρ. If $r = d_A$ we say that ρ has *full rank*. With these definitions, we have the following:

Corollary 12.15 (Characterization of Completely Mixed States) *A state is completely mixed if and only if it has full rank.*

Proof Necessity: if $p_i = 0$ for some i, then ρ is perfectly discriminable from φ_i. Hence, it cannot be completely mixed. Sufficiency follows immediately from Theorem 10.11. □

In particular, for two-dimensional systems we have the result:

Corollary 12.16 *For $d_A = 2$ any state on the border of* $\mathsf{St}_1(A)$ *is pure.*

12.6 Diagonalizing Effects

Using the machinery of the maps $\dagger, \sharp$, and $\flat$, the diagonalization of states can be turned into a similar result for effects:

Corollary 12.17 *For every system* A*, every effect* $a \in \mathsf{Eff}(A)$ *can be decomposed as*

$$a = \sum_{i=1}^{r} c_i \, a_i \tag{12.11}$$

where $\{a_i\}_{i=1}^r$ *are coexisting atomic effects belonging to a maximally discriminating measurement and* $c_i \in [0, 1]$ *for every* $i \in \{1, \dots, r\}$.

Proof Applying the map $\flat$, we can turn a into multiple of a state of the conjugate system $\overline{A}$. Defining $\overline{\rho} := a^\flat/(e_{\overline{A}}|a^\flat)$ and diagonalizing it as $\overline{\rho} = \sum_{i=1}^r p_i \overline{\alpha}_i$ we obtain the decomposition

$$a^\flat = \sum_{i=1}^{r} c_i \overline{\alpha}_i, \qquad c_i := \left(e_{\overline{A}}|a^\flat\right) p_i.$$

where $\{\overline{\alpha}_i\}_{i=1}^r$ are perfectly discriminable pure states of $\overline{A}$. Applying the map $\sharp$ on both sides, we then obtain

$$a = \left(a^\flat\right)^\sharp = \sum_{i=1}^{r} c_i \, (\overline{\alpha}_i)^\sharp .$$

Viewing the state $\overline{\alpha}_i$ as the conjugate of a state $\alpha_i \in \mathsf{PurSt}_1(A)$, we finally get

$$a = \sum_{i=1}^{r} c_i \left[\left(\alpha_i^\dagger\right)^\flat\right]^\sharp$$

$$= \sum_{i=1}^{r} c_i \, \alpha_i^\dagger .$$

Hence, a is a linear combination of the atomic effects $\left\{\alpha_i^\dagger\right\}_{i=1}^r$, which can coexist in a measurement due to Corollary 11.9. Moreover, we have $c_i = (a|\alpha_i)$, which implies that c_i belongs to the interval $[0, 1]$. □

12.7 Operational Versions of the Spectral Theorem

In finite-dimensional quantum theory, the diagonalization of states is a special case of the spectral theorem, which states that every Hermitian matrix can be decomposed into a linear combination of orthogonal projectors. The Hermitian matrices can be interpreted in two different ways, either as elements of the vector space $St_\mathbb{R}(A)$ or as elements of the vector space $Eff_\mathbb{R}(A)$ – for the two vector spaces coincide in quantum theory. At the operational level, we can prove two analogs of the spectral theorem, one for $St_\mathbb{R}(A)$ and one for $Eff_\mathbb{R}(A)$.

Theorem 12.18 (Spectral Theorem for the Span of States) *For every system A, every element of the vector space $St_\mathbb{R}(A)$ can be decomposed into a real linear combination of perfectly discriminable pure states.*

The proof of the theorem is based on the following:

Lemma 12.19 *If ξ is an element of the vector space $St_\mathbb{R}(A)$, it can be written as $\xi = a\,\chi_A - b\rho$, where ρ is a normalized state and a, b are two non-negative coefficients.*

Proof Let us write ξ as $\xi = c_+\rho_+ - c_-\rho_-$, where $c_\pm$ are non-negative coefficients and $\rho_\pm$ are normalized states. Diagonalize ρ_+ as $\rho_+ = \sum_{i=1}^{r} p_i\alpha_i$ where $\{\alpha_i\}_{i=1}^{r}$ are perfectly discriminable, define $p := \max\{p_i\}$, and consider the vector ζ defined by

$$\zeta := \chi_A - \frac{1}{d_A\,c_+ + p}\,\xi \tag{12.12}$$

$$= \frac{1}{d_A}\sum_{i=1}^{d_A}\left(1 - \frac{p_i}{p}\right)\alpha_i + \frac{c_-}{c_+}\,\rho_- \,,$$

having set $p_i = 0$ for $i > r$. Note that ζ is a linear combination of the states $\{\alpha_i\}_{i=1}^{d_A}$ and ρ_-, and that the coefficients of the combination are non-negative. Hence, ζ is proportional to a state, which we denote by $\rho := \zeta/(e_A|\zeta)$. The desired decomposition then follows from Eq. (12.12), setting $a := d_A\,c_+ + p_*$ and $b := d_A\,c_+ + p_*\,(e_A|\zeta)$. □

Proof of Theorem 12.18 Let ξ be a generic element of $St_\mathbb{R}(A)$, decomposed as $\xi = a\,\chi_A - b\,\rho$. Diagonalizing ρ as $\rho = \sum_{i=1}^{k} p_i\,\alpha_i$ and extending the set $\{\alpha_i\}_{i=1}^{k}$ to a pure maximal set we then obtain

$$\xi = \sum_{i=1}^{d_A}\left(\frac{a}{d_A} - bp_i\right)\alpha_i \,,$$

having defined $p_i = 0$ for $i > k$. □

An analog decomposition can be proved for the elements of the vector space spanned by the effects:

Corollary 12.20 (Spectral Theorem for the Span of Effects) *For every system* A, *every element of the vector space* $\mathsf{Eff}_{\mathbb{R}}(A)$ *can be decomposed as a linear combination of coexisting atomic effects belonging to a maximally discriminating measurement.*

The proof is identical to the proof of diagonalization for effects.

Using the spectral theorem, we can now prove a result that you may have already suspected:

Proposition 12.21 *Every system has a continuum of pure states.*

Proof Thanks to ideal compression, it is enough to prove the result for two-dimensional systems: if every two-dimensional system has a continuum of pure states, then every two-dimensional face of a generic system should have a continuum of pure states.

Let A be a two-dimensional system. From Exercise 12.3 we know that A must have at least three pure states $\{\alpha_0, \alpha_1, \alpha_2\}$. Of course, the three states must be linearly independent, otherwise one of them would have to be mixed. This implies that the subspace

$$\mathsf{K} = \left\{ \xi \in \mathsf{St}_{\mathbb{R}}(A) \mid (e_A|\xi) = 0 \right\} \subset \mathsf{St}_{\mathbb{R}}(A)$$

is at least two-dimensional, because it contains the linearly independent vectors $\alpha_0 - \alpha_1$ and $\alpha_0 - \alpha_2$. Now, let ξ be an arbitrary element of K. Using the spectral theorem, ξ can be decomposed as

$$\xi = c \left(\varphi_\xi - \varphi_\xi^\perp \right),$$

where $\{\varphi_\xi, \varphi_\xi^\perp\}$ are two perfectly discriminable pure states and c is a constant. Inserting the relation $\chi_A = \left(\varphi_\xi + \varphi_\xi^\perp \right)/2$, we also have

$$\xi = 2c \left(\varphi_\xi - \chi \right).$$

This equation shows that the map $\xi \mapsto \varphi_\xi$ is injective on the rays of K, that is $\varphi_{\xi_1} = \varphi_{\xi_2}$ only if $\xi_2 = t\xi_1$ for some constant t. Now, since K has a continuous infinity of rays, there must be a continuous set of pure states. □

Exercise 12.3 Combining purification, perfect discriminability, and local discriminability, show that every system must have more than two pure states.

12.8 Operational Version of the Schmidt Decomposition

In addition to the operational versions of the spectral theorem, we can also prove an operational version of the Schmidt decomposition in quantum theory:

Proposition 12.22 *Let* Ψ *be a pure state of* AB *and let* ρ_A *and* ρ_B *be its marginals on systems* A *and* B, *respectively. For every diagonalization of* ρ_A, *say*

$$\rho_A = \sum_{i=1}^{r} p_i \alpha_i,$$

the state ρ_B can be diagonalized as

$$\rho_B = \sum_{i=1}^{r} p_i\, \beta_i$$

for a suitable set of perfectly discriminable pure states $\{\beta_i\}_{i=1}^{r}$. Moreover, there exist two maximally discriminating measurements $\{a_i\}_{i=1}^{d_A}$ and $\{b_j\}_{j=1}^{d_B}$ such that

$$\Psi \begin{array}{c} A - a_i \\ B - b_j \end{array} = p_i\, \delta_{ij} \qquad \forall i \le r, \forall j \le r. \tag{12.13}$$

Proof   Let $\{\alpha_i\}_{i=1}^{d_A}$ be a pure maximal set extending the set $\{\alpha_i\}_{i=1}^{r}$, and let $\left\{\alpha_i^{\dagger}\right\}_{i=1}^{d_A}$ be the corresponding measurement. Applying the effect $\alpha_i^{\dagger}$ on one side of the pure state Ψ, we obtain

$$\Psi \begin{array}{c} A - \alpha_i^{\dagger} \\ B \end{array} = q_i\ \left(\beta_i \vdash B \right), \tag{12.14}$$

where β_i is a normalized pure state and q_i is a probability. Note that, in fact, $q_i = p_i$; indeed, we have

$$q_i = \Psi \begin{array}{c} A - \alpha_i^{\dagger} \\ B - e \end{array} = \left(\rho_A \right) A - \alpha_i^{\dagger} = p_i,$$

the last equality coming from the diagonalization of ρ_A. On the other hand, steering implies the relation

$$p_i\ \left(\alpha_i \vdash A \right) = \Psi \begin{array}{c} A \\ B - b_j \end{array}, \tag{12.15}$$

for some suitable measurement $\{b_j\}_{j=1}^{r}$. Combining Eqs. (12.14) and (12.15) we then obtain the relation

$$p_i\ \left(\beta_i \vdash B - b_j \right) = \Psi \begin{array}{c} A - \alpha_i^{\dagger} \\ B - b_j \end{array} = p_i\ \left(\alpha_j \vdash B - \alpha_i^{\dagger} \right) = p_i\, \delta_{ij}, \tag{12.16}$$

valid for every $i, i \le r$. Since p_i is non-zero for every $i \le r$, the above relation implies that the states $\{\beta_i\}_{i=1}^{r}$ are perfectly discriminable. Then, the desired decomposition of ρ_B comes from Eq. (12.14), summing over i and recalling that $q_i = p_i$. Finally, extending the set $\{\beta_i\}_{i=1}^{r}$ to a pure maximal set $\{\beta_i\}_{i=1}^{d_B}$, we have

$$\Psi \begin{array}{c} A - \alpha_i^{\dagger} \\ B - \beta_j^{\dagger} \end{array} = p_i\ \left(\beta_i \vdash B - \beta_j^{\dagger} \right) = p_i\, \delta_{ij} \qquad \forall i \le r, \forall j \le r. \tag{12.17}$$

thus proving Eq. (12.13) with $a_i := \alpha_i^{\dagger}$ and $b_j := \beta_j^{\dagger}$. $\square$

Exercise 12.4 Given an arbitrary purification of the state ρ, show that the dimension of the purifying system cannot be smaller than the rank of ρ.

12.9 Summary

In this chapter we defined conjugate systems as systems that can be maximally correlated, i.e. they can be in a pure state that has invariant marginals on both sides. We showed that two conjugate systems are mirror images of one another, with the states and measurements on one system in one-to-one correspondence with the states and measurements of the other. Using this fact, we proved two important results:

1. All maximal sets of perfectly discriminable pure states of a system have the same cardinality, which we call *informational dimension*. This number quantifies the maximum number of states that can be perfectly discriminated in the system. In quantum theory, it corresponds to the dimension of the system's Hilbert space.
2. The barycenter of a maximal set of perfectly discriminable pure states is the invariant state.

In turn, these two results allowed us to prove that every state can be diagonalized, i.e. decomposed into a mixture of perfectly discriminable pure states. This provided operational versions of the spectral theorem for Hermitian matrices, as well as an operational version of the Schmidt decomposition of pure bipartite quantum states.

Solutions to Selected Problems and Exercises

Exercise 12.2

Suppose that the maximal set of perfectly discriminable pure states $\{\varphi_i\}_{i=1}^{d_A}$ is contained in F. Then the completely mixed state $\chi_A = 1/d_A \sum_{i=1}^{d_A} \varphi_i$ belongs to F. Conversely, let $\rho \in F$ be completely mixed. Then, $\mathsf{RefSet}_1(\rho) = \mathsf{RefSet}_1(\chi_A) \equiv \mathsf{St}_1(A)$. Since every maximal set of pure states $\{\varphi_i\}_{i=1}^{d_A}$ belongs to $\mathsf{RefSet}_1(\chi_A)$, then it also belongs to $\mathsf{RefSet}_1(\rho) = F$.

Exercise 12.3

By contradiction, suppose that system A has only two pure states α_0 and α_1. Then, every mixed state can be written as $\rho = p\alpha_0 + (1-p)\alpha_1$. Moreover, the states α_0 and α_1 must be perfectly discriminable (by the perfect discriminability axiom). Denoting by $\{a_0, a_1\}$ the measurement that discriminates among them, we have

$$\rho = \sum_{i=0,1} (a_i|\rho)\,\alpha_i \qquad \forall \rho \in \mathsf{St}(A)\,.$$

By local discriminability, this means that the measurement $\{a_0, a_1\}$ followed by the preparation of the states $\{\alpha_0, \alpha_1\}$ has no disturbance. Clearly, this would be in contradiction with the no information without disturbance, which follows from purification. $\square$

PART IV

QUANTUM THEORY FROM
THE PRINCIPLES

Conclusive Teleportation   13

In quantum theory, states are represented by Hermitian operators on complex Hilbert spaces. This fact leads to beautiful mathematical structures, but also to puzzling questions: Why *Hermitian operators*? And why a *complex* Hilbert space? In this chapter we attack these questions, establishing a surprising link between the Hilbert-space representation of states and the task of *conclusive teleportation*, where the sender attempts to transfer an unknown state to the receiver without communicating any classical message. We will see that conclusive teleportation cannot be achieved deterministically, unless causality and local discriminability are satisfied. Quantitatively, we will establish two achievable bounds on the maximum probability of conclusive teleportation for a given system A. The first bound is in terms of d_A, the maximum number of perfectly discriminable states. The second bound is in terms of D_A, the dimension of the vector space spanned by the states of the system. Combining the two bounds, we will prove the fundamental equality

$$D_A = d_A^2 \,,$$

universally valid for every system A. Thanks to this equality, every state of system A can be represented as a $d_A \times d_A$ real matrix, or, equivalently, as a Hermitian operator on a complex d_A-dimensional Hilbert space. Being able to represent states as Hermitian operators is an important milestone towards the full derivation of quantum theory, achieved in the final chapter of this book.

13.1 The Task

In Chapter 7, we saw that our principles imply the existence of a teleportation protocol, capable of transferring an arbitrary state from a sender to a receiver using only pre-existing correlations and a finite amount of classical communication. Here we focus on a variant of teleportation where the sender and the receiver do not communicate at all: the input state at the sender's end is transferred to the receiver only *probabilistically*, provided that a suitable measurement performed by the sender gives the right outcome.

As in the usual teleportation scenario, the goal is to transfer a generic state of a system A from the sender to the receiver. As a resource, the sender has a system B correlated with another copy of system A in the receiver's laboratory. Her strategy to transfer the state will be to perform a binary measurement on the input system A together with system B, obtaining one of the outcomes "yes" or "no." The protocol is designed in such a way that, conditionally on the outcome "yes," the state will be transferred to the receiver. The outcome "no" will result in a failure.

In pictures, the condition of successful state transfer is

$$
\left(\Psi \begin{array}{c} A \\ B \end{array} B_{\text{yes}} \; (\rho\!-\!A \right) = p_{\text{yes}} \; (\rho\!-\!A \qquad \forall \rho \in \mathrm{St}(A), \tag{13.1}
$$

where Ψ is the resource state used for teleportation, ρ is the state to be teleported, B_{yes} is the effect corresponding to the successful outcome, and p_{yes} is the probability that the successful outcome occurs.

In principle, the probability p_{yes} could depend on the input state. However, it is easy to see that it does not: indeed, for every pair of input states ρ_1 and ρ_2 the condition (13.1) implies the requirement

$$
\left(\Psi \begin{array}{c} A \\ B \end{array} B_{\text{yes}} \; (\rho_i\!-\!A \right) = p_{\text{yes}}^{(i)} \; (\rho_i\!-\!A \qquad \forall i \in \{1, 2, 3\},
$$

with $\rho_3 := (\rho_1 + \rho_2)/2$. Combining the three relations, one obtains

$$
p_{\text{yes}}^{(3)} \; (\rho_3\!-\!A = \frac{1}{2} \left(\Psi \begin{array}{c} A \\ B \end{array} B_{\text{yes}} \; (\rho_1\!-\!A \right) + \frac{1}{2} \left(\Psi \begin{array}{c} A \\ B \end{array} B_{\text{yes}} \; (\rho_2\!-\!A \right)
$$

$$
= \frac{p_{\text{yes}}^{(1)}}{2} \; (\rho_1\!-\!A + \frac{p_{\text{yes}}^{(2)}}{2} \; (\rho_2\!-\!A ,
$$

which in turn implies

$$
\left(\frac{p_{\text{yes}}^{(3)} - p_{\text{yes}}^{(2)}}{2}\right) \; (\rho_2\!-\!A = \left(\frac{p_{\text{yes}}^{(1)} - p_{\text{yes}}^{(3)}}{2}\right) \; (\rho_1\!-\!A .
$$

As long as ρ_1 and ρ_2 are distinct, the above equation forces the equality $p_{\text{yes}}^{(1)} = p_{\text{yes}}^{(2)} = p_{\text{yes}}^{(3)}$. This proves that the probability p_{yes} in Eq. (13.1) is independent of the input state ρ. We will call p_{yes} the *probability of conclusive teleportation*.

Since p_{yes} is independent of ρ, local discriminability allows one to simplify Eq. (13.1) into

$$
\left(\Psi \begin{array}{c} A \\ B \end{array} B_{\text{yes}} \right) = p_{\text{yes}} \; -\!A\!-\boxed{I}\!-\!A\!- . \tag{13.2}
$$

Besides being more elegant, this formulation will become useful in the next sections.

13.2 The Causality Bound

Considering conclusive teleportation, the natural question is: *How large is the probability of success?* Intuitively, causality implies that the probability cannot be 1, otherwise the sender would be able to transmit a signal to the receiver without sending any physical system. Let us make this intuition precise: denoting by B_{no} the effect corresponding to the unsuccessful outcome, the normalization condition $B_{yes} + B_{no} = e_B \otimes e_A$, implied by causality, implies the relation

(13.3)

where ρ_0 is the marginal of Ψ on system A. Now, if the outcome "yes" occurred with probability $p_{yes} = 1$, the term corresponding to the outcome "no" should vanish, and we would have

Combined with the teleportation condition (13.2), the above equation would imply the relation

meaning that the identity on system A is an *erasure channel*, that is, it produces a fixed state independently of the input. As a consequence, system A must be *trivial*: all of its states must be multiples of a single state. In summary: the probability of conclusive teleportation must be smaller than 1 for every system that could be used to send a signal.

We now make the argument quantitative, using causality to derive an upper bound to the probability of conclusive teleportation in terms of the number of perfectly discriminable states. The upper bound is as follows:

Theorem 13.1 (Causality Bound) *In every conclusive teleportation protocol for system A, the probability of success satisfies the bound*

$$p_{yes} \leq \frac{1}{d_A^2},$$

(13.4)

where d_A is the maximum number of perfectly discriminable states of system A.

Proof Let $\mathcal{T}$ be the twirling channel for system A, decomposed as $\mathcal{T} = \sum_{x \in X} p_x \mathcal{U}_x$ for suitable probabilities $\{p_x\}_{x \in X}$ and suitable reversible transformations $\{\mathcal{U}_x\}_{x \in X}$ – see Eq. (7.6) for the proof of this decomposition. Clearly, the condition of conclusive teleportation (13.2) implies that one has

for every $x \in \mathsf{X}$. Applying the transformations $\mathcal{U}_x$ and $\mathcal{U}_x^{-1}$ on both sides of Eq. (13.3), multiplying by p_x, and summing over x we obtain

Here the third equality follows from the relation $(e_{\mathrm{A}}|\,\mathcal{U}_x^{-1} = (e_{\mathrm{A}}|\quad \forall x \in \mathsf{X}$ and from the decomposition $\mathcal{T} = \sum_{x \in \mathsf{X}} p_x \mathcal{U}_x$. The fourth equality follows from the fact that the twirling channel transforms every state into the invariant state χ_{A}. The above chain of equalities, combined with the condition of conclusive teleportation (13.2), yields the relation

$$p_{\mathrm{yes}} \quad \text{---A---}\boxed{\mathcal{I}}\text{---A---} \;+\; \text{---A---}\boxed{\mathcal{A}_{\mathrm{no}}}\text{---A---} \;=\; \text{---A---}(e)\;\; (\chi)\text{---A---}, \tag{13.5}$$

where $\mathcal{A}_{\mathrm{no}}$ is the transformation defined by $\mathcal{A}_{\mathrm{no}} := \sum_{x \in \mathsf{X}} p_x\,(\mathcal{U}_x \otimes \mathcal{B}_{\mathrm{no}})(\Psi \otimes \mathcal{U}_x^{-1})$. Now, let $\Phi \in \mathsf{PurSt}(\mathrm{A\bar{A}})$ be a minimal purification of the invariant state χ_{A}. Applying both members of Eq. (13.5) on system A, we obtain

the right-hand side following from the fact that the marginal of Φ on system $\bar{\mathrm{A}}$ is the invariant state $\chi_{\bar{\mathrm{A}}}$ (cf. Lemma 11.17). Finally, applying the effect $\Phi^{\dagger}$ on both sides yields the inequality

$$p_{yes} \quad \leq \quad \left(\begin{array}{c} \boxed{\chi} \!-\! A \\ \boxed{\chi} \!-\! \overline{A} \end{array} \!\!\boxed{\Phi^\dagger} \right).$$ (13.6)

In order to obtain the desired bound, it is enough to decompose the invariant state $\chi_A \otimes \chi_{\overline{A}}$ as $\chi_A \otimes \chi_{\overline{A}} = \sum_{i=1}^{d_A^2} \Phi_i / d_A^2$, where $\{\Phi_i\}_{i=1}^{d_A^2}$ is a maximal set of perfectly discriminable pure states containing the state Φ. □

The name "causality bound" emphasizes that the origin of the bound is Eq. (13.3), which follows directly from causality. However, the bound does not follow *only* from causality. In fact, the right-hand side of Eq. (13.4) depends on a number of operational features that are specific to our theories. This is clear from our proof, which used:

1. the existence of a twirling channel and the existence of the invariant state χ;
2. the fact that the invariant state has a minimal purification Φ;
3. the fact that the marginal of Φ on the purifying system is invariant;
4. the fact that the state $\chi_A \otimes \chi_{\overline{A}}$ can be decomposed into a mixture of d_A^2 perfectly discriminable states, one of which is Φ;
5. the fact that the state χ_{AB} is of the form $\chi_A \otimes \chi_B$.

A theory that fails to satisfy one of these properties could have a different value in the r.h.s. of the causality bound. This is the case, for example, of classical theory, in which the causality bound reads $p_{yes} \leq 1/d_A$:

Exercise 13.1 Show that the probability of conclusive teleportation for a classical system of dimension d_A is upper bounded by $1/d_A$ and provide a protocol that achieves the bound.

13.3 Achieving the Causality Bound

Here we now show that the causality bound is attainable, thanks to steering:

Theorem 13.2 *For every system A, there exists a conclusive teleportation procotol with success probability*

$$p_{yes} = \frac{1}{d_A^2} .$$ (13.7)

Proof Let $\Phi \in \mathsf{PurSt}(A\overline{A})$ be a minimal purification of the invariant state χ_A. Then, since the marginal of Φ on system $\overline{A}$ is the invariant state (Proposition 12.2), we have the identity

$$\begin{array}{c} \boxed{\Phi} \begin{array}{l} \!-\! A \\ \!-\! \overline{A} \!\boxed{e} \end{array} \\ \boxed{\Phi} \begin{array}{l} \!-\! A \!\boxed{e} \\ \!-\! \overline{A} \end{array} \end{array} = \begin{array}{c} \boxed{\chi} \!-\! A \\ \boxed{\chi} \!-\! \overline{A} \end{array} = \boxed{\chi} \!-\! A\overline{A} \;,$$

where the last equality is proved in Exercise 7.2. Then, the state $|\Phi\rangle_{A\overline{A}}|\Phi\rangle_{A\overline{A}}$ is a minimal purification of the invariant state $\chi_{A\overline{A}}$. On the other hand, by Theorem 12.8 the maximum probability of a pure state in the convex decomposition of $\chi_{A\overline{A}}$ is $p_{\max}^{A\overline{A}} = 1/d_{A\overline{A}}$, and by Corollaries 12.10 and 12.11 one has $p_{\max}^{A\overline{A}} = 1/(d_A d_{\overline{A}}) = 1/d_A^2$. By steering, there must exist an effect $\Phi^\sharp$ that induces the pure state Φ with probability $p_{\max}^{A\overline{A}} = 1/d_A^2$: precisely, one has

$$
\begin{array}{c}
\Phi \\
\Phi
\end{array}
\quad \Phi^\sharp = \frac{1}{d_A^2} \quad \Phi \qquad (13.8)
$$

cf. Eq. (11.10) for the definition $\Phi^\sharp$. Now, since Φ is faithful, the above equation implies

$$
\Phi \quad \Phi^\sharp = \frac{1}{d_A^2} \quad A \boxed{\mathcal{I}} A . \qquad (13.9)
$$

Hence, conclusive teleportation is achieved with probability $p_{\text{yes}} = 1/d_A^2$.   □

In summary, the maximum probability of conclusive teleportation for system A, denoted by $p_{\text{tele}}^{(A)}$, is given by

$$
p_{\text{tele}}^{(A)} = \frac{1}{d_A^2} . \qquad (13.10)
$$

This is a deep and remarkable equality. It links in a non-trivial way two distinct operational tasks – teleporting information and discriminating states. Moreover, it plays a crucial role in our derivation of quantum theory. In the next sections we will prove an alternative expression for the maximum probability of conclusive teleportation: the alternative expression is in terms of the D_A, the dimension of the vector space spanned by the states of system A.

13.4 The Local Discriminability Bound

We now provide a new upper bound on the probability of conclusive teleportation. Remarkably, the bound relies *only* on local discriminability:

Theorem 13.3 (Local Discriminability Bound)   *In every conclusive teleportation protocol for system A, the probability of success satisfies the bound*

$$
p_{\text{yes}} \le \frac{1}{D_A} . \qquad (13.11)
$$

Proof Thanks to local discriminability, every bipartite state $\Psi \in \mathrm{St}(AB)$ can be written as

$$
\Psi \begin{array}{c} A \\ B \end{array} = \sum_{i=1}^{D_A} \sum_{j=1}^{D_B} \Psi_{ij} \quad \begin{array}{c} \alpha_i \;\text{---}\; A \\ \beta_j \;\text{---}\; B \end{array} , \tag{13.12}
$$

where $\{\alpha_i\}$ and $\{\beta_j\}$ are bases for the vector spaces $\mathrm{St}_{\mathbb{R}}(A)$ and $\mathrm{St}_{\mathbb{R}}(B)$, respectively. Similarly, every bipartite effect $F \in \mathrm{Eff}(BA)$ can be written as

$$
\begin{array}{c} B \;\text{---} \\ A \;\text{---} \end{array} F = \sum_{k=1}^{D_A} \sum_{l=1}^{D_B} F_{kl} \quad \begin{array}{c} B \;\text{---}\; \beta_k^* \\ A \;\text{---}\; \alpha_l^* \end{array} , \tag{13.13}
$$

where $\{\beta_k^*\}$ and $\{\alpha_l^*\}$ are the dual bases of $\{\beta_j\}$ and $\{\alpha_i\}$, respectively – that is, they satisfy the relations $(\alpha_k^* | \alpha_i) = \delta_{ik}$ and $(\beta_l^* | \beta_j) = \delta_{jl}$ for arbitrary i, j, k, l. As a consequence, we have the relations

$$
\Psi \begin{array}{c} A \;\text{---} \\ B \end{array}\!\!\!\begin{array}{c} \\ A \end{array} F = \sum_{i,l} [\Psi F]_{il} \quad \begin{array}{c} \alpha_i \;\text{---}\; A \\ A \;\text{---}\; \alpha_l^* \end{array} , \tag{13.14}
$$

and

$$
\Psi \begin{array}{c} A \\ B \end{array} S_{A,B} \begin{array}{c} B \\ A \end{array} F = \mathrm{Tr}\,[\Psi F] , \tag{13.15}
$$

where $S_{A,B}$ is the reversible transformation that swaps A and B.

Now, consider a protocol consisting of a resource state Ψ and a successful effect B_{yes}. Using Eq. (13.14), we can rewrite the condition for successful teleportation (13.2) as

$$
[\Psi B_{\mathrm{yes}}]_{il} = p_{\mathrm{yes}} \delta_{il} \qquad \forall i, l \in \{1, \dots, D_A\} . \tag{13.16}
$$

Taking the trace on both sides and using Eq. (13.15) we then obtain the bound

$$
p_{\mathrm{yes}} D_A = \mathrm{Tr}\,[\Psi B_{\mathrm{yes}}]
$$

$$
= \Psi \begin{array}{c} A \\ B \end{array} S_{A,B} \begin{array}{c} B \\ A \end{array} B_{\mathrm{yes}}
$$

$$
\leq 1 ,
$$

and therefore $p_{\mathrm{yes}} \leq 1/D_A$. □

Note that the success probability is strictly smaller than 1 for every non-trivial system: only a system with $D_A = 1$ can have success probability $p_{\mathrm{yes}} = 1$. It is worth stressing that the bound $p_{\mathrm{yes}} \leq 1/D_A$ does not require any assumption other than local discriminability. For example, causality does not play any role here: *even in a non-causal world, a sender cannot use conclusive teleportation to transfer the state of a system without sending any message to the receiver!*

13.5 Achieving the Local Discriminability Bound

We now show that the bound $p_{\text{yes}} \leq 1/D_A$ can be achieved. This result requires much more than just local discriminability – most prominently, it requires the fact that pure states can be identified by atomic effects.

The key to the achievability proof is to consider the atomic effect $\Phi^\dagger$ that identifies the entangled state Φ, namely $(\Phi|\Phi^\dagger) = 1$. The first ingredient used in the proof is the observation that the effect $\Phi^\dagger$ can be expressed in terms of the effect $\Phi^\sharp$, previously used to construct a conclusive teleportation protocol. Essentially, the two effects $\Phi^\dagger$ and $\Phi^\sharp$ coincide, up to a swap of systems A and $\overline{\text{A}}$ and to a local transformation on A:

Lemma 13.4 *For every system A, one has*

$$\begin{array}{c}\text{A}\\ \overline{\text{A}}\end{array}\!\!\Phi^\dagger \quad = \quad \begin{array}{c}\text{A}\;\boxed{\mathcal{U}}\;\text{A}\quad\quad\overline{\text{A}}\\ \overline{\text{A}}\quad\boxed{S_{\text{A},\overline{\text{A}}}}\quad\text{A}\;\Phi^\sharp\end{array}, \tag{13.17}$$

where $\mathcal{U}$ is a reversible transformation and $S_{\text{A},\overline{\text{A}}}$ is the reversible transformation that swaps A with A'.

The proof is quite elaborate and we postpone it to the final section of this chapter. The second step towards proving achievability of the local discriminability bound is a simple group-theoretic result:

Lemma 13.5 *For every system A, there exists a basis for $\text{St}_\mathbb{R}(\text{A})$, denoted by $\{\alpha_i\}$, such that the matrix $(O_\mathcal{U})_{ij} := (\alpha_i^*|\mathcal{U}|\alpha_j)$ is orthogonal for every reversible transformation $\mathcal{U}$.*

The proof is standard and can be found in the Appendix to this chapter. Using Lemmas 13.4 and 13.5, we can now show the desired result:

Theorem 13.6 (State Identification Bound) *For every system A there exists a conclusive teleportation protocol with probability*

$$p_{\text{yes}} = \frac{1}{D_A}. \tag{13.18}$$

Proof of Theorem 13.6 Let $\mathcal{U}$ be the reversible transformation in Lemma 13.4 and $\{\alpha_i\}$ be the basis in Lemma 13.5. By definition, we have

$$\boxed{\alpha_i}\;\text{A}\;\boxed{\mathcal{U}}\;\text{A} \quad = \quad \sum_k \mathcal{U}_{ki}\;\boxed{\alpha_k}\;\text{A}.$$

Expanding the state Φ as

$$\Phi\!\begin{array}{c}\text{A}\\ \overline{\text{A}}\end{array} \quad = \quad \sum_{i,j=1}^{D_A} \Phi_{ij}\;\begin{array}{c}\boxed{\alpha_i}\;\text{A}\\ \boxed{\tilde{\alpha}_j}\;\overline{\text{A}}\end{array}, \tag{13.19}$$

we then obtain

$$(13.20)$$

Now, we have the chain of relations

$$= \mathrm{Tr}\left[\mathcal{U}\Phi\,\Phi^{\sharp}\right]$$
$$= p_{\mathrm{yes}}\,\mathrm{Tr}[\mathcal{U}] \qquad (13.21)$$
$$\leq p_{\mathrm{yes}}\,D_{\mathrm{A}}\,.$$

Here the second equality follows from Eq. (13.15) with $\Psi \equiv (\mathcal{U} \otimes \mathcal{I}_{\overline{A}})\,\Phi$ and $F \equiv \Phi^{\dagger}$. The third equality follows from the fact that Φ and $\Phi^{\sharp}$ allow for conclusive teleportation, and therefore satisfy the relation $[\Phi\,\Phi^{\sharp}]_{ij} = p_{\mathrm{yes}}\,\delta_{ij}$, as in Eq. (13.16). The last inequality comes from the fact $\mathcal{U}_{ij}$ is a $D_{\mathrm{A}} \times D_{\mathrm{A}}$ orthogonal matrix (Lemma 13.5) and consequently its trace cannot be larger than D_{A}. In conclusion, we have proved the bound $p_{\mathrm{yes}} \geq 1/D_{\mathrm{A}}$. Since $1/D_{\mathrm{A}}$ is also an upper bound (by the discriminability bound), we have the equality $p_{\mathrm{yes}} = 1/D_{\mathrm{A}}$. $\qquad\square$

An interesting byproduct of the above proof is that the reversible transformation $\mathcal{U}$ appearing in Lemma 13.4 must be the identity:

Exercise 13.2 Show that the effects $\Phi^{\dagger}$ and $\Phi^{\sharp}$ coincide up to swap of system A with system $\overline{\mathrm{A}}$.

In summary, we obtained that the maximum success probability over all protocols of conclusive teleportation for system A is given by

$$p_{\mathrm{tele}}^{(\mathrm{A})} = \frac{1}{D_{\mathrm{A}}} \qquad (13.22)$$

and is achieved by a protocol that uses the resource state Φ and the successful effect $\Phi^{\dagger}$ (up to swap of systems A and $\overline{\mathrm{A}}$).

13.6 The Origin of the Hilbert Space

We have seen two alternative expressions for the maximum probability of conclusive teleportation: one in terms of the informational dimension d_{A} [Eq. (13.10)] and one in terms of the state space dimension D_A [Eq. (13.22)]. Combining them, we obtain the fundamental equality

$$D_A = d_A^2 .$$ (13.23)

This equality implies that the states of system A can be represented as real $d_A \times d_A$ matrices. Indeed, we can pick any basis for the real vector space $St_\mathbb{R}(A)$ and expand every state $\rho \in St(A)$ in terms of D_A real coefficients. Since $D_A = d_A^2$, the expansion coefficients can be arranged in a square $d_A \times d_A$ matrix – let us denote it by M_ρ. Equivalently, the matrix M_ρ can be replaced by a complex Hermitian matrix S_ρ, defined by

$$S_\rho := f\left(M_\rho\right) \qquad f(M) := \frac{M + M^T}{2} + i\frac{M - M^T}{2} ,$$ (13.24)

M^T denoting the transpose. Note that the two representations M_ρ and S_ρ are completely equivalent: indeed, the function f is an invertible linear map, with inverse f^{-1} given by $f^{-1}(S) = (S + S^*)/2 - i(S - S^*)/2$, S^* denoting the complex conjugate.

Equation (13.24) is the first point where the complex numbers become visible in our reconstruction: it tells us that states can be represented as complex Hermitian matrices – or equivalently, as Hermitian operators acting on a complex Hilbert space. In addition, we know that the operator corresponding to states must span the space of *all* Hermitian operators, as required by Eq. (13.23). For example, a theory where only the real Hermitian matrices represent states would not be compatible with Eq. (13.23). This observation excludes not only quantum theory on real Hilbert spaces (which, in fact, was already excluded by local discriminability), but also every other theory that describes states using only real Hermitian matrices.

The above observations bring us quite close to the quantum formalism, although there is still work to do before we get there; for example, we still need to show that the Hermitian operators representing states are positive semidefinite. Also, we still need to show that every unit vector in the Hilbert space is associated to a pure state: while the Hilbert space is already here, we still have to give operational meaning to its vectors. All this will be the subject of the last chapter of the book.

13.7 Isotropic States and Effects

In this section we provide the proof of Lemma 13.4. The proof is interesting in itself, as it is based on the notions of *isotropic state* and *isotropic effect*, which in turn rely on the notions of *transpose* and *conjugate* of a reversible transformation.

Let us start by defining the transpose. Note that for any reversible transformation $\mathcal{U}$ on system A one has

Hence, the uniqueness of purification [Eq. (7.2)] implies that there exists a reversible transformation $\mathcal{U}^{\mathrm{T}}$ on $\overline{\mathrm{A}}$ such that

$$(13.25)$$

Since Φ is faithful on $\overline{\mathrm{A}}$, the transformation $\mathcal{U}^{\mathrm{T}}$ is uniquely determined by the above equation. We call the unique transformation $\mathcal{U}^{\mathrm{T}}$ satisfying Eq. (13.25) the *transpose of $\mathcal{U}$ relative to the state* Φ.

Note that the map $\mathcal{U} \mapsto \mathcal{U}^{\mathrm{T}}$ is injective. This follows easily from the fact that the state Φ is faithful on A. Furthermore, since the marginal of Φ on system $\overline{\mathrm{A}}$ is the invariant state, the roles of A and $\overline{\mathrm{A}}$ can be exchanged: we can define the transpose of a reversible transformation $\mathcal{V} \in \mathbf{G}_{\mathrm{A}}$ as the unique transformation $\mathcal{V}^{\widetilde{\mathrm{T}}}$ satisfying the relation

$$(13.26)$$

Again, the map $\mathcal{V} \mapsto \mathcal{V}^{\widetilde{\mathrm{T}}}$ is injective. Moreover, combining Eqs. (13.25) and (13.26) shows that the map yields the relations T and $\widetilde{\mathrm{T}}$, which are the inverse of each other

$$\left(\mathcal{U}^{\tau}\right)^{\widetilde{\tau}} = \mathcal{U} \quad \forall \mathcal{U} \in \mathbf{G}_{\mathrm{A}} \qquad \text{and} \qquad \left(\mathcal{V}^{\widetilde{\tau}}\right)^{\tau} = \mathcal{V} \quad \forall \mathcal{V} \in \mathbf{G}_{\overline{\mathrm{A}}}.$$

In particular, this means that T and $\widetilde{\mathrm{T}}$ are also surjective.

Once transposition is defined, one can define conjugation:

Definition 13.7 The *conjugate* of a reversible gate $\mathcal{U} \in \mathbf{G}_{\mathrm{A}}$ is the reversible gate $\mathcal{U}^* \in \mathbf{G}_{\overline{\mathrm{A}}}$ defined by $\mathcal{U}^* := \left(\mathcal{U}^{\mathrm{T}}\right)^{-1} \equiv \left(\mathcal{U}^{-1}\right)^{\mathrm{T}}$.

It is easy to see that conjugation is a group isomorphism from $\mathbf{G}_{\mathrm{A}}$ to $\mathbf{G}_{\overline{\mathrm{A}}}$, meaning that one has

$$(\mathcal{U}\mathcal{V})^* = \mathcal{U}^*\mathcal{V}^* \quad \forall \mathcal{U}, \mathcal{V} \in \mathbf{G}_{\mathrm{A}}.$$

The conjugation map allows us to define *isotropic* states and effects:

Definition 13.8 A state $\Psi \in \mathsf{St}(A\overline{A})$ is *isotropic* if it is invariant under all reversible transformations $\mathcal{U} \otimes \mathcal{U}^*$, namely

$$
\Psi \!\!\begin{array}{c} \text{A} \,\boxed{\mathcal{U}}\, \text{A} \\ \overline{\text{A}} \,\boxed{\mathcal{U}^*}\, \overline{\text{A}} \end{array} \quad = \quad \Psi \!\!\begin{array}{c} \text{A} \\ \overline{\text{A}} \end{array} \qquad \forall \mathcal{U} \in \mathbf{G}_{\text{A}} . \tag{13.27}
$$

Similarly, an effect $F \in \mathsf{Eff}(A\overline{A})$ is *isotropic* if it is invariant under all reversible transformations $\mathcal{U} \otimes \mathcal{U}^*$, namely

$$
\begin{array}{c} \text{A} \,\boxed{\mathcal{U}}\, \text{A} \\ \overline{\text{A}} \,\boxed{\mathcal{U}^*}\, \overline{\text{A}} \end{array} \!\! F \quad = \quad \begin{array}{c} \text{A} \\ \overline{\text{A}} \end{array} \!\! F \qquad \forall \mathcal{U} \in \mathbf{G}_{\text{A}} . \tag{13.28}
$$

By definition, the state Φ is isotropic; indeed, we have

$$
\Phi \!\!\begin{array}{c} \text{A} \,\boxed{\mathcal{U}}\, \text{A} \\ \overline{\text{A}} \,\boxed{\mathcal{U}^*}\, \overline{\text{A}} \end{array} = \Phi \!\!\begin{array}{c} \text{A} \hspace{3cm} \text{A} \\ \overline{\text{A}} \,\boxed{\mathcal{U}^\tau}\,\boxed{(\mathcal{U}^\tau)^{-1}}\, \overline{\text{A}} \end{array}
$$

$$
= \quad \Phi \!\!\begin{array}{c} \text{A} \\ \overline{\text{A}} \end{array} \qquad \forall \mathcal{U} \in \mathbf{G}_{\text{A}} .
$$

Similarly, the effect $\Phi^\dagger$ is isotropic; indeed, for every $\mathcal{U} \in \mathbf{G}_{\text{A}}$ we have

$$
\Phi \!\!\begin{array}{c} \text{A} \,\boxed{\mathcal{U}}\, \text{A} \\ \overline{\text{A}} \,\boxed{\mathcal{U}^*}\, \overline{\text{A}} \end{array} \!\! \Phi^\dagger \quad = \quad \Phi \!\!\begin{array}{c} \text{A} \\ \overline{\text{A}} \end{array} \!\! \Phi^\dagger \quad = \quad 1 .
$$

Since $\Phi^\dagger$ is the *only* atomic effect satisfying $(\Phi^\dagger | \Phi) = 1$, the above equality implies $(\Phi^\dagger | (\mathcal{U} \otimes \mathcal{U}^*) = (\Phi^\dagger | - $ that is, $\Phi^\dagger$ is isotropic.

The same line of argument allows one to prove the more general statement:

Lemma 13.9 *A pure state Ψ is isotropic if and only if the corresponding effect $\Psi^\dagger$ is isotropic.*

Now, *pure* isotropic states have an important property: they are all equivalent to the state Φ, up to local reversible transformations.

Lemma 13.10 *Every isotropic pure state $\Psi \in \mathsf{PurSt}_1(A\overline{A})$ is of the form*

$$
\Psi \!\!\begin{array}{c} \text{A} \\ \overline{\text{A}} \end{array} \quad = \quad \Phi \!\!\begin{array}{c} \text{A} \,\boxed{\mathcal{U}}\, \text{A} \\ \overline{\text{A}} \end{array}
$$

for some reversible transformation $\mathcal{U} \in \mathbf{G}_{\text{A}}$.

Proof Since Ψ satisfies the isotropy condition (13.27), its marginal on system $\overline{A}$ is the invariant state $\chi_{\overline{\text{A}}}$. Hence, the states Ψ and Φ are purifications of the same state. The uniqueness of the purification then yields the desired result. $\qquad\square$

By the duality between states and effects, it is immediate to obtain the following:

Lemma 13.11 *For every normalized isotropic pure state* $\Psi \in \mathsf{PurSt}_1(A\overline{A})$, *the effect* $\Psi^\dagger$ *is of the form*

$$
\begin{array}{c}
\overline{}A\phantom{} \\
\overline{}\overline{A}\phantom{}
\end{array}
\Psi^\dagger
\Big)
=
\begin{array}{c}
A\ \boxed{\mathcal{U}^{-1}}\ A \\
\overline{A}
\end{array}
\Phi^\dagger
\Big)
$$

for some reversible transformation $\mathcal{U} \in \mathbf{G}_A$.

The above result allows us to prove Lemma 13.4. To this purpose, it is enough to recognize that the teleportation effect $\Phi^\sharp$ is an isotropic effect of $\overline{A}A$:

Lemma 13.12 *For every system* A, *the teleportation effect* $\Phi^\sharp \in \mathsf{Eff}(\overline{A}A)$ *is isotropic.*

Proof For every reversible transformation $\mathcal{U} \in \mathbf{G}_A$, one has

$$
\Phi\!\!\begin{array}{c}A\\ \overline{A}\ \boxed{\mathcal{U}^*}\end{array}\!\!\begin{array}{c}\\ \boxed{\mathcal{U}}\end{array}\!\!\Phi^\sharp
\ \ \Phi\!\!\begin{array}{c}A\\ \overline{A}\end{array}
=
\Phi\!\!\begin{array}{c}A\ \boxed{\mathcal{U}^{-1}}\ A\\ \overline{A}\end{array}\!\!\Phi^\sharp
\ \ \Phi\!\!\begin{array}{c}A\\ \overline{A}\ \boxed{\mathcal{U}^\tau}\ \overline{A}\end{array}
$$

$$
= \frac{1}{d_A^2}\ \Phi\!\!\begin{array}{c}A\ \boxed{\mathcal{U}^{-1}}\ A\\ \overline{A}\ \boxed{\mathcal{U}^\tau}\ \overline{A}\end{array}
$$

$$
= \frac{1}{d_A^2}\ \Phi\!\!\begin{array}{c}A\\ \overline{A}\end{array}
$$

$$
=
\begin{array}{c}\Phi\!\!\begin{array}{c}A\\ \overline{A}\end{array}\\ \Phi\!\!\begin{array}{c}A\\ \overline{A}\end{array}\Phi^\sharp\end{array}\Big).
$$

Here, the second and fourth equalities follow from the teleportation condition (13.9), while the third equality follows from the fact that Φ is isotropic. Since the state $\Phi \otimes \Phi$ is faithful for $\overline{A}A$, the above relation implies $(\Phi^\sharp|(\mathcal{U}^* \otimes \mathcal{U}) = (\Phi^\sharp|$ for every $\mathcal{U} \in \mathbf{G}_A$. Hence, $\Phi^\sharp$ is an isotropic effect of $\overline{A}A$. $\square$

Proof of Lemma 13.4 We know that $\Phi^\sharp$ is an isotropic effect of $\overline{A}A$ (Lemma 13.12) and that it is normalized (Theorem 11.6). Hence, the swapped effect $(\Phi^\sharp|\,\mathcal{S}_{A,\overline{A}}$ is a normalized isotropic effect of $A\overline{A}$. Since $(\Phi^\sharp|\,\mathcal{S}_{A,\overline{A}}$ is normalized, it is of the form $(\Phi^\sharp|\,\mathcal{S}_{A,\overline{A}} = (\Psi^\dagger|$ for some suitable pure state Ψ. We are then in condition to apply Lemma 13.11, obtaining the relation

$$\frac{\text{A}}{\overline{\text{A}}} \boxed{S_{\text{A},\overline{\text{A}}}} \genfrac{}{}{0pt}{}{\overline{\text{A}}}{\text{A}} \Phi^{\sharp}\rangle = \genfrac{}{}{0pt}{}{\text{A}}{} \boxed{\mathcal{U}^{-1}} \genfrac{}{}{0pt}{}{\text{A}}{\overline{\text{A}}} \Phi^{\dagger}\rangle ,$$

where $\mathcal{U}$ is a suitable reversible transformation. Multiplying by $\mathcal{U}$ on both sides we obtain the desired result.   □

13.8 Summary

In this chapter we studied the task of conclusive teleportation. We quantified the maximum probability of success, providing two alternative expressions, one in terms of the informational dimension and one in terms of the state space dimension. Comparing the two expressions we obtained the fundamental relation $D_{\text{A}} = d_{\text{A}}^2$, which allows us to represent the states in our theory as Hermitian operators on a complex Hilbert space. This brings us quite close to quantum theory, but more work has still to be done in order to achieve a full reconstruction: it remains to show that the matrices representing states are positive and that all the projectors on unit vectors in the Hilbert space correspond to pure states. These results will be established in the final chapter of the book.

Appendix 13.1 Unitary and Orthogonal Representations

In this appendix we prove an important result of group theory that allows us to prove Lemma 13.5. This result plays a crucial role in our subsequent derivation.

Lemma 13.13 *Let M_g be a real representation of a compact group $\mathbf{G}$ on a Hilbert space $\mathcal{H}$. Then the representation M_g is similar to an orthogonal representation O_g.*

Proof Consider the positive definite matrix P defined by the integral

$$P := \int_{\mathbf{G}} \mathrm{d}g M_g^{\mathsf{T}} M_g,$$

where $\mathrm{d}g$ is the Haar measure on the group $\mathbf{G}$, that exists because $\mathbf{G}$ is compact. By definition one has $P^{\mathsf{T}} = P$, and exploiting the invariance of the Haar measure one can easily verify that $M_g^{\mathsf{T}} P M_g = P$ for every $g \in \mathbf{G}$. Let us now define the new representation

$$O_g := P^{\frac{1}{2}} M_g P^{-\frac{1}{2}},$$

similar to M_g by the change of basis $P^{-\frac{1}{2}}$ on $\mathcal{H}$. The representation O_g satisfies the following identity:

$$\begin{aligned}
O_g^{\mathsf{T}} O_g &= \left(P^{\frac{1}{2}} M_g P^{-\frac{1}{2}} \right)^{\mathsf{T}} \left(P^{\frac{1}{2}} M_g P^{-\frac{1}{2}} \right) \\
&= P^{-\frac{1}{2}} \left(M_g^{\mathsf{T}} P M_g \right) P^{-\frac{1}{2}} \\
&= I.
\end{aligned}$$

Finally, this means that the representation O_g is orthogonal.   □

By a straightforwardly analogous argument, one can prove the following lemma, whose proof we omit.

Lemma 13.14   *Let M_g be a complex representation of the compact group G on a Hilbert space $\mathcal{H}$. Then M_g is similar to a unitary representation U_g.*

We are now in position to prove Lemma 13.5:

Proof of Lemma 13.5   The group $\mathbf{G}_A$ of reversible transformations acting on a given system is compact. Indeed, by local discriminability every set of physical transformations is represented by a set of matrices with bounded matrix elements. Moreover, the group of reversible transformations is closed: every converging sequence of reversible transformations must converge to a transformation (by the closure of the set of transformations), which is easily checked to be reversible. Being a closed and bounded set in finite dimension, the group is compact. Let $\{\beta_i\}$ be a basis in $\mathsf{St}_{\mathbb{R}}(A)$, and $M_{\mathcal{U}}$ denote the square matrix with matrix elements $(M_{\mathcal{U}})_{ij} = (\beta_i^* | \mathcal{U} | \beta_j)$. The representation $M_{\mathcal{U}}$ of $\mathbf{G}_A$ fulfils the hypotheses of Lemma 13.14, and thus one can find a similar representation $O_{\mathcal{U}}$ which is orthogonal, with $(O_{\mathcal{U}})_{ij} = (\alpha_i^* | \mathcal{U} | \alpha_j)$ and the basis $\{\alpha_i\}$ defined as $|\alpha_i) := \sum_j P_{ij}^{-\frac{1}{2}} |\beta_j)$.   □

14

The Qubit

Until now, our reconstruction of quantum theory from the principles consisted in proving some general, conceptually deep properties of the theory. In this chapter, we finally start seizing the mathematics of the theory on more concrete grounds. Indeed, we prove that the most basic systems of our theory must be *qubits*, the elementary systems of quantum information theory. The proof is constructive, and it will explicitly show the way in which the mathematics of Hilbert spaces materializes, unfolding its structure out of the seeds of the general framework and the principles that we formulated.

14.1 Two-dimensional Systems

The first step in our construction is to prove that the set of normalized states $St_1(A)$ of a system A with $d_A = 2$ is a sphere. For this purpose, we prove the following lemma that holds for a generic system.

Lemma 14.1 *With a suitable choice of basis for the vector space $St_{\mathbb{R}}(A)$, every reversible transformation $\mathcal{U} \in \mathbf{G}_A$ is represented by a matrix $M_T(\mathcal{U})$ of the form*

$$M_T(\mathcal{U}) = \left(\begin{array}{c|c} 1 & 0 \\ \hline 0 & O_{\mathcal{U}} \end{array} \right), \tag{14.1}$$

where $O_{\mathcal{U}}$ is an orthogonal $(D_A - 1) \times (D_A - 1)$ matrix.

Proof Let $\{\xi_i\}$ be a basis for $St_{\mathbb{R}}(A)$, chosen in such a way that the first basis vector is χ, while the remaining vectors satisfy $(e|\xi_i) = 0, \forall i = 2, \ldots, D_A$. Such a choice is always possible since every vector $v \in St_{\mathbb{R}}(A)$ can be written as $v = (e|v) \chi + \xi$, where ξ satisfies $(e|\xi) = 0$. Now, since $\mathcal{U}\chi = \chi$, the first column of the matrix representation $M_T(\mathcal{U})$ of $\mathcal{U}$ on the chosen basis must be $(1, 0, \ldots, 0)^\mathsf{T}$. Moreover, since $(e|\mathcal{U} = (e|$, one must have $(e|\mathcal{U}|\xi) = 0$ for every ξ such that $(e|\xi) = 0$. Hence, the first row of $M_T(\mathcal{U})$ must be $(1, 0, \ldots, 0)$, namely $M_T(\mathcal{U})$ has the block form of Eq. (14.1). Applying the procedure of the proof of Lemma 13.13 to the representation $M_T(\mathcal{U})$, one obtains a similar representation that is orthogonal, while preserving the block structure of Eq. (14.1). □

Recalling that the group $\mathbf{G}_A$ is compact, we have then the following.

Corollary 14.2 *For every system A, the group of reversible transformations $\mathbf{G}_A$ is (isomorphic to) a compact subgroup of $O(D_A - 1)$.*

Exercise 14.1   Let us write an arbitrary state $\rho \in \mathsf{St}_1(A)$ as $\rho = \chi_A + \xi$, with $(e|\xi) = 0$, as in the proof of Lemma 14.1. Prove that the linear map $\mathcal{N}$ defined by $\mathcal{N}(\rho) = \chi_A - \xi$ is not a physical transformation.

From here to the end of the chapter we will assume that system A has dimension $d_A = 2$, and thus $D_A = 4$. The idea of the proof that the set of normalized states $\mathsf{St}_1(A)$ is a sphere is now a simple geometric observation: in the ordinary three-dimensional space the sphere is the only compact convex set that has an infinite number of pure states connected by orthogonal transformations. The complete proof is given in the following theorem.

Theorem 14.3 (The Bloch Sphere)   *The normalized pure states of a system A with $d_A = 2$ form a three-dimensional sphere and the group G_A is* SO(3).

Proof   According to equation (13.23) and Corollary 14.2, the group of reversible transformations G_A is a compact subgroup of the orthogonal group O(3). It cannot be the whole O(3) because, as we saw in Exercise 14.1, the inversion $-I$ cannot represent a physical transformation. We now show that G_A must be SO(3) by excluding all the other possibilities. From Corollary 12.21 we know that the cardinality of the set of pure states of system A is continuous. Therefore, transitivity of reversible transformations on pure states implies that the group G_A must be continuous. Now, from the classification of subgroups of O(3) we know that there are only two possibilities: (i) G_A is SO(3) and (ii) G_A is the subgroup generated by SO(2), the group of rotations around a fixed axis, say the z-axis, and possibly the binary subgroups of reflections with respect to planes containing the z-axis, and with respect to the xy-plane. As to possibility (ii), it is excluded because in this case the action of the group G_A cannot be transitive on $\mathsf{PurSt}_1(A)$. Indeed, by the SO(2) symmetry, the set of pure states must contain at least a circle in the xy-plane. This circle will be necessarily invariant under all operations in the group. However, since the convex set of states is three dimensional, there is at least a pure state outside the circle. Thus, in this case the group G_A cannot be transitive on pure states, thus contradicting the uniqueness of purification in the purification axiom. The only remaining alternative is then case (i), namely $G_A = $ SO(3). The set of pure states then corresponds to a sphere.   □

As a consequence of Theorem 14.3, the convex set of normalized states of a system A with $d_A = 2$ is a ball. Since the convex set of density matrices on a two-dimensional Hilbert space is also a ball – the Bloch ball – we can represent the states in $\mathsf{St}_1(A)$ as density matrices. Precisely, we can construct the representation as follows. First, we choose three orthogonal axes passing through the center of the ball, and call them the x, y, z axes. We then take $\varphi_{+,k}, \varphi_{-,k}, k = x, y, z$ as the two perfectly discriminable pure states lying on the k-axis,[1] and define $\sigma_k := \varphi_{k,+} - \varphi_{k,-}$. From the geometry of the ball we know that any state $\rho \in \mathsf{St}_1(A)$ can be written as

$$\rho = \chi + \frac{1}{2} \sum_{k=x,y,z} n_k \sigma_k \qquad \sum_{k=x,y,z} n_k^2 \leq 1, \qquad (14.2)$$

---

[1]   Since $d_A = 2$, also each pure state can be perfectly discriminable from exactly another one. In the representation of states in terms of points on the sphere every discriminable pair must lie on a diameter, otherwise the discriminating effect would not be bounded by 1.

where the pure states are those for which $\sum_{k=x,y,z} n_k^2 = 1$. The Bloch representation S_ρ of state ρ is then obtained by associating the basis vectors $\chi, \sigma_x, \sigma_y, \sigma_z$ to the matrices

$$S_\chi = \frac{1}{2} \begin{pmatrix} 1 & 0 \\ 0 & 1 \end{pmatrix} \quad S_{\sigma_x} = \begin{pmatrix} 0 & 1 \\ 1 & 0 \end{pmatrix}$$

$$S_{\sigma_y} = \begin{pmatrix} 0 & -i \\ i & 0 \end{pmatrix} \quad S_{\sigma_z} = \begin{pmatrix} 1 & 0 \\ 0 & -1 \end{pmatrix} \tag{14.3}$$

and defining S_ρ by linearity from Eq. (14.2). Clearly, in this way we obtain

$$S_\rho := \frac{1}{2} \begin{pmatrix} 1 + n_z & n_x - i n_y \\ n_x + i n_y & 1 - n_z \end{pmatrix}, \tag{14.4}$$

which is the expression of a generic density matrix. Denoting by $M_2(\mathbb{C})$ the set of complex two-by-two matrices we have the following.

Corollary 14.4 (Qubit Density Matrices)   *For a system A with $d_A = 2$ the set of states $\mathsf{St}_1(A)$ is isomorphic to the set of density matrices in $M_2(\mathbb{C})$ through the isomorphism $\rho \mapsto S_\rho$ defined in equation (14.3).*

Once we decide to represent the states in $\mathsf{St}_1(A)$ as matrices, the effects in $\mathsf{Eff}(A)$ are necessarily represented by matrices too. The matrix representation of an effect, given by the map $a \in \mathsf{Eff}(A) \mapsto E_a \in M_2(\mathbb{C})$ is defined uniquely by the relation

$$\mathrm{Tr}[E_a S_\rho] := (a|\rho) \quad \forall \rho \in \mathsf{St}(A). \tag{14.5}$$

Furthermore, we have the following.

Corollary 14.5   *For $d_A = 2$ the set of effects $\mathsf{Eff}(A)$ is isomorphic to the set of positive Hermitian matrices $P \in M_2(\mathbb{C})$ such that $P \leq I$.*

Proof   Since the set of deterministic pure states is in one-to-one correspondence with the set of rank-one projections, the matrix E_a must be non-negative definite for every effect a, since we have $\mathrm{Tr}[E_a S_\rho] = (a|\rho) \geq 0$ for every density matrix S_ρ. Moreover, since $1 = (e_A|\rho) = \mathrm{Tr}[S_\rho E_{e_A}]$ for every density matrix S_ρ, the deterministic effect e_A must correspond to the identity matrix $E_{e_A} = I_2$. Finally, since we have $\mathrm{Tr}[E_a S_\rho] = (a|\rho) \leq 1$ for every density matrix S_ρ, we have $0 \leq E_a \leq I$ for every effect $a \in \mathsf{Eff}(A)$. On the other hand, we know that for every couple of perfectly discriminable pure states $\varphi, \varphi_\perp$ there exists an atomic effect $\varphi^\dagger$ such that $(\varphi^\dagger|\varphi) = 1$ and $(\varphi^\dagger|\varphi_\perp) = 0$. Since the two pure states $\varphi, \varphi_\perp$ are represented by orthogonal rank-one projections S_φ and $S_{\varphi_\perp}$, we must have

$$E_{\varphi^\dagger} = S_\varphi. \tag{14.6}$$

This proves that the atomic effects are the whole set of positive rank-one projections. Since the set of effects $\mathsf{Eff}(A)$ is closed under convex combinations, every matrix P with $0 \leq P \leq I$ represents some effect a.   □

Finally, the reversible transformations are represented as conjugation by unitary matrices in SU(2):

Corollary 14.6 *For every reversible transformation* $\mathcal{U} \in \mathbf{G}_A$ *with* $d_A = 2$ *there exists a unitary matrix* $U \in SU(2)$ *such that*

$$S_{\mathcal{U}\rho} = US_\rho U^\dagger \qquad \rho \in \mathsf{St}(A). \tag{14.7}$$

Conversely, for every $U \in SU(2)$ *there exists a reversible transformation* $\mathcal{U} \in \mathbf{G}_A$ *such that Eq. (14.7) holds.*

Proof Every rotation of the Bloch sphere is represented by conjugation by some $SU(2)$ matrix. Conversely, every conjugation by an $SU(2)$ matrix represents some rotation of the Bloch sphere. On the other hand, we know that $\mathbf{G}_A$ is the group of all rotations of the Bloch sphere (Theorem 14.3). $\qquad\qquad\square$

Remark The choice of representation of the Bloch ball is not uniquely defined. Indeed, one could re-define the representation performing any transformation in the symmetry group of the ball, namely a rotation in $SO(3)$ or a reflection $-I$. Notice that the reflection can also be thought of as a reflection with respect to the xz-plane followed by a π-rotation around the y-axis. This type of representation, call it S', differs from the original one S by a sign before S_{σ_y}, namely

$$S'_\chi = S_\chi, \qquad S'_{\sigma_x} = S_{\sigma_x} \qquad S'_{\sigma_z} = S_{\sigma_z} \qquad S'_{\sigma_y} = -S_{\sigma_y}, \tag{14.8}$$

modulo rotations of the x, y, and z directions. Since S_{σ_y} is completely imaginary, while the other matrices in equation (14.3) are real, one has $S'_\rho = S^*_\rho$, and thus

$$\begin{aligned} S'_{\mathcal{U}\rho} &= (US_\rho U^\dagger)^* \\ &= U^* S'_\rho U^T. \end{aligned}$$

This change of representation can then be performed by the simple exchange

$$S'_{\varphi_{y,\pm}} = S_{\varphi_{y,\mp}}.$$

Finally, it is straightforwardly proved that also in the representation S' the representation of effects enjoys the property that $E'_{\varphi^\dagger} = S'_\varphi$.

Remark We proved that all two-dimensional systems A and B in our theory have the same sets of states ($\mathsf{St}_1(A) \simeq \mathsf{St}_1(B)$), effects ($\mathsf{Eff}(A) \simeq \mathsf{Eff}(B)$), and reversible transformations ($\mathbf{G}_A \simeq \mathbf{G}_B$), but we did not show that A and B are operationally equivalent. For example, A and B could behave differently when they are composed with a third system C, i.e. AC $\not\simeq$ BC. We will prove in the next chapters that, actually, all two-dimensional systems are operationally equivalent.

We conclude this section with a simple fact that will be very useful later:

Corollary 14.7 (Superposition Principle for Qubits) *Let* $\{\varphi_1, \varphi_2\} \subset \mathsf{St}_1(A)$ *be two perfectly discriminable pure states of a system* A *with* $d_A = 2$. *Let* $\{\varphi_1^\dagger, \varphi_2^\dagger\}$ *be the observation* *test such that* $(\varphi_i^\dagger|\varphi_j) = \delta_{ij}$. *Then, for every probability* $0 \le p \le 1$ *there exists a pure state* $\psi_p \in \mathsf{PurSt}_1(A)$ *such that*

$$(\varphi_1^\dagger|\psi_p) = p \qquad (\varphi_2^\dagger|\psi_p) = 1 - p. \tag{14.9}$$

Precisely, the set of pure states $\psi_p \in \mathsf{St}_1(A)$ *satisfying Eq. (14.9) is a circle in the Bloch sphere.*

Proof Elementary property of density matrices on $\mathbb{C}^2$. □

Exercise 14.2 Prove that for a qubit system Q one has

$$(\varphi_{k,s}^\dagger|\varphi_{k',s'}) = \frac{1}{2}(1 + ss'\delta_{kk'}), \tag{14.10}$$

for every $s = \pm$ and for every $k = x, y, z$.

14.2 Summary

In this chapter we showed that every two-dimensional system in our theory is a *qubit*. With this expression we mean that normalized states of the system can be represented as density matrices for a quantum system with two-dimensional Hilbert space. This choice of representation also allowed us to show that effects of two-dimensional systems correspond to positive Hermitian matrices bounded by the identity, and that reversible transformations act on the states by conjugation with unitary matrices in $\mathrm{SU}(2)$.

Solutions to Selected Problems and Exercises

Exercise 14.1

Let us write the state Φ as $\Phi = \chi_A \otimes \chi_{\overline{A}} + \Xi$. Since $(e|_A|\Phi)_{A\overline{A}} = |\chi)_{\overline{A}}$ one must have $(e|_A|\Xi)_{A\overline{A}} = 0$. Therefore, Ξ must be of the form $\Xi = \sum_i \alpha_i \otimes \beta_i$ with $(e|\alpha_i) = 0$ for all i. Applying the transformation $\mathcal{N}$ one then obtains $(\mathcal{N} \otimes \mathcal{I}_{\overline{A}})\Phi = \chi_A \otimes \chi_{\overline{A}} - \Xi$. We now prove that this is not a state, and therefore, $\mathcal{N}$ cannot be a physical transformation. Let $\Phi^\dagger$ be the atomic effect such that $(\Phi^\dagger|\Phi) = 1$. Then, we have $1 = (\Phi^\dagger|\chi_A \otimes \chi_{\overline{A}}) + (\Phi^\dagger|\Xi) = 1/d_A^2 + (\Phi^\dagger|\Xi)$. Now, we have

$$(\Phi^\dagger|(\mathcal{N} \otimes \mathcal{I}_{\overline{A}})|\Phi) = \frac{1}{d_A^2} - (\Phi^\dagger|\Xi) = \frac{2}{d_A^2} - 1.$$

Since this quantity is negative for every $d_A > 1$, the map $\mathcal{N}$ cannot be a physical transformation. As a consequence, the matrix $[N]$ defined as

$$[N] = \left(\begin{array}{c|c} 1 & 0 \\ \hline 0 & -I_{D_A-1} \end{array}\right), \tag{14.11}$$

cannot represent a transformation of system A.

Exercise 14.2

We recall that $\chi = \frac{1}{2}(\varphi_{k,+} + \varphi_{k,-})$ for every $\sigma_k = \varphi_{k,+} - \varphi_{k,-}$ and for every $k = x, y, z$. Thus, one has

$$\varphi_{k,s} = \chi + \frac{s}{2}\sigma_k, \qquad s = \pm 1.$$

Now, by Corollary 14.5 one has $E_{\varphi^\dagger} = S_\varphi$, and then

$$
\begin{aligned}
(\varphi_{k,s}^\dagger | \varphi_{k',s'}) &= \mathrm{Tr}[E_{\varphi_{k,s}^\dagger} S_{\varphi_{k',s'}}] \\
&= \mathrm{Tr}[S_{\varphi_{k,s}^\dagger} S_{\varphi_{k',s'}}] \\
&= \mathrm{Tr}[(S_\chi + \tfrac{s}{2}S_{\sigma_k})(S_\chi + \tfrac{s'}{2}S_{\sigma_{k'}'})] \\
&= \frac{1}{2}(1 + ss'\delta_{kk'}),
\end{aligned}
$$

where we used the explicit form of S_χ and S_{σ_k} in Eq. (14.3), which gives

$$\mathrm{Tr}[S_\chi^2] = \frac{1}{2}, \quad \mathrm{Tr}[S_\chi S_{\sigma_k}] = 0, \quad \mathrm{Tr}[S_{\sigma_k} S_{\sigma_{k'}}] = 2\delta_{kk'}.$$

15

Projections

So far, we know that systems of a theory satisfying our principles, with dimension $d_A = 2$, have the same set of states, which is a four-dimensional cone with a ball as a basis. Now, we want to extend our result to higher-dimensional systems, proving that all the systems with the same dimension are operationally equivalent, and most importantly their cone of states coincides with the cone of quantum states for some quantum system.

We already have some information about the geometric structure of cones of states, though a very partial one: indeed, the ideal compression axiom tells us that any face of the cone of states of some system is itself the cone of states of a smaller system. However, to reconstruct the sets of quantum states we need more information about the structure of cones of states.

The purpose of the present chapter is to provide a tool to enrich the information that we have about sets of states, and in particular about their faces. Indeed, we will associate a projection to every face of the set of states, and the properties of projections will allow us to derive important geometric information.

15.1 Orthogonal Complements

The main purpose of the present section is to show that we can canonically associate a state to a face of a convex set, in such a way that we can equivalently define the notions related to discriminability in terms of states and faces. In particular, given the state canonically associated with a face, we can define its orthogonal complement. In the next section, we will then show that orthogonal complements identify perfectly discriminable faces.

We start by showing a canonical way to associate a state ω_F with a face F.

Lemma 15.1 (State Associated with a Face) *Let F be a face of the convex set $\mathsf{St}_1(A)$ and let $\{\varphi_i\}_{i=1}^{|F|}$ be a maximal set of perfectly discriminable pure states in F. Then the state $\omega_F :=$ $\frac{1}{|F|}\sum_{i=1}^{|F|}\varphi_i$ depends only on the face F and not on the particular set $\{\varphi_i\}_{i=1}^{|F|}$. Morever, F is the face identified by ω_F.*

Proof Let $(C, \mathcal{E}, \mathcal{D})$ be the ideal encoding scheme for the face F. By Corollary 12.13, $\{\mathcal{E}\varphi_i\}_{i=1}^{|F|}$ is a maximal set of perfectly discriminable pure states of C and by Theorem 12.8 one has $\chi_C = \frac{1}{|F|}\sum_{i=1}^{|F|}\mathcal{E}\varphi_i$. Hence, $\omega_F = \frac{1}{|F|}\sum_{i=1}^{|F|}\varphi_i = \frac{1}{|F|}\sum_{i=1}^{|F|}\mathcal{D}\mathcal{E}\varphi_i = \mathcal{D}\chi_C$. Since the right-hand side of the equality is independent of the particular set $\{\varphi_i\}_{i=1}^{|F|}$, the state ω_F on the left-hand side is independent too. To prove that F is the face identified by ω_F it is enough to prove that $\mathsf{RefSet}_1(\omega_F) = F$. This fact follows from the relation $\omega_F = \mathcal{D}\chi_C$ and from Exercise 8.6   □

We now define the *orthogonal complement* of the state ω_F.

Definition 15.2 The *orthogonal complement of the state* ω_F associated with the face F is the state $\omega_F^\perp \in \mathsf{St}_1(A) \cup \{0\}$ defined as follows:

1. if $|F| = d_A$, then $\omega_F^\perp = 0$;
2. if $F < d_A$, then $\omega_F^\perp$ is defined by the relation

$$\chi_A = \frac{|F|}{d_A}\omega_F + \frac{d_A - |F|}{d_A}\omega_F^\perp. \tag{15.1}$$

An easy way to write the orthogonal complement of a state ω_F is provided by the following lemma.

Lemma 15.3 *Take a maximal set* $\{\varphi_i\}_{i=1}^{|F|}$ *of perfectly discriminable pure states in* F *and extend it to a maximal set* $\{\varphi_i\}_{i=1}^{d_A}$ *of perfectly discriminable pure states in* $\mathsf{St}_1(A)$, *then for* $|F| < d_A$ *we have*

$$\omega_F^\perp = \frac{1}{d_A - |F|} \sum_{i=|F|+1}^{d_A} \varphi_i.$$

Proof By definition, for $|F| < d_A$ we have $\omega_F^\perp = \frac{1}{d_A - |F|}(d_A\chi_A - |F|\omega_F)$. Substituting the expressions $\chi_A = \frac{1}{d_A}\sum_{i=1}^{d_A}\varphi_i$ and $\omega_F = \frac{1}{|F|}\sum_{i=1}^{|F|}\varphi_i$ we then obtain the thesis.   □

Notice that, by definition, the orthogonal complement $\omega_F^\perp$ only depends on the face F — and not on the choice of the maximal set $\{\varphi\}_{i=1}^{d_A}$. An obvious consequence of Lemma 15.3 is the following corollary.

Corollary 15.4 *The states* ω_F *and* $\omega_F^\perp$ *are perfectly discriminable.*

Proof Take a maximal set $\{\varphi_i\}_{i=1}^{|F|}$ of perfectly discriminable pure states in F, extend it to a maximal set $\{\varphi_i\}_{i=1}^{d_A}$, and take the observation test $\{\varphi_i^\dagger\}_{i=1}^{d_A}$. Since $(\varphi_i^\dagger|\varphi_j) = \delta_{ij}$, the binary test $\{a_F, e - a_F\}$ defined by $a_F := \sum_{i=1}^{|F|}\varphi_i^\dagger$ discriminates perfectly between ω_F and $\omega_F^\perp$.   □

Corollary 15.5 *Let AB be a composite system. Define the face* $\tilde{F}$ *associated with the state* $\omega_F \otimes \chi_B$. *Then one has*

$$\omega_{\tilde{F}} = \omega_F \otimes \chi_B, \tag{15.2}$$

$$\omega_{\tilde{F}}^\perp = \omega_F^\perp \otimes \chi_B. \tag{15.3}$$

The proof is left as an exericse for the reader.

Exercise 15.1 Prove Corollary 15.5.

We say that a state $\tau \in \mathsf{St}_1(A)$ is *perfectly discriminable from the face* F if τ is perfectly discriminable from every state σ in the face F. With this definition we have the following.

Lemma 15.6   *The following statements are equivalent:*

1. *τ is perfectly discriminable from the face F;*
2. *τ is perfectly discriminable from ω_F;*
3. *τ belongs to the face identified by $\omega_F^\perp$, i.e. $\tau \in F_{\omega_F^\perp}$.*

Proof $(1 \Leftrightarrow 2)$ τ is perfectly discriminable from ω_F if and only if then there exists a binary test $\{a, e - a\}$ such that $(a|\tau) = 1$ and $(a|\omega_F) = 0$. By Exercise 8.1 the latter equality is equivalent to $a =_{\omega_F} 0$, and since $\mathsf{RefSet}_1(\omega_F) = F$, statement 2 is equivalent to the condition $(a|\sigma) = 0$ for every $\sigma \in F$. As a result, statement 2 is equivalent to the requirement that τ is discriminable from any state σ in the face F. $(2 \Rightarrow 3)$ Let $\{\varphi_i\}_{i=1}^{|F|}$ be a maximal set of perfectly discriminable states in F, so that $\omega_F = \frac{1}{|F|} \sum_{i=1}^{|F|} \varphi_i$. Let $\{\psi_i\}_{i=1}^{k}$ be the set of perfectly discriminable pure states in the spectral decomposition $\tau = \sum_{i=1}^{r} p_i \psi_i$, with $r \leq k$. Since τ is perfectly discriminable from ω_F, all states $\{\psi_i\}_{i=1}^{r}$ must be perfectly discriminable from ω_F. By Proposition 10.5 this implies that the states $\{\varphi_i\}_{i=1}^{|F|} \cup \{\psi_i\}_{i=1}^{r}$ are perfectly discriminable. Let us set $\varphi_{|F|+i} := \psi_i$, $i = 1, \ldots, r$, so that the states $\{\varphi_i\}_{i=1}^{|F|+r}$ are perfectly discriminable. Let us extend this set to a maximal set $\{\varphi_i\}_{i=1}^{d_A}$. By Lemma 15.3 we have $\omega_F^\perp = \frac{1}{d_A - |F|} \sum_{i=|F|+1}^{d_A} \varphi_i$. Hence, all the states $\{\varphi_i\}_{i=|F|+1}^{d_A}$ are in the face $F_{\omega_F^\perp}$. Since τ is a mixture of these states, it also belongs to the face $F_{\omega_F^\perp}$. $(3 \Rightarrow 2)$ Since ω_F and $\omega_F^\perp$ are perfectly discriminable, if τ belongs to the face identified by $\omega_F^\perp$, then by Proposition 10.1 τ is perfectly discriminable from ω_F. $\square$

Corollary 15.7   *If ρ is perfectly discriminable from σ and from τ, then ρ is perfectly discriminable from any convex mixture of σ and τ.*

Proof   Let F be the face identified by ρ. Then by Lemma 15.6 we have $\sigma, \tau \in F_{\omega_F^\perp}$. Since $F_{\omega_F^\perp}$ is a convex set, any mixture of σ and τ belongs to it. By Lemma 15.6, this means that any mixture of σ and τ is perfectly discriminable from ρ. $\square$

15.2 Orthogonal Faces

We now introduce the notion of *orthogonal faces*.

Definition 15.8 (Orthogonal Face)   Given a face $F \subseteq \mathsf{St}_1(A)$, the *orthogonal face $F^\perp$* is the set of all states that are perfectly discriminable from the face F.

By Lemma 15.6 it is clear that $F^\perp$ is the face identified by $\omega_F^\perp$, that is $F^\perp = F_{\omega_F^\perp}$. In the following lemma we list a few elementary facts about orthogonal faces.

Lemma 15.9   *The following properties hold:*

1. $|F^\perp| = d_A - |F|$;
2. $\chi_A = \frac{|F|}{d_A} \omega_F + \frac{|F^\perp|}{d_A} \omega_{F^\perp}$;

3. $\omega_{F^\perp} = \omega_F^\perp$;

4. $\omega_{F^\perp}^\perp = \omega_F$;

5. $\left(F^\perp\right)^\perp = F$.

Proof (1) If $|F| = d_A$ the thesis is obvious. If $|F| < d_A$, take a maximal set $\{\varphi_i\}_{i=1}^{|F|}$ of perfectly discriminable pure states in F, and analogously take the maximal set $\{\varphi_j\}_{j=|F|+1}^{|F|+|F^\perp|}$ in $F^\perp$. Hence we have

$$\omega_F = \frac{1}{|F|} \sum_{i=1}^{|F|} \varphi_i \qquad \omega_{F^\perp} = \frac{1}{|F^\perp|} \sum_{j=|F|+1}^{|F|+|F^\perp|} \varphi_j.$$

By Corollary 15.4, the states ω_F and $\omega_{F^\perp}$ are perfectly discriminable. Consequently, by the refined discriminability property of Proposition 10.5, the pure states $\{\varphi_i\}_{i=1}^{|F|+|F^\perp|}$ are jointly perfectly discriminable. Now, we must have $|F| + |F^\perp| = d_A$, otherwise there would be a pure state ψ that is perfectly discriminable from the states $\{\varphi_i\}_{i=1}^{|F|+|F^\perp|}$. This would imply that ψ belongs to $F^\perp$, and that states $\{\psi\} \cup \{\varphi_j\}_{j=|F|+1}^{|F|+|F^\perp|}$ are perfectly discriminable in $F^\perp$, in contradiction with the hypotheses that the set $\{\varphi_j\}_{j=|F|+1}^{|F|+|F^\perp|}$ is maximal in $F^\perp$. (2) Immediate from (1) and Definition 15.2. (3 and 4) Both items follow by comparison of (2) with Eq. (15.1). (5) By condition 3 of Lemma 15.6, $\left(F^\perp\right)^\perp$ is the face identified by the state $\omega_{F^\perp}^\perp$, which, by (4), is ω_F. Since the face identified by ω_F is F, we have $\left(F^\perp\right)^\perp = F$. □

We now show that there is a canonical way to associate an effect a_F with a face F:

Definition 15.10 (Effect Associated with a Face) We say that $a_F \in \mathsf{Eff}(A)$ is the *effect associated with the face* $F \subseteq \mathsf{St}_1(A)$ if and only if $a_F =_F e$ and $a_F =_{F^\perp} 0$.

In other words, the definition imposes that $(a_F|\rho) = 1$ for every $\rho \in F$ and $(a_F|\sigma) = 0$ for every $\sigma \in F^\perp$. We now prove that the notion of effect associated with a face is well defined, as such an effect exists and is unique.

Lemma 15.11 Let $\{\varphi_i\}_{i=1}^{|F|}$ be a maximal set of perfectly discriminable pure states in F, and $\{\varphi_i\}_{i=1}^{d_A}$ be its extension to a maximal set in $\mathsf{St}_1(A)$. Then the effect $a = \sum_{i=1}^{|F|} \varphi_i^\dagger$ is associated with the face F.

Proof We can write ω_F and $\omega_{F^\perp}$ as

$$\omega_F = \frac{1}{|F|} \sum_{j=1}^{|F|} \varphi_j, \qquad \omega_{F^\perp} = \frac{1}{|F^\perp|} \sum_{j=|F|+1}^{d_A} \varphi_j.$$

Thus, one can easily verify that $(a|\omega_F) = 1$ and $(a|\omega_{F^\perp}) = 0$, which implies $a =_{\omega_F} e$ and $a =_{\omega_{F^\perp}} 0$, by Exercise 8.1, namely $a =_F e$ and $a =_{F^\perp} 0$. □

We then prove uniqueness of the effect associated with a face.

Lemma 15.12 *The effect a_F associated with the face F is unique.*

Proof Let a_F be an effect associated with the face F. Let us consider the spectral decomposition $a_F = \sum_{i=1}^{d_A} a_i \psi_i^\dagger$ (Corollary 12.17). Since the atomic effects $\{\psi_i^\dagger\}_{i=1}^{d_A}$ make a perfectly discriminating measurement for some maximal set of perfectly discriminable pure states $\{\psi_i\}_{i=1}^{d_A}$, we have $0 \le (a_F|\psi_i) = a_i \le 1$ for every i. Moreover we have

$$\sum_{i=1}^{d_A} a_i(\psi_i^\dagger|\omega_F) = (a_F|\omega_F) = 1 = (e|\omega_F) = \sum_{i=1}^{d_A}(\psi_i^\dagger|\omega_F),$$

which implies

$$\sum_{i=1}^{d_A}(1 - a_i)(\psi_i^\dagger|\omega_F) = 0.$$

Since $(1 - a_i) \ge 0$ and $(\psi_i^\dagger|\omega_F) \ge 0$ for every i, we must have $(1 - a_i)(\psi_i^\dagger|\omega_F) = 0$ for every i. Thus, either $(\psi_i^\dagger|\omega_F) = 0$ or $a_i = 1$. This fact means that, for those i for which $a_i < 1$ the state ψ_i belongs to the face $F^\perp$, because the effect $\psi_i^\dagger$ perfectly discriminates ω_F from ψ_i. Thus, for $a_i < 1$ one has

$$a_i = \sum_{j=1}^{d_A} a_j(\psi_j^\dagger|\psi_i) = (a_F|\psi_i) = 0.$$

Consequently, either $a_i = 1$ and $(\psi_i^\dagger|\omega_{F^\perp}) = 0$, or $a_i = 0$ and $(\psi_i^\dagger|\omega_F) = 0$. Upon suitable reordering of the indices, the states $\{\psi_i\}_{i=1}^{d_A}$ then split in two subsets $\mathsf{A} := \{\psi_i\}_{i=1}^{k}$ and $\mathsf{B} := \{\psi_i\}_{i=k+1}^{d_A}$, such that $\mathsf{A} \subseteq F$ and $\mathsf{B} \subseteq F^\perp$. The two sets A and B are maximal in F and $F^\perp$, respectively. Indeed, suppose that there exists $\varphi \in F$ such that $\mathsf{A} \cup \{\varphi\} \subseteq F$ is perfectly discriminable. By hypothesis, φ is perfectly discriminable also from the set B. Then, the set $\{\psi_i\}_{i=1}^{d_A} \cup \{\varphi\}$ would be perfectly discriminable, in contradiction to the fact that $\{\psi_i\}_{i=1}^{d_A}$ is maximal. The same argument applies to the set B. Then we have $k = |F|$, and

$$a_F = \sum_{i=1}^{|F|} \psi_i^\dagger.$$

Suppose now that there exists another effect a_F' associated with the face F. Then, there must exist a maximal set $\{\varphi_i\}_{i=1}^{d_A}$ with discriminating measurement $\{\varphi_i^\dagger\}_{i=1}^{d_A}$, such that $a_F' = \sum_{i=1}^{|F|} \varphi_i^\dagger$. Now, the set $\{\varphi_i\}_{i=1}^{|F|} \cup \{\psi_i\}_{i=|F|+1}^{d_A}$ is perfectly discriminable by Proposition 10.5, and the discriminating measurement is $\{\varphi_i^\dagger\}_{i=1}^{|F|} \cup \{\psi_i^\dagger\}_{i=|F|+1}^{d_A}$, thanks to Lemma 11.3 and Theorem 11.5. This implies that

$$a_F = e - \sum_{i=|F|+1}^{d_A} \psi_i^\dagger = a_F'.$$

Then the effect a_F associated with the face F is unique. $\square$

Two immediate consequences of the last result are summarized in the following corollaries.

Corollary 15.13 *Let F be a face in* $\mathsf{St}_1(A)$. *Then*

$$a_{F^\perp} = e - a_F. \tag{15.4}$$

Corollary 15.14 *Let F be the face* $\{\varphi\} \subseteq \mathsf{St}_1(A)$. *Then* $a_F = \varphi^\dagger$ *is the unique atomic effect that identifies* φ.

The main property of the effect associated with a face is that it "identifies the face," in the following sense:

Lemma 15.15 *A state* $\rho \in \mathsf{St}_1(A)$ *belongs to the face F if and only if* $(a_F|\rho) = 1$.

Proof By definition, if ρ belongs to F, then $(a_F|\rho) = 1$. Conversely, if $(a_F|\rho) = 1$, then ρ is perfectly discriminable from $\omega_{F^\perp}$, because $(a_F|\omega_{F^\perp}) = 0$. By Lemma 15.6 the fact that ρ is perfectly discriminable from $\omega_{F^\perp}$ implies that ρ belongs to $(F^\perp)^\perp$, which is just F (item 5 of Lemma 15.9). $\qquad\square$

Exercise 15.2 Show that for a qubit state ρ the condition $(\varphi_0^\dagger|\rho) = 0$ implies $\rho \propto \varphi_1$, where $\{\varphi_0, \varphi_1\}$ is a maximal perfectly discriminable set of pure states.

15.3 Projections

We are now in position to define the central object of this chapter, namely the projection on a face.

Definition 15.16 (Projection) Let F be a face of $\mathsf{St}_1(A)$. A *projection* on the face F is a transformation Π_F such that:

1. $\Pi_F =_F \mathcal{I}_A$;
2. $\Pi_F =_{F^\perp} 0$.

When F is the face identified by a pure state $\varphi \in \mathsf{St}_1(A)$, we have $F = \{\varphi\}$, and the corresponding projection $\Pi_{\{\varphi\}}$ is called *projection on the pure state* φ.

The first condition in Definition 15.16 means that the projection Π_F does not disturb the states in the face F. The second condition means that Π_F annihilates all states in the orthogonal face $F^\perp$. Consistently with Definition 15.16, we will indicate with $\Pi_F^\perp$ the projection on the face $F^\perp$, that is, we will use the definition $\Pi_F^\perp := \Pi_{F^\perp}$.

An equivalent condition for Π_F to be a projection on the face F is the following:

Lemma 15.17 *The transformation* Π_F *in* $\mathsf{Transf}(A)$ *is a projection on F if and only if for every maximal set of perfectly discriminable pure states* $\{\varphi_i\}_{i=1}^{d_A}$ *for system A such that* $\{\varphi_i\}_{i=1}^{|F|}$ *is maximal in the face F one has:*

1. $\Pi_F|\varphi_j) = |\varphi_j)$ for all $j \leq |F|$;
2. $\Pi_F|\varphi_l) = 0$ for all $l > |F|$.

Proof   The condition is clearly necessary, by Definition 15.16. However, if $\Pi_F|\varphi_j) = |\varphi_j)$ for $j \le |F|$ and $\Pi_F|\varphi_l) = 0$ for $l > |F|$, then by definition of $\omega_{F\perp}$ we have $\Pi_F|\omega_{F\perp}) = 0$, and, therefore $\Pi_F =_{F\perp} 0$. Moreover, by the spectral decomposition Theorem 12.14, for every state $\tau \in F$ there exists a set of perfectly discriminable pure states $\{\varphi_i\}_{i=1}^{|F|}$ in the face F such that $\tau = \sum_{i=1}^{|F|} p_i \varphi_i$. Thus, by hypothesis,

$$\Pi_F|\tau) = \sum_{i=1}^{|F|} p_i \Pi_F|\varphi_i)$$

$$= \sum_{i=1}^{|F|} p_i|\varphi_i)$$

$$= |\tau).$$

This implies that $\Pi_F =_F \mathcal{I}_\mathrm{A}$.   □

A result that will be useful later is:

Lemma 15.18   *The transformation $\Pi_F \otimes \mathcal{I}_\mathrm{B}$ is a projection on the face $\tilde{F}$ identified by the state $\omega_F \otimes \chi_\mathrm{B}$.*

Proof   We first show that $\Pi_F \otimes \mathcal{I}_\mathrm{B} =_{\omega_F \otimes \chi_\mathrm{B}} \mathcal{I}_\mathrm{A} \otimes \mathcal{I}_\mathrm{B}$: Indeed, by local discriminability, it is easy to see that every state $\sigma \in F_{\omega_F \otimes \chi_\mathrm{B}}$ can be written as $|\sigma) = \sum_{i=1}^{r} \sum_{j=1}^{d_\mathrm{B}} \sigma_{ij}|\alpha_i)|\beta_j)$, where $\{\alpha_i\}_{i=1}^{r}$ is a basis for $\mathsf{Span}(F)$ and $\{\beta_j\}_{j=1}^{d_\mathrm{B}}$ is a basis for $\mathsf{St}_1(\mathrm{B})$. Since $\Pi_F =_F \mathcal{I}_\mathrm{A}$, we have

$$|\sigma) = (\Pi_F \otimes \mathcal{I}_\mathrm{B})|\sigma)$$

$$= \sum_{i=1}^{r} \sum_{j=1}^{d_\mathrm{B}} \sigma_{ij} \Pi_F|\alpha_i)|\beta_j)$$

$$= \sum_{i=1}^{r} \sum_{j=1}^{d_\mathrm{B}} \sigma_{ij}|\alpha_i)|\beta_j)$$

$$= |\sigma),$$

which implies $\Pi_F \otimes \mathcal{I}_\mathrm{B} =_{\omega_F \otimes \chi_\mathrm{B}} \mathcal{I}_\mathrm{A} \otimes \mathcal{I}_\mathrm{B}$. Finally, note that by definition of the face $\tilde{F}$, one has $\omega_{\tilde{F}} = \omega_F \otimes \chi_\mathrm{B}$, and by Corollary 15.5 $\omega_{\tilde{F}\perp} = \omega_{F\perp} \otimes \chi_\mathrm{B}$. Since we have $(\Pi_F \otimes \mathcal{I}_\mathrm{B})|\omega_{\tilde{F}\perp}) = \Pi_F|\omega_{F\perp}) \otimes |\chi_\mathrm{B}) = 0$, we can conclude $\Pi_F \otimes \mathcal{I}_\mathrm{B} =_{\tilde{F}\perp} 0$. Hence $\Pi_F \otimes \mathcal{I}_\mathrm{B}$ is a projection on $\tilde{F}$.   □

In the following we will show that for every face F there exists a unique projection Π_F and we will prove several properties of projections. Let us start from an elementary observation.

Theorem 15.19   *If Π_F is a projection on the face F, then one has $(e_\mathrm{A}|\Pi_F = (a_F|$.*

Proof   Since $\Pi_F|\omega_F) = |\omega_F)$ and $\Pi_F|\omega_{F\perp}) = 0$, we have

$$(e|\Pi_F|\omega_F) = 1, \quad (e|\Pi_F|\omega_{F\perp}) = 0,$$

which by Exercise 8.1 implies

$$(e|\Pi_F =_F (e|, \quad (e|\Pi_F =_{F^\perp} 0,$$

namely $(e|\Pi_F$ satisfies the defining properties of the effect associated with the face F. Finally, by the uniqueness Lemma 15.12, we have $(e|\Pi_F = (a_F|$. □

Corollary 15.20 *The transformation $\Pi_F + \Pi_F^\perp$ is a channel.*

Proof This is a consequence of Theorem 15.19 and Corollary 15.13, since they provide $(e|(\Pi_F + \Pi_{F^\perp}) = (e|$. □

We can now prove the existence of projections.

Lemma 15.21 (Existence of Projections) *For every face F of $\mathsf{St}_1(A)$ there exists an atomic projection Π_F.*

Proof By Lemma 10.3, there exists a system B and an atomic transformation $\mathcal{A} \in \mathsf{Transf}(A \to B)$ with $(e|_B \mathcal{A} = (a_F|$. Then, if $\Psi_{\omega_F} \in \mathsf{St}(A\tilde{A})$ is a minimal purification of ω_F, we can define the state $|\Sigma)_{B\tilde{A}} := (\mathcal{A} \otimes \mathcal{I}_{\tilde{A}})|\Psi_{\omega_F})_{A\tilde{A}}$. By the atomicity postulate, Σ is a pure state. Moreover, the pure states Σ and Ψ_{ω_F} have the same marginal on system $\tilde{A}$: indeed, we have

and, by definition, $a_F =_F e_A$, which by Theorem 8.3 implies

Then, the uniqueness of purification implies that for every pair of pure states ψ_0 and φ_0, there exists a reversible channel $\mathcal{U}$ such that

$$(15.5)$$

Now, take the atomic effect $b \in \mathsf{Eff}(B)$ such that $(b|\psi_0) = 1$, and define the transformation Π_F as

Applying b on both sides of Eq. (15.5) we then obtain

$$\Psi_{\omega F}\!\!\begin{array}{c} \text{A} \;\boxed{\Pi_F}\; \text{A} \\ \tilde{\text{A}} \end{array} = \Psi_{\omega F}\!\!\begin{array}{c} \text{A} \\ \tilde{\text{A}} \end{array}$$

and, thanks to Theorem 8.3, $\Pi_F =_F \mathcal{I}_{\text{A}}$. Moreover, we have $\Pi_F =_{F^\perp} 0$: indeed, by construction of Π_F we have

$$\left(\rho\right)\!-\!\text{A}\!-\!\boxed{\Pi_F}\!-\!\text{A}\!-\!\left(e\right) = \begin{array}{c} \boxed{\varphi_0}\text{A} \quad \text{B}\;\boxed{b} \\ \boxed{\mathcal{U}} \\ \left(\rho\right)\text{A}\boxed{\mathcal{A}}\text{B}\quad\text{A}\left(e\right) \end{array}$$

$$\leq \begin{array}{c} \boxed{\varphi_0}\text{A} \quad \text{B}\left(e\right) \\ \boxed{\mathcal{U}} \\ \left(\rho\right)\text{A}\boxed{\mathcal{A}}\text{B}\quad\text{A}\left(e\right) \end{array}$$

$$= \left(\rho\right)\!-\!\text{A}\!-\!\boxed{\mathcal{A}}\!-\!\text{B}\!-\!\left(e\right)$$

$$= \left(\rho\right)\!-\!\text{A}\!-\!\left(a_F\right).$$

This implies $(e_{\text{A}}|\Pi_F|\omega_{F^\perp}) \leq (a_F|\omega_{F^\perp}) = 0$, and, therefore, by Exercise 8.1, $\Pi_F =_{F^\perp} 0$. Moreover, the transformation Π_F is atomic, being the composition of atomic transformations. In conclusion, Π_F is the desired atomic projection. □

To prove the uniqueness of the projection Π_F we need two auxiliary lemmas, given in the following.

Lemma 15.22 *Let $\Phi \in \mathsf{St}_1(\text{A}\tilde{\text{A}})$ be a minimal purification of the invariant state χ_{A}, and let $\Pi_F \in \mathsf{Transf}(\text{A})$ be an atomic projection on the face $F \subseteq \mathsf{St}_1(\text{A})$. Then, the pure state $\Phi_F \in \mathsf{St}_1(\text{A}\tilde{\text{A}})$ defined by*

$$\Phi_F\!\!\begin{array}{c} \text{A} \\ \tilde{\text{A}} \end{array} := \frac{d_{\text{A}}}{|F|} \; \Phi\!\!\begin{array}{c} \text{A}\;\boxed{\Pi_F}\;\text{A} \\ \tilde{\text{A}} \end{array} \qquad (15.6)$$

is a purification of ω_F.

Proof The state Φ_F is pure by the atomicity principle. Now, we have

$$\Phi_F\!\!\begin{array}{c} \text{A} \\ \tilde{\text{A}}\left(e\right) \end{array} = \frac{d_{\text{A}}}{|F|} \; \Phi\!\!\begin{array}{c} \text{A}\;\boxed{\Pi_F}\;\text{A} \\ \tilde{\text{A}}\left(e\right) \end{array}$$

$$= \frac{d_{\text{A}}}{|F|} \left(\chi\right)\!-\!\text{A}\!-\!\boxed{\Pi_F}\!-\!\text{A}.$$

Using the fact that $\chi_{\text{A}} = \frac{|F|}{d_{\text{A}}}\omega_F + \frac{|F^\perp|}{d_{\text{A}}}\omega_{F^\perp}$, and the definition of Π_F, we then obtain

$$\frac{d_{\text{A}}}{|F|} \left(\chi\right)\!-\!\text{A}\!-\!\boxed{\Pi_F}\!-\!\text{A} = \left(\omega_F\right)\!-\!\text{A},$$

and finally

$$\Phi_F \!\begin{array}{l} A \\ \tilde{A}\,\boxed{e} \end{array} = \omega_F \!-\!\! A \;,$$

which is the thesis. □

Lemma 15.23 *Let* $\Pi_F \in \mathsf{Transf}(A)$ *be an atomic projection on the face F. A transformation* $\mathcal{C} \in \mathsf{Transf}(A)$ *satisfies* $\mathcal{C} =_F \mathcal{I}_A$ *if and only if*

$$\mathcal{C}\Pi_F = \Pi_F. \qquad (15.7)$$

Proof Let Φ_F be the purification of ω_F defined in Lemma 15.22. Since $\mathcal{C} =_F \mathcal{I}_A$, we have $(\mathcal{C} \otimes \mathcal{I})|\Phi_F) = |\Phi_F)$. In other words, we have $(\mathcal{C}\Pi_F \otimes \mathcal{I})|\Phi) = (\Pi_F \otimes \mathcal{I})|\Phi)$. Since Φ is faithful, this implies that $\mathcal{C}\Pi_F = \Pi_F$. Conversely, Eq. (15.7) implies that for $\sigma \in F$, $\mathcal{C}|\sigma) = \mathcal{C}\Pi_F|\sigma) = \Pi_F|\sigma) = |\sigma)$, namely $\mathcal{C} =_F \mathcal{I}_A$. □

Theorem 15.24 (Uniqueness of Projections) *The projection* Π_F *satisfying Definition 15.16 is unique.*

Proof Let Π_F be the atomic projection on F derived in Lemma 15.21. Let Π'_F be another (possibly non-atomic) projection on the same face F. Define the pure state Φ_F as in Lemma 15.22, and define the (possibly mixed) state $\Phi'_F := (\Pi'_F \otimes \mathcal{I}_{\tilde{A}})|\Phi)$. Now, Φ_F and Φ'_F are both extensions of the same state $\tilde{\omega}_F \in \tilde{A}$: indeed, one has

$$\Phi_F \!\begin{array}{l} A\,\boxed{e} \\ \tilde{A} \end{array} = \frac{d_A}{|F|}\; \Phi \!\begin{array}{l} A\,\boxed{\Pi_F}\,A\,\boxed{e} \\ \tilde{A} \end{array}$$

$$= \frac{d_A}{|F|}\; \Phi \!\begin{array}{l} A\,\boxed{\Pi'_F}\,A\,\boxed{e} \\ \tilde{A} \end{array}$$

$$= \Phi'_F \!\begin{array}{l} A\,\boxed{e} \\ \tilde{A} \end{array}\;,$$

having used the relation $(e_A|\Pi_F = (a_F| = (e_A|\Pi'_F$ proved in Theorem 15.19, and uniqueness of the effect a_F of Lemma 15.12. Let now $\Sigma \in \mathsf{St}(A\tilde{A}B)$ be a purification of Φ'_F. By Proposition 7.1, we have

$$\Phi \!\begin{array}{l} A\,\boxed{\Pi'_F}\,A \\ \tilde{A} \end{array} = \Phi'_F \!\begin{array}{l} A \\ \tilde{A} \end{array}$$

$$= \Phi_F \!\begin{array}{l} A\,\boxed{\mathcal{C}}\,A \\ \tilde{A} \end{array}$$

$$= \Phi \!\begin{array}{l} A\,\boxed{\Pi_F}\,A\,\boxed{\mathcal{C}}\,A \\ \tilde{A} \end{array}\;,$$

for some channel $\mathcal{C} \in \mathsf{Transf}(A)$. Now, since Φ is faithful, $\Pi'_F = \mathcal{C}\Pi_F$. By Definition 15.16 we have $\Pi'_F =_F \mathcal{I}_A$ and $\Pi_F =_F \mathcal{I}_A$, and we can then conclude that $\mathcal{C} =_F \mathcal{I}_A$. Finally, using Lemma 15.23 we obtain $\Pi'_F = \mathcal{C}\Pi_F = \Pi_F$. □

As a consequence, we have the following important property of projections.

Corollary 15.25 (Atomicity of Projections)   *The projection Π_F on a face F is atomic.*

We now show a few simple properties of projections. In the following, we will consider an arbitrary maximal set of perfectly discriminable pure states $\{\varphi_i\}_{i=1}^{d_A}$. Given then any subset $V \subseteq \{1, \ldots, d_A\}$ we define (with a slight abuse of notation) $\omega_V := \sum_{i \in V} \varphi_i / |V|$, and Π_V as the projection on the face $F_V := F_{\omega_V}$. We refer to F_V as *the face generated by V*.

Lemma 15.26   *For two arbitrary subsets $V, W \subseteq \{1, \ldots, d_A\}$ one has*

$$\Pi_V \Pi_W = \Pi_{V \cap W}.$$

In particular, if $V \cap W = \emptyset$ one has $\Pi_V \Pi_W = 0$.

Proof   Since the face $F_{V \cap W}$ is contained in the faces F_V and F_W, we have $\Pi_V \Pi_W |\rho) = \Pi_V |\rho) = |\rho)$ for every $\rho \in F_{V \cap W}$. In other words, $\Pi_V \Pi_W =_{F_{V \cap W}} \mathcal{I}_A$. Moreover, if $l \notin V \cap W$ we have $\Pi_V \Pi_W |\varphi_l) = 0$. By Lemma 15.17 and by the uniqueness of projections (Theorem 15.24) we then obtain that $\Pi_V \Pi_W$ is the projection on the face generated by $V \cap W$. □

Corollary 15.27 (Idempotence)   *Every projection Π_F satisfies the identity $\Pi_F^2 = \Pi_F$.*

Proof   Consider a maximal set of perfectly discriminable pure states $\{\varphi_i\}_{i=1}^{d_A}$ such that $\{\varphi_i\}_{i \in V}$ is maximal in F. In this way F is the face generated by V, and, therefore $\Pi_F = \Pi_V$. The thesis follows by taking $V = W$ in Lemma 15.26. □

Corollary 15.28   *For every state $\rho \in \mathsf{St}_1(A)$ such that $\rho \notin F^\perp$, the normalized state ρ' defined by*

$$|\rho') = \frac{\Pi_F |\rho)}{(e|\Pi_F|\rho)} \tag{15.8}$$

belongs to the face F.

Proof   By Theorem 15.19, we have $(e|\Pi_F = (a_F|$. Since $\rho \notin F^\perp$, we must have $(e|\Pi_F|\rho) = (a_F|\rho) > 0$, and, therefore, the state ρ' in Eq. (15.8) is well defined. Considering that $(a_F|\Pi_F|\rho) = (e|\Pi_F^2|\rho)$, by Theorem 15.19, and using idempotence from Corollary 15.27, we have $(a_F|\Pi_F|\rho) = (e|\Pi_F|\rho)$. Now, using the definition of ρ' we obtain

$$(a_F|\rho') = \frac{(a_F|\Pi_F|\rho)}{(e|\Pi_F|\rho)} = 1,$$

Finally, Lemma 15.15 implies that ρ' belongs to the face F. □

Corollary 15.29 *Let $\Pi_{\{\varphi\}}$ be the projection on the pure state $\varphi \in \mathsf{St}_1(A)$. Then for every state $\rho \in \mathsf{St}_1(A)$ one has $\Pi_{\{\varphi\}}|\rho) = p|\varphi)$ where $p = (\varphi^\dagger|\rho)$.*

Proof Recall that, by Corollary 15.14 and Theorem 15.19, we have $(\varphi^\dagger| = (e|\Pi_{\{\varphi\}}$. If $(\varphi^\dagger|\rho) = 0$ then clearly $\Pi_{\{\varphi\}}|\rho) = 0$. Otherwise, the proof is a straightforward application of Corollary 15.28. □

Corollary 15.30 *Let $\Pi_{\{\varphi\}}$ be the projection on the pure state $\varphi \in \mathsf{PurSt}_1(A)$. Then $\Pi_{\{\varphi\}} = |\varphi)(\varphi^\dagger|$.*

Corollary 15.31 *Let $\varphi \in \mathsf{PurSt}_1(A)$, and $\tilde{F}$ be the face of $\mathsf{St}_1(AB)$ associated with $\varphi \otimes \chi_B$. Then $\tilde{F} = \{\varphi \otimes \rho | \rho \in \mathsf{St}_1(B)\} = \{\varphi\} \otimes \mathsf{St}_1(B)$.*

Proof By Lemma 15.18, one has $\Pi_{\tilde{F}} = \Pi_\varphi \otimes \mathcal{I}_B$, and by Corollary 15.30

$$\Pi_{\tilde{F}} = |\varphi)(\varphi^\dagger| \otimes \mathcal{I}_B.$$

Let $\sigma \in \tilde{F}$, then $|\sigma) = \Pi_{\tilde{F}}|\sigma) = |\varphi) \otimes (\varphi^\dagger|_A|\sigma) = |\varphi) \otimes |\rho)$, where $|\rho) = (\varphi^\dagger|_A|\sigma)$. □

We conclude the present subsection with a result that will be useful in the following. The result requires two lemmas, the first of which regards a useful property of the effect a_F:

Lemma 15.32 *Let φ be a pure state in the face $F \subseteq \mathsf{St}_1(A)$. If $\mathcal{A} \in \mathsf{Transf}(A)$ is an atomic transformation such that $\mathcal{A} =_F \mathcal{I}_A$, then $(\varphi^\dagger|\mathcal{A} = (\varphi^\dagger|$. Moreover, if a_F is the effect associated with the face F, then we have $(a_F|\mathcal{A} = (a_F|$.*

Proof By the atomicity axiom, the effect $(\varphi^\dagger|\mathcal{A}$ is atomic. Now, since $\mathcal{A}|\varphi) = |\varphi)$, we have $(\varphi^\dagger|\mathcal{A}|\varphi) = (\varphi^\dagger|\varphi) = 1$. On the other hand, $\varphi^\dagger$ is the unique atomic effect such that $(\varphi^\dagger|\varphi) = 1$ (Theorem 11.5). Hence, $(\varphi^\dagger|\mathcal{A} = (\varphi^\dagger|$. Now, since $a_F = \sum_{i=1}^{|F|} \varphi_i^\dagger$, for perfectly discriminable states $\varphi_i \in F$ we have $(a_F|\mathcal{A} = \sum_{i=1}^{|F|}(\varphi_i^\dagger|\mathcal{A} = \sum_{i=1}^{|F|}(\varphi_i^\dagger| = (a_F|$ (Lemma 15.11). □

Lemma 15.33 *An atomic transformation $\mathcal{A} \in \mathsf{Transf}(A)$ satisfies $\mathcal{A} =_F \mathcal{I}_A$ if and only if*

$$\Pi_F \mathcal{A} = \Pi_F. \tag{15.9}$$

Proof Suppose that $\mathcal{A} =_F \mathcal{I}_A$. Let $\Phi \in \mathsf{St}_1(A\tilde{A})$ be a minimal purification of the invariant state χ_A and define the two pure states

$$|\Phi_F) := \frac{d_A}{|F|}(\Pi_F \otimes \mathcal{I}_{\tilde{A}})|\Phi)$$

$$|\Phi'_F) := \frac{d_A}{|F|}(\Pi_F \mathcal{A} \otimes \mathcal{I}_{\tilde{A}})|\Phi).$$

Then we have

$$
\Phi'_F \!\!\begin{array}{c} A \,\boxed{e} \\ \tilde{A} \end{array}
=
\Phi \!\!\begin{array}{c} A \,\boxed{\mathcal{A}}\, A \,\boxed{a_F} \\ \tilde{A} \end{array}
$$

$$
=
\Phi \!\!\begin{array}{c} A \,\boxed{\Pi_F}\, A \,\boxed{e} \\ \tilde{A} \end{array}
$$

$$
=
\Phi_F \!\!\begin{array}{c} A \,\boxed{e} \\ \tilde{A} \end{array},
$$

having used the condition $(a_F | \mathcal{A} = (a_F |$ of Lemma 15.32. Now, we proved that Φ_F and Φ'_F have the same marginal on system $\tilde{A}$. By the uniqueness of purification, there exists a reversible transformation $\mathcal{V} \in \mathbf{G}_A$ such that $|\Phi'_F) = (\mathcal{V} \otimes \mathcal{I}_{\tilde{A}})|\Phi_F)$. Since Φ is faithful, this implies $\Pi_F \mathcal{A} = \mathcal{V} \Pi_F$. Now, for every ρ in F one has $\mathcal{V}|\rho) = \mathcal{V}\Pi_F|\rho) = \Pi_F \mathcal{A}|\rho) = |\rho)$, namely $\mathcal{V} =_F \mathcal{I}_A$. Then,

$$
\Pi_F \mathcal{A} = \mathcal{V} \Pi_F
$$
$$
= \Pi_F.
$$

Conversely, suppose that Eq. (15.9) is satisfied. Let $\varphi \in F$ be a pure state in F. Then, we have

$$
(\varphi^\dagger | \mathcal{A} | \varphi) = (\varphi^\dagger | \Pi_F \mathcal{A} | \varphi) = (\varphi^\dagger | \Pi_F | \varphi) = (\varphi^\dagger | \varphi) = 1,
$$

having used that $(\varphi^\dagger | \Pi_F = (\varphi^\dagger |$ by Lemma 15.32. Since $\varphi^\dagger$ identifies the state φ, we must have $\mathcal{A}\varphi = \varphi$. Finally, since $\varphi \in F$ is arbitrary, the last identity implies $\mathcal{A} =_F \mathcal{I}_A$. □

15.4 Projection of a Pure State on Two Orthogonal Faces

Some properties of two-dimensional systems will be extended to the case of generic systems using projections on orthogonal faces. In this section we will prove some of them, starting from the following lemmas.

Lemma 15.34 *Consider a pure state $\varphi \in \mathsf{St}_1(A)$ and two complementary projections $\Pi_1 := \Pi_F$ and $\Pi_2 := \Pi_F^\perp$ such that $\Pi_i | \varphi) \neq 0$. Then one has*

$$
a_{F_\theta} = \varphi_1^\dagger + \varphi_2^\dagger, \tag{15.10}
$$

with $|\theta) := (\Pi_F + \Pi_F^\perp)|\varphi)$ and $|\varphi_i) := \Pi_i|\varphi)/(e|\Pi_i|\varphi)$.

Proof By the atomicity axiom, the states $|\varphi_i)$ are pure. We can define the two probabilities $p_i := (e|\Pi_i|\varphi)$, $i = 1, 2$. In this way we have $\Pi_i|\varphi) = p_i|\varphi_i)$ for $i = 1, 2$, and

$\theta = p_1\varphi_1 + p_2\varphi_2$. Using the effects a_F and $a_{F\perp}$, it is trivial to verify that the pure states φ_1 and φ_2 are perfectly discriminable. Consider the atomic effects $\varphi_i^\dagger$, and define $\tilde{a} := \varphi_1^\dagger + \varphi_2^\dagger$. Now, $(\tilde{a}|\theta) = p_1(\varphi_1^\dagger|\varphi_1) + p_2(\varphi_2^\dagger|\varphi_2) = 1$, thus $\tilde{a} =_{F_\theta} e$, while if we extend $\{\varphi_1, \varphi_2\}$ to a maximal set, we have $(\varphi_1^\dagger|\varphi_j) = (\varphi_2^\dagger|\varphi_j) = 0$ for $j > 2$, and then $\tilde{a} =_{F_\theta^\perp} 0$ by Lemma 15.17. Thus, $\tilde{a} = a_{F_\theta}$ is the effect associated with the face $F_\theta = \mathsf{RefSet}_1(\theta)$. $\square$

Lemma 15.35 *Consider a pure state $\varphi \in \mathsf{St}_1(A)$ and two complementary projections Π_F and $\Pi_F^\perp$. Then, φ belongs to the face identified by the state*

$$|\theta) := (\Pi_F + \Pi_F^\perp)|\varphi). \tag{15.11}$$

Proof If $\Pi_F|\varphi) = 0$, or $\Pi_F^\perp|\varphi) = 0$, then there is nothing to prove: this means that $\Pi_F^\perp|\varphi) = |\varphi)$, or $\Pi_F|\varphi) = |\varphi)$, respectively, and the thesis is trivially true. By Corollary 15.20 one cannot have both $\Pi_F|\varphi) = 0$ and $\Pi_F^\perp|\varphi) = 0$. Suppose then that $\Pi_F|\varphi) \neq 0$ and $\Pi_F^\perp|\varphi) \neq 0$. By Lemma (15.34), we have $a_{F_\theta} = \varphi_1^\dagger + \varphi_2^\dagger$. Recalling that $(\varphi_i^\dagger|\Pi_i = (\varphi_i^\dagger|$ for $i = 1, 2$ by virtue of Lemma 15.32, we then conclude the following:

$$\begin{aligned} (a_{F_\theta}|\varphi) &= [(\varphi_1^\dagger| + (\varphi_2^\dagger|]|\varphi) \\ &= (\varphi_1^\dagger|\Pi_1|\varphi) + (\varphi_2^\dagger|\Pi_2|\varphi) \\ &= \sum_{i=1,2} p_i(\varphi_i^\dagger|\varphi_i) = 1. \end{aligned}$$

Finally, Lemma 15.15 yields $\varphi \in F_\theta$. $\square$

A consequence of Lemma 15.35 is the following.

Lemma 15.36 *Let $\varphi \in \mathsf{St}_1(A)$ be a pure state, and F be a face in $\mathsf{St}_1(A)$. If ρ is perfectly discriminable from $\Pi_F|\varphi)$ and from $\Pi_F^\perp|\varphi)$ then ρ is perfectly discriminable from $|\varphi)$. In particular, one has $(\varphi^\dagger|\rho) = 0$.*

Proof Since ρ is perfectly discriminable from $\Pi_F|\varphi)$ and $\Pi_F^\perp|\varphi)$, it is also perfectly discriminable from any convex combination of them (Corollary 15.7). Equivalently, ρ is perfectly discriminable from the face F_θ identified by $|\theta) := \Pi_F|\varphi) + \Pi_F^\perp|\varphi)$. In particular, ρ must be perfectly discriminable from φ, which belongs to F_θ by virtue of Lemma 15.35. Then by Exercise 11.3 we have $(\varphi^\dagger|\rho) = 0$. $\square$

We now provide a technical result that will be useful in the following.

Lemma 15.37 *Let $\varphi \in \mathsf{St}_1(A)$ be a pure state such that $\Pi_F|\varphi) \neq 0$ and $\Pi_F^\perp|\varphi) \neq 0$. Define the pure states $|\varphi_1) := \Pi_F|\varphi)/(e|\Pi_F|\varphi)$ and $|\varphi_2) := \Pi_F^\perp|\varphi)/(e|\Pi_F^\perp|\varphi)$ and the mixed state $|\theta) := (\Pi_F + \Pi_F^\perp)|\varphi)$. Then, we have*

$$\Pi_F \Pi_{F_\theta} = \Pi_{\{\varphi_1\}}$$
$$\Pi_F^\perp \Pi_{F_\theta} = \Pi_{\{\varphi_2\}}.$$

Proof   Let $\{\psi_i\}_{i=1}^{|F|}$ be a maximal set of perfectly discriminable pure states in F, chosen in such a way that $\psi_1 = \varphi_1$, and let $\{\psi_i\}_{i=|F|+1}^{d_A}$ be a maximal set of perfectly discriminable pure states in $F^\perp$, chosen in such a way that $\psi_{|F|+1} = \varphi_2$. Defining the sets $V_1 := \{1, \ldots, |F|\}$, $V_2 := \{|F| + 1, \ldots, d_A\}$, and $U := \{1, |F| + 1\}$ we then have $\Pi_{V_1} = \Pi_F$, $\Pi_{V_2} = \Pi_F^\perp$, and $\Pi_U = \Pi_{F_\theta}$, where the last equality follows from Lemma 15.34. Using Lemma 15.26 we obtain

$$\Pi_{V_i}\Pi_U = \Pi_{V_i \cap U}$$
$$= \Pi_{\{\varphi_i\}},$$

namely the thesis.   □

We conclude this subsection with an important observation about the group of reversible transformations that act as the identity on two orthogonal faces F and $F^\perp$, denoted as

$$\mathbf{G}_{F,F^\perp} := \{\mathcal{U} \in \mathbf{G}_A | \mathcal{U} =_F \mathcal{I}_A, \mathcal{U} =_{F^\perp} \mathcal{I}_A\}.$$

Then we have the following:

Theorem 15.38   *For every face $F \subset \mathsf{St}_1(A)$ such that $F \neq \emptyset$ and $F \neq \mathsf{St}_1(A)$, the group $\mathbf{G}_{F,F^\perp}$ is topologically equivalent to a circle.*

Proof   Let $\mathcal{U}$ be a transformation in $\mathbf{G}_{F,F^\perp}$, $\Phi \in \mathsf{St}(A\tilde{A})$ be a minimal purification of the invariant state χ_A, and $|\Phi_{\mathcal{U}}) := (\mathcal{U} \otimes \mathcal{I}_{\tilde{A}})|\Phi)$ be the Choi–Jamiołkowski state of $\mathcal{U}$. Define the face $\tilde{F} := F_{\omega_F \otimes \chi_{\tilde{A}}}$. By Corollary 15.5 one has $\tilde{F}^\perp = F_{\omega_{F^\perp} \otimes \chi_{\tilde{A}}}$, and the projections $\Pi_{\tilde{F}} := \Pi_F \otimes \mathcal{I}_{\tilde{A}}$ and $\Pi_{\tilde{F}}^\perp := \Pi_F^\perp \otimes \mathcal{I}_{\tilde{A}}$ (see Lemma 15.18). Using Lemma 15.33 we then obtain

$$\Pi_{\tilde{F}}|\Phi_{\mathcal{U}}) = (\Pi_F \otimes \mathcal{I}_{\tilde{A}})|\Phi_{\mathcal{U}})$$
$$= (\Pi_F\mathcal{U} \otimes \mathcal{I}_{\tilde{A}})|\Phi)$$
$$= (\Pi_F \otimes \mathcal{I}_{\tilde{A}})|\Phi)$$
$$= \frac{|F|}{d_A}|\Phi_F),$$

and, similarly,

$$\Pi_{\tilde{F}}^\perp|\Phi_{\mathcal{U}}) = (\Pi_F^\perp \otimes \mathcal{I}_{\tilde{A}})|\Phi_{\mathcal{U}})$$
$$= (\Pi_F^\perp\mathcal{U} \otimes \mathcal{I}_{\tilde{A}})|\Phi)$$
$$= (\Pi_F^\perp \otimes \mathcal{I}_{\tilde{A}})|\Phi)$$
$$= \frac{|F^\perp|}{d_A}|\Phi_{F^\perp}).$$

This means that the projections of $\Phi_{\mathcal{U}}$ on the faces $\tilde{F}$ and $\tilde{F}^\perp$ are independent of $\mathcal{U}$. Also, it means that $\Phi_{\mathcal{U}}$ belongs to the face F_θ identified by the state $|\theta) := \frac{|F|}{d_A}|\Phi_F) + \frac{|F^\perp|}{d_A}|\Phi_{F^\perp})$ (Lemma 15.35). Now, by the compression axiom, F_θ is equivalent to the state space of a qubit via some ideal encoding scheme $(\mathcal{E}, \mathcal{D}, B)$. The two perfectly discriminable states $\mathcal{E}|\Phi_F)$ and $\mathcal{E}|\Phi_{F^\perp})$ can then be represented as the North and South poles of the Bloch sphere, respectively. Moreover, by the properties of ideal compression, the transformations

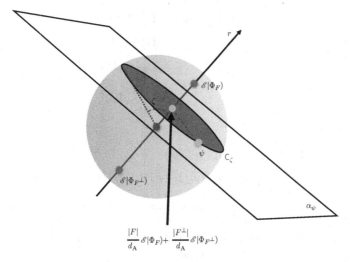

The representation of the circle C_ζ in the Bloch ball. The axis r joins the opposite points $\mathcal{E}|\Phi_F)$ and $\mathcal{E}|\Phi_{F^\perp})$. The point $\frac{|F|}{d_A}\mathcal{E}|\Phi_F) + \frac{|F^\perp|}{d_A}\mathcal{E}|\Phi_{F^\perp})$, coinciding with $\mathcal{E}\Pi_F \mathcal{D}|\psi) + \mathcal{E}\Pi_{F^\perp}\mathcal{D}|\psi)$, belongs to the intersection of the axis r and the plane α_ψ orthogonal to r and containing $\psi \in \mathsf{PurSt}_1(A)$. The latitude ζ of the circle C_ζ has $\cos\zeta$ coinciding with the distance between the center of the sphere and the point $\frac{|F|}{d_A}\mathcal{E}|\Phi_F) + \frac{|F^\perp|}{d_A}\mathcal{E}|\Phi_{F^\perp})$, which amounts to $\cos\zeta = \frac{|F|}{d_A} - \frac{|F^\perp|}{d_A}$.

Fig. 15.1

$\mathcal{E}\Pi_F \mathcal{D}$ and $\mathcal{E}\Pi_F^\perp \mathcal{D}$ are projections on the two perfectly discriminable pure states $\mathcal{E}|\Phi_F)$ and $\mathcal{E}|\Phi_{F^\perp})$. Thus, all the compressed Choi–Jamiołkowski states $\{\mathcal{E}|\Phi_\mathcal{U})\}_{\mathcal{U} \in \mathbf{G}_{F,F^\perp}}$ have fixed projections $\frac{|F|}{d_A}\mathcal{E}|\Phi_F)$ and $\frac{|F^\perp|}{d_A}\mathcal{E}|\Phi_{F^\perp})$ on the perfectly discriminable pure states corresponding to the poles of the Bloch sphere. By the geometry of the Bloch ball we then know that all the compressed Choi–Jamiołkowski states $\{\mathcal{E}|\Phi_\mathcal{U})\}_{\mathcal{U} \in \mathbf{G}_{F,F^\perp}}$ belong to the same horizontal plane intersecting the vertical axis of the poles at latitude ζ given by $\cos\zeta = (|F| - |F^\perp|)/d_A$ (see Fig. 15.1). Since these states are pure, this implies that the set $\{\mathcal{E}|\Phi_\mathcal{U})\}_{\mathcal{U} \in \mathbf{G}_{F,F^\perp}}$ is a subset of a circle C_ζ on the Bloch sphere. Precisely, the circle C_ζ is given by

$$C_\zeta := \{\psi \in \mathsf{St}_1(B)| \, \mathcal{E}\Pi_{\{\Phi_F\}}\mathcal{D}|\psi) = |F|/d_A\mathcal{E}|\Phi_F), \, \mathcal{E}\Pi_{\{\Phi_{F^\perp}\}}\mathcal{D}|\psi) = |F^\perp|/d_A\mathcal{E}|\Phi_{F^\perp})\}.$$

We now prove that in fact the states $\{\mathcal{E}\phi_\mathcal{U}\}_{\mathcal{U} \in \mathbf{G}_{F,F^\perp}}$ are the whole circle. Let ψ be a state in C_ζ. Since $\mathcal{D}|\psi)$ belongs to the face F_θ, we obtain

$$(\Pi_F \otimes \mathcal{I}_{\tilde{A}})\mathcal{D}|\psi) = \Pi_{\tilde{F}}\mathcal{D}|\psi)$$
$$= \Pi_{\tilde{F}}\Pi_{F_\theta}\mathcal{D}|\psi)$$
$$= \Pi_{\{\Phi_F\}}\mathcal{D}|\psi)$$
$$= \mathcal{D}\mathcal{E}\Pi_{\{\Phi_F\}}\mathcal{D}|\psi)$$
$$= \frac{|F|}{d_A}|\Phi_F),$$

where the third equality comes from Lemma 15.37 with the substitutions $F \to \tilde{F}$, $\varphi \to \psi$, $\varphi_1 \to \Phi_F$, and $\varphi_2 \to \Phi_{F^\perp}$. Similarly,

$$(\Pi_F^\perp \otimes \mathcal{I}_{\tilde{A}})\mathcal{D}|\psi) = \Pi_{\tilde{F}}^\perp \mathcal{D}|\psi)$$

$$= \Pi_{\tilde{F}}^\perp \Pi_{F_\theta} \mathcal{D}|\psi)$$

$$= \Pi_{\{\Phi_{F^\perp}\}} \mathcal{D}|\psi)$$

$$= \frac{|F^\perp|}{d_A}|\Phi_{F^\perp}).$$

Therefore, we have

$$(e|_A \mathcal{D}|\psi) = (a_{\tilde{F}} + a_{\tilde{F}^\perp}|_A \mathcal{D}|\psi)$$

$$= (e|_A(\Pi_F \otimes \mathcal{I}_{\tilde{A}})\mathcal{D}|\psi) + (e|_A(\Pi_F^\perp \otimes \mathcal{I}_{\tilde{A}})\mathcal{D}|\psi)$$

$$= \frac{|F|}{d_A}(e|_A|\Phi_F) + \frac{|F^\perp|}{d_A}(e|_A|\Phi_{F^\perp})$$

$$= (e|_A(\Pi_F \otimes \mathcal{I}_{\tilde{A}})|\Phi) + (e|_A(\Pi_F^\perp \otimes \mathcal{I}_{\tilde{A}})|\Phi)$$

$$= (a_{\tilde{F}} + a_{\tilde{F}^\perp}|_A|\Phi)$$

$$= (e|_A|\Phi)$$

$$= |\chi_{\tilde{A}}).$$

Since $\mathcal{D}|\psi)$ and Φ are both purifications of the invariant state $\chi_{\tilde{A}}$, by the uniqueness of purification there must be a reversible transformation $\mathcal{U} \in \mathbf{G}_A$ such that $\mathcal{D}|\psi) = (\mathcal{U} \otimes \mathcal{I}_{\tilde{A}})|\Phi)$. Finally, it is easy to check that $\Pi_F \mathcal{U} = \Pi_F$ and $\Pi_F^\perp \mathcal{U} = \Pi_F^\perp$, which, by Lemma 15.33, implies $\mathcal{U} =_F \mathcal{I}_A$ and $\mathcal{U} =_{F^\perp} \mathcal{I}_A$. This proves that the Choi–Jamiołkowski states $\{\Phi_{\mathcal{U}}\}_{\mathcal{U} \in \mathbf{G}_{F,F^\perp}}$ are the image of the whole circle $\mathbf{C}_\zeta$ under the reversible linear map $\mathcal{D}$. Since the Choi–Jamiołkowski isomorphism is continuous in the operational norm, the group $\mathbf{G}_{F,F^\perp}$ is topologically equivalent to a circle.   □

15.5 Summary

In this chapter we defined projections and proved several properties thereof. The projection on the face F is defined as a transformation $\Pi_F \in \mathsf{Transf}(A)$ that acts as the identity on states in the face F and that annihilates the states on the *orthogonal face $F^\perp$*. We proved existence and uniqueness of the projection on a face F, along with its atomicity. Finally, we gave some useful results about the projections of a pure state on two orthogonal faces.

Solutions to Selected Problems and Exercises

Exercise 15.2

Define the normalized state $\bar{\rho} := \rho/(e|\rho)$. Since $e = \varphi_0^\dagger + \varphi_1^\dagger$, one has $\bar{\rho} = \rho/(\varphi_1^\dagger|\rho)$. This implies that $(\varphi_1^\dagger|\bar{\rho}) = 1$, and using Lemma 15.15 one has $\rho = \varphi_1(e|\rho)$.

16

The Superposition Principle

What is the operational meaning of the superposition principle? This question can be answered by recalling that every pure state belongs to a maximal perfectly discriminable set. One can think of a maximal set of perfectly discriminable pure states as a *quantity* that can be measured by the corresponding perfectly discriminating measurement. If a system is prepared in one of the pure states making a maximal set, the information gained via the discriminating measurement can be thought of as ascertaining the value of the corresponding quantity.

A strong property of quantum theory is that for every quantity corresponding to a perfectly discriminating measurement, and for every conceivable probability distribution over its values, there exists a *pure state* in which the considered quantity has values distributed according to that specific probability distribution. This property is what we call the *superposition principle*.

16.1 The Superposition Principle

We now show that in a theory satisfying our postulates the superposition principle holds.

Theorem 16.1 (Superposition Principle for General Systems) *Let* $\{\varphi_i\}_{i=1}^{d_A} \subseteq \mathsf{St}_1(A)$ *be a maximal set of perfectly discriminable pure states of system* A. *Then, for every choice of probabilities* $\{p_i\}_{i=1}^{d_A}$, $p_i \geq 0$, $\sum_{i=1}^{d_A} p_i = 1$ *there exists a pure state* $\varphi_p \in \mathsf{St}_1(A)$ *such that*

$$p_i = (\varphi_i^\dagger | \varphi_p) \qquad \forall i = 1, \ldots, d_A. \tag{16.1}$$

or, equivalently,

$$\Pi_{\{\varphi_i\}} | \varphi_p) = p_i | \varphi_i) \qquad \forall i = 1, \ldots, d_A, \tag{16.2}$$

where $\Pi_{\{\varphi_i\}}$ *is the projection on* φ_i.

Proof Let us first prove the equivalence between Eqs. (16.1) and (16.2). From Eq. (16.2) we obtain Eq. (16.1) using the relation $(e|\Pi_{\{i\}} = (\varphi_i^\dagger|$, which follows from Theorem 15.19. Conversely, from Eq. (16.1) we obtain Eq. (16.2) using Corollary 15.29. Now, we will prove Eq. (16.1) by induction. The statement for $N = 2$ is proved by Corollary 14.7. Assume that the statement holds for every system B of dimension $d_B = N$ and suppose that $d_A = N + 1$. Let F be the face of $\mathsf{St}_1(A)$ identified by $\omega_F = 1/N \sum_{i=1}^{N} \varphi_i$ and $F^\perp$ be the orthogonal face, identified by the state φ_{N+1}. Now there are two cases: either $p_{N+1} = 1$

or $p_{N+1} \neq 1$. If $p_{N+1} = 1$, then there is nothing to prove: the desired state is φ_{N+1}. Then, suppose that $p_{N+1} \neq 1$. Using the induction hypothesis and the compression axiom in Section 4.3 we can find a state $\psi_{\mathbf{q}} \in F$ such that $(\varphi_i^\dagger | \psi_{\mathbf{q}}) = q_i$, with $q_i = p_i/(1 - p_{N+1})$, $i = 1, \ldots, N$. Let us then define a new maximal set of perfectly discriminable pure states $\{\varphi_i'\}_{i=1}^{N+1}$, with $\varphi_1' = \psi_{\mathbf{q}}$ and $\varphi_{N+1}' = \varphi_{N+1}$. Note that, since $\varphi_{N+1}' = \varphi_{N+1} = \omega_{F\perp}$, by Lemma 15.9 one has $\omega_F = 1/N \sum_{i=1}^N \varphi_i'$, that is, F is the face identified by the states $\{\varphi_i'\}_{i=1}^N$. Now consider the two-dimensional face F' identified by $\theta = \frac{1}{2}(\varphi_1' + \varphi_{N+1}')$. By Corollary 14.7 (superposition principle for qubits) we know that there exists a pure state $\varphi \in F'$ with $(\varphi_1'^\dagger | \varphi) = 1 - p_{N+1}$ and $(\varphi_{N+1}'^\dagger | \varphi) = p_{N+1}$. Let us define $V := \{1, \ldots, N\}$ and $W := \{1, N+1\}$. Then, we have $\Pi_F = \Pi_V$ and $\Pi_{F'} = \Pi_W$, and by Lemma 15.26,

$$\Pi_F | \varphi) = \Pi_F \Pi_{F'} | \varphi)$$
$$= \Pi_{V \cap W} | \varphi)$$
$$= \Pi_{\{\varphi_1'\}} | \varphi)$$
$$= \Pi_{\{\psi_{\mathbf{q}}\}} | \varphi)$$
$$= (1 - p_{N+1}) | \psi_{\mathbf{q}}),$$

having used Corollary 15.29 for the last equality. Finally, for $i = 1, \ldots, N$ we have

$$(\varphi_i^\dagger | \varphi) = (\varphi_i^\dagger | \Pi_F | \varphi)$$
$$= (1 - p_{N+1})(\varphi_i^\dagger | \psi_{\mathbf{q}})$$
$$= (1 - p_{N+1}) q_i$$
$$= p_i.$$

On the other hand, we have $(\varphi_{N+1}^\dagger | \varphi) = (\varphi_{N+1}'^\dagger | \varphi) = p_{N+1}$. $\qquad \qquad \square$

16.2 Completeness for Purification

Using the superposition principle and the spectral decomposition of Theorem 12.14 we can now show that every state of system A has a purification in AB provided $d_B \geq d_A$.

Lemma 16.2 *For every state $\rho \in \mathsf{St}_1(A)$ and for every system B with $d_B \geq d_A$ there exists a purification of ρ in $\mathsf{St}_1(AB)$.*

Proof Take the spectral decomposition of ρ, given by $\rho = \sum_{i=1}^{d_A} p_i \varphi_i$, where $\{p_i\}$ are probabilities and $\{\varphi_i\}_{i=1}^{d_A}$ for system A is a maximal set of perfectly discriminable pure states. Let $\{\psi_i\}_{i=1}^{d_B}$ be a maximal set of perfectly discriminable pure states for system B. Clearly, $\{\varphi_i \otimes \psi_j\}$ is a maximal set of perfectly discriminable pure states for AB. By the superposition principle (Theorem 16.1) there exists a pure state $\Psi_\rho \in \mathsf{St}_1(AB)$ such that $(\varphi_i^\dagger \otimes \psi_j^\dagger | \Psi_\rho) = p_i \delta_{ij}$. Equivalently, we have $(\psi_i^\dagger |_B | \Psi_\rho)_{AB} = p_i | \varphi_i)_A$ for every $i = 1, \ldots, d_A$ and $(\psi_i^\dagger |_B | \Psi_\rho)_{AB} = 0$ for $i > d_A$. Summing over i we then obtain $(e|_B | \Psi_\rho)_{AB} = \sum_{i=1}^{d_B} (\psi_i^\dagger |_B | \Psi_\rho)_{AB} = \sum_{i=1}^{d_A} p_i | \varphi_i)_A = | \rho)_A$. $\qquad \square$

As a consequence of Lemma 16.2 we have the following:

Corollary 16.3 *Every system* B *with* $d_B = d_A$ *is operationally equivalent to the conjugate system* $\tilde{A}$.

Proof By Lemma 16.2, the invariant state $\chi_A \in St_1(A)$ has a purification Ψ in $St_1(AB)$. Moreover, by Proposition 12.22, the marginal of Ψ on B is the invariant state χ_B. By definition, this means that B is a conjugate system of A. Since the conjugate system $\tilde{A}$ is unique up to operational equivalence (Theorem 8.11), this implies the thesis. $\quad\square$

16.3 Equivalence of Systems with Equal Dimension

We are now in position to prove that two systems A and B with the same dimension are operationally equivalent, namely that there is a reversible transformation from A to B. In other words, we prove that the informational dimension classifies the systems of our theory up to operational equivalence. Notice the remarkable fact that this property is derived from the principles, rather than being assumed from the start.

Corollary 16.4 (Operational Equivalence of Systems with Equal Dimension) *Every two systems* A *and* B *with* $d_A = d_B$ *are operationally equivalent.*

Proof By Corollary 16.3, A and B are both operationally equivalent to the conjugate system $\tilde{A}$. Hence, they are operationally equivalent to each other. $\quad\square$

16.4 Reversible Operations of Perfectly Discriminable Pure States

Another important consequence of the superposition principle is the possibility of transforming any arbitrary maximal set of perfectly discriminable pure states into any other via a reversible transformation.

Corollary 16.5 *Let* A *and* B *be two systems with* $d_A = d_B =: d$ *and let* $\{\varphi_i\}_{i=1}^d$ *(resp.* $\{\psi_i\}_{i=1}^d$*) be a maximal set of perfectly discriminable pure states in* A *(resp.* B*). Then, there exists a reversible transformation* $\mathcal{U} \in \mathsf{Transf}(A \rightarrow B)$ *such that* $\mathcal{U}|\varphi_i) = |\psi_i)$.

Proof Let $\Phi \in St(A\tilde{A})$ be a purification of the invariant state χ_A. Although we know that A and $\tilde{A}$ are operationally equivalent (Corollary 16.3) we use the notation A and $\tilde{A}$ to distinguish between the two subsystems of $A\tilde{A}$. Define the pure state $\tilde{\varphi}_i$ via the relation $(\varphi_i^\dagger|_A|\Phi)_{A\tilde{A}} = \frac{1}{d}|\tilde{\varphi}_i)_{\tilde{A}}$. Then, by Lemma 11.6 we have

$$(\tilde{\varphi}_i^\dagger|_{\tilde{A}}|\Phi)_{A\tilde{A}} = \frac{1}{d}|\varphi_i)_A. \tag{16.3}$$

On the other hand, using the superposition principle (Theorem 16.1) we can construct a state $\Psi \in \mathsf{St}_1(\mathrm{B\tilde{A}})$ such that $(\psi_i^\dagger \otimes \tilde{\varphi}_j^\dagger | \Psi) = \delta_{ij}/d$, or, equivalently,

$$(\tilde{\varphi}_i^\dagger |_{\tilde{A}} | \Psi)_{\mathrm{B\tilde{A}}} = \frac{1}{d} |\psi_i)_{\mathrm{B}}. \tag{16.4}$$

Now, Φ and Ψ have the same marginal on system $\tilde{A}$: they are both purifications of the invariant state $\chi_{\tilde{A}}$. Moreover, A and B are operationally equivalent because they have the same dimension (Corollary 16.4). Hence, by the uniqueness of purification, there must be a reversible transformation $\mathcal{U} \in \mathsf{Transf}(A \to B)$ such that

$$|\Psi)_{\mathrm{B\tilde{A}}} = (\mathcal{U} \otimes \mathcal{I}_{\tilde{A}})|\Phi)_{\mathrm{A\tilde{A}}}. \tag{16.5}$$

Combining Eqs. (16.3), (16.4), and (16.5) we finally obtain

$$\frac{1}{d}\mathcal{U}|\varphi_i)_A = \left[\mathcal{U} \otimes (\tilde{\varphi}_i^\dagger |_{\tilde{A}}\right] |\Phi)_{\mathrm{A\tilde{A}}}$$
$$= (\tilde{\varphi}_i^\dagger |_{\tilde{A}} |\Psi)_{\mathrm{B\tilde{A}}}$$
$$= \frac{1}{d}|\psi_i)_{\mathrm{B}},$$

that is, $\mathcal{U}|\varphi_i) = |\psi_i)$ for every $i = 1, \ldots, d$. $\square$

16.5 Summary

In this chapter we proved one of the distinctive features of quantum theory, namely the superposition principle. We phrased superposition in terms of pure states and atomic effects. We then used the superposition principle to show that two systems with the same dimension are operationally equivalent.

In this chapter we finally conclude the reconstruction of quantum theory from the principles. To this end, we prove that states of a system with dimension d_A are in one-to-one correspondence with density matrices on the Hilbert space $\mathcal{H}_A \simeq \mathbb{C}^{d_A}$, and effects are in one-to-one correspondence with positive operators bounded by the identity operator on $\mathcal{H}_A$.

In principle, this does not characterize the whole theory, since all the sets of transformations are still missing from the picture. However, as a consequence of the fact that "everything not forbidden is allowed" (Section 7.12, and in particular Proposition 7.10), we know that in a causal theory with local discriminability and unique purification, the set of states completely specify the theory, namely uniquely determine the set of transformations, thanks to the isomorphism between states and effects.

Thus, the characterization of the sets $\mathsf{St}(A)$ and $\mathsf{Eff}(A)$ for every system A completely specifies the theory, which in our case will be quantum theory.

17.1 The Basis

In order to specify the correspondence between states and matrices we choose a particular basis for the vector space $\mathsf{St}_{\mathbb{R}}(A)$. The basis is constructed as follows:[1]

1. Let us first choose a maximal set of d_A perfectly discriminable states $\{\varphi_m\}_{m=1}^{d_A}$, and declare that they are the first d_A basis vectors.
2. Now, for every $m < n$ the face F_{mn} generated by $\{\varphi_m, \varphi_n\}$ can be ideally encoded in a two-dimensional system by the ideal compression scheme $(Q, \mathcal{E}^{(mn)}, \mathcal{D}^{(mn)})$. Let

$$\eta_0 := \mathcal{E}^{(mn)}\varphi_m, \quad \eta_1 := \mathcal{E}^{(mn)}\varphi_n. \tag{17.1}$$

As we know from Corollary 14.4, the convex set of states of a two-dimensional system is the Bloch sphere, and we can choose the z-axis to be the line joining the two states $\{\mathcal{E}^{(mn)}\varphi_m, \mathcal{E}^{(mn)}\varphi_n\}$, e.g. with the positive direction of the z-axis being the direction from η_0 to η_1. Once direction of the z-axis has been specified, we can choose the x and y axes. Note that any couple of orthogonal directions in the plane orthogonal to z-axis is a valid choice for the x- and y-axes.[2] Let $\varphi_{x,+}^{mn}, \varphi_{x,-}^{mn} \in F_{mn}$ and $\varphi_{y,+}^{mn}, \varphi_{y,-}^{mn} \in F_{mn}$ be the

---

[1] Hardy (2011).

[2] At the moment there is no relation among the different choices of axes made for different values of m and n. However, to prove that the states are represented by positive matrices, later we will have to find a suitable way of connecting all these choices of axes.

decoding of the two pairs of perfectly discriminable states $\eta_{x,\pm}, \eta_{y,\pm} \in \mathsf{St}_1(Q)$ in the direction of the x-axis and y-axis, respectively, namely

$$\varphi_{k,\pm}^{mn} := \mathcal{D}^{(mn)}\eta_{k,\pm}, \quad k = x, y. \tag{17.2}$$

Let us also define

$$\sigma_k^Q := \eta_{k,+} - \eta_{k,-},$$
$$\sigma_k^{mn} := \mathcal{D}^{(mn)}\sigma_k^Q$$
$$= \varphi_{k,+}^{mn} - \varphi_{k,-}^{mn} \quad k = x, y. \tag{17.3}$$

It is clear that the two elements $\sigma_x^{mn}, \sigma_y^{mn}$ of $\mathsf{St}_\mathbb{R}(A)$ are the decoding of the two elements σ_x^Q, σ_y^Q in $\mathsf{St}_\mathbb{R}(Q)$ that are represented by the Pauli matrices in the matrix representation $S : \mathsf{St}(Q) \to \mathsf{Herm}(\mathbb{C}^2)$ of Eq. (14.4).

An immediate observation is the following.

Lemma 17.1 *The four vectors $\{\varphi_m, \varphi_n, \sigma_x^{mn}, \sigma_y^{mn}\} \subseteq \mathsf{St}_\mathbb{R}(A)$ are linearly independent for any choice of $1 \le m < n \le d_A$.*

Proof   If we consider the matrix representation S of Eq. (14.4), we have

$$S_{\mathcal{E}^{(mn)}\varphi_m} = \begin{pmatrix} 1 & 0 \\ 0 & 0 \end{pmatrix}, \qquad S_{\mathcal{E}^{(mn)}\varphi_n} = \begin{pmatrix} 0 & 0 \\ 0 & 1 \end{pmatrix},$$
$$S_{\mathcal{E}^{(mn)}\sigma_x^{(mn)}} = \begin{pmatrix} 0 & 1 \\ 1 & 0 \end{pmatrix}, \qquad S_{\mathcal{E}^{(mn)}\sigma_y^{(mn)}} = \begin{pmatrix} 0 & -i \\ i & 0 \end{pmatrix}. \tag{17.4}$$

Since both maps $S^{-1} : \mathsf{Herm}(\mathbb{C}^2) \to \mathsf{St}_\mathbb{R}(Q)$ and $\mathcal{D}^{(mn)} : \mathsf{St}_\mathbb{R}(Q) \to \mathsf{St}_\mathbb{R}(A)$ are injective, linear independence is a consequence of linear independence of the four matrices in Eq. (17.4). $\square$

We now show that the collection of vectors $\{\varphi_n\}_{m=1}^{d_A} \cup \{\sigma_k^{mn}\}_{n>m=1,...,d_A\ k=x,y}$ constructed in this way is a basis for $\mathsf{St}_\mathbb{R}(A)$. To this purpose we use the following result.

Lemma 17.2 *Let $V \subset \{1, \ldots, d_A\}$, and consider the projection Π_V. Then, if $m \notin V$ or $n \notin V$, one has $\Pi_V|\sigma_k^{mn}) = 0$ for $k = x, y$.*

Proof   The case when $m, n \notin V$ is simply treated, because in this case all the states $\varphi_{k,\pm}^{mn}$ involved in the definition of σ_k^{mn} belong to $F_V^\perp$, and thus $\Pi_V|\sigma_k^{mn}) = 0$. Let us then consider the case where $m \in V$ and $n \notin V$. Using Lemma 15.26 and Corollary 15.30 we obtain

$$\Pi_V|\varphi_{k,\pm}^{mn}) = \Pi_V\Pi_{\{m,n\}}|\varphi_{k,\pm}^{mn})$$
$$= \Pi_{\{m\}}|\varphi_{k,\pm}^{mn})$$
$$= |\varphi_m) (\varphi_m^\dagger|\varphi_{k,\pm}^{mn}).$$

Since the face F_{mn} is isomorphic to the Bloch sphere and the states $\varphi_{k,\pm}^{mn}$, $k = x, y$ lie on the equator of the Bloch sphere, we have

$$(\varphi_m^\dagger|\varphi_{k,\pm}^{mn}) = (\varphi_m^\dagger|\mathcal{D}^{(mn)}\mathcal{E}^{(mn)}|\varphi_{k,\pm}^{mn}) = (\eta_m^\dagger|\eta_{k\pm}^{mn}) = \frac{1}{2}, \qquad (17.5)$$

where we used the qubit algebra, along with the fact that, since $(\varphi^\dagger|\mathcal{D}^{(mn)}$ is atomic by Lemma 8.9, and

$$(\varphi_m^\dagger|\mathcal{D}^{(mn)}|\eta_n) = (\varphi_m^\dagger|\mathcal{D}^{(mn)}\mathcal{E}^{(mn)}|\varphi_n) = (\varphi_m^\dagger|\varphi_n) = \delta_{mn},$$

one has $(\varphi_m^\dagger|\mathcal{D}^{(mn)} = (\eta_m^\dagger|$. Equation (17.5) then implies

$$\Pi_V|\sigma_k^{mn}) = \Pi_V\left(|\varphi_{k,+}^{mn}) - |\varphi_{k,-}^{mn})\right)$$

$$= |\varphi_m)\left(\frac{1}{2} - \frac{1}{2}\right) = 0.$$

The case where $m \notin V$ and $n \in V$ can be treated exactly in the same way. □

Thanks to this result, we can now prove the following statement.

Lemma 17.3 *The vectors $\{\varphi_n\}_{m=1}^{d_A} \cup \{\sigma_k^{mn}\}_{n>m=1,\dots,d_A\ k=x,y}$ form a basis for $\mathsf{St}_\mathbb{R}(A)$.*

Proof Since the number of vectors is exactly d_A^2, to prove that they form a basis it is enough to show that they are linearly independent. Suppose that there exist coefficients $\{c_m\}_{m=1}^{d_A} \cup \{c_k^{mn}\}_{n>m=1,\dots,d_A\ k=x,y}$ such that

$$\sum_{m=1}^{d_A} c_m\varphi_m + \sum_{m=1}^{d_A-1}\sum_{n=m+1}^{d_A}\sum_{k=x,y} c_k^{mn}\sigma_k^{mn} = 0.$$

Applying the projection $\Pi_{\{m,n\}}$ on both sides and using Lemma 17.2 we obtain

$$c_m|\varphi_m) + c_n|\varphi_n) + c_x^{mn}|\sigma_x^{mn}) + c_y^{mn}|\sigma_y^{mn}) = 0.$$

However, from Lemma 17.1 we know that the vectors $\{\varphi_m, \varphi_n, \sigma_x^{mn}, \sigma_y^{mn}\}$ are linearly independent. Consequently, $c_m = c_n = c_k^{mn} = 0$ for all m, n, k. □

Analogously to the construction of the basis $\{\varphi_n\}_{m=1}^{d_A} \cup \{\sigma_k^{mn}\}_{n>m=1,\dots,d_A\ k=x,y}$ for $\mathsf{St}_\mathbb{R}(A)$, one can construct a basis $\{\varphi_n^\dagger\}_{m=1}^{d_A} \cup \{\sigma_k^{mn\dagger}\}_{n>m=1,\dots,d_A\ k=x,y}$ in $\mathsf{Eff}_\mathbb{R}(A)$, where $\sigma_k^{mn\dagger} := \varphi_{k,+}^{mn\dagger} - \varphi_{k,-}^{mn\dagger}$. Notice that by construction one has $(\varphi_{k,\pm}^{mn\dagger}| = (\eta_{k,\pm}|\mathcal{D}^{(mn)}$, and thus

$$(\sigma_k^{mn\dagger}| = (\sigma_k^{Q\dagger}|\mathcal{D}^{(mn)}, \qquad (17.6)$$

where

$$\sigma_k^{Q\dagger} = \eta_{k,+}^\dagger - \eta_{k,-}^\dagger.$$

Exercise 17.1 Show that $(\sigma_k^{mn\dagger}|\varphi_m) = (\sigma_k^{mn\dagger}|\varphi_n) = 0$, for $k = x, y$.

Exercise 17.2 Show that one can adopt the basis $\{\varphi_n^\dagger\}_{m=1}^{d_A} \cup \{\sigma_k^{mn\dagger}\}_{n>m=1,\dots,d_A\ k=x,y}$ in the space of effects $\mathsf{Eff}_\mathbb{R}(A)$.

Exercise 17.3 Show that $(\sigma_k^{mn\dagger}|\sigma_{k'}^{mn}) = 2\delta_{kk'}$.

17.2 Matrix Representation of States and Effects

Since the state space $St(A)$ for system A spans a real vector space of dimension $D_A = d_A^2$, we can decide to represent the vectors $\{\varphi_m\}_{m=1}^{d_A} \cup \{\sigma_k^{mn}\}_{n>m=1,\ldots,d_A\ k=x,y}$ as Hermitian $d_A \times d_A$ matrices. Precisely, we associate the vector φ_m to the matrix S_{φ_m} defined by

$$\left[S_{\varphi_m} \right]_{rs} = \delta_{rm}\delta_{sm}, \tag{17.7}$$

the vector σ_x^{mn} to the matrix

$$\left[S_{\sigma_x^{mn}} \right]_{rs} = \delta_{rm}\delta_{sn} + \delta_{rn}\delta_{sm}, \tag{17.8}$$

and the vector σ_y^{mn} to the matrix

$$\left[S_{\sigma_y^{mn}} \right]_{rs} = i\lambda^{mn} \left(\delta_{rm}\delta_{sn} - \delta_{rn}\delta_{sm} \right), \tag{17.9}$$

where λ^{mn} can take values $+1$ or -1. The freedom in the choice of λ^{mn} will be useful in Section 17.3, where we will introduce the representation of composite systems of two qubits. However, this choice of sign plays no role in the present section, and for simplicity we will adopt the positive sign.

Recall that in principle any orthogonal direction in the plane orthogonal to the z-axis can be chosen to be the x-axis. In general, once a reference pair of axes is chosen for the above representation, switching to the other possible choices for the x-axis leads to matrices of the form

$$\left[S_{\sigma_{x,\phi}^{mn}} \right]_{rs} = \delta_{rm}\delta_{sn}e^{i\phi_{mn}} + \delta_{rn}\delta_{sm}e^{-i\phi_{mn}} \qquad \phi_{mn} \in [0, 2\pi), \tag{17.10}$$

and the corresponding choice for the y-axis will lead to matrices of the form

$$\left[S_{\sigma_{y,\phi}^{mn}} \right]_{rs} = i\lambda^{mn} \left(\delta_{rm}\delta_{sn}e^{i\phi_{mn}} - \delta_{rn}\delta_{sm}e^{-i\phi_{mn}} \right) \qquad \phi_{mn} \in [0, 2\pi). \tag{17.11}$$

Since the vectors $\{\varphi_m\}_{m=1}^{d_A} \cup \{\sigma_k^{mn}\}_{n>m=1,\ldots,d_A;\ k=x,y}$ are a basis for the real vector space $St_{\mathbb{R}}(A)$, we can expand any state $\rho \in St(A)$ on them:

$$|\rho) = \sum_{m=1}^{d_A} \rho_m |\varphi_m) + \sum_{m=1}^{d_A-1} \sum_{n=m+1}^{d_A} \sum_{k=x,y} \rho_k^{mn} |\sigma_k^{mn}) \tag{17.12}$$

and the expansion coefficients $\{\rho_m\}_{m=1}^{d_A} \cup \{\rho_k^{mn}\}_{n>m=1,\ldots,d_A;\ k=x,y}$ are all real. Hence, each state ρ is in one-to-one correspondence with a Hermitian matrix, given by

$$S_\rho = \sum_{m=1}^{d_A} \rho_m S_{\varphi_m} + \sum_{m=1}^{d_A-1} \sum_{n=m+1}^{d_A} \sum_{k=x,y} \rho_k^{mn} S_{\sigma_k^{mn}}. \tag{17.13}$$

Extending by linearity the mapping defined by Eqs. (17.12) and (17.13) to the whole $St_\mathbb{R}(A)$, we can define a linear mapping S as follows:

$$S : St_\mathbb{R}(A) \to Herm(\mathbb{C}^{d_A}),$$

$$S : \rho \mapsto S_\rho := \sum_{m=1}^{d_A} \rho_m S_{\varphi_m} + \sum_{m=1}^{d_A-1} \sum_{n=m+1}^{d_A} \sum_{k=x,y} \rho_k^{mn} S_{\sigma_k^{mn}}, \qquad (17.14)$$

where $Herm(\mathbb{C}^d)$ denotes the real vector space of Hermitian matrices over $\mathbb{C}^d$. Notice that the map S is invertible, and thus equation (17.14) defines an isomorphism between $St_\mathbb{R}(A)$ and $Herm(\mathbb{C}^{d_A})$.

The representation S naturally induces a representation $E : Eff_\mathbb{R}(A) \to Herm(\mathbb{C}^{d_A})$ through the identity

$$(a|\rho) = Tr[E_a S_\rho].$$

The representation E enjoys the following property.

Lemma 17.4 *Let $\varphi_m \in Eff(A)$ be maximal set of perfectly discriminable pure states used to define the matrix representation S_ρ. Then, the effect $\varphi_m^\dagger$ has matrix representation $E_{\varphi_m^\dagger}$ such that $E_{\varphi_m^\dagger} = S_{\varphi_m}$ and the generalized effects $\sigma_k^{mn\dagger}$ for $k = x,y$ have matrix representation $E_{\sigma_k^{mn\dagger}} = S_{\sigma_k^{mn}}$.*

Proof Let $\rho \in St_1(A)$ be an arbitrary state. Expanding ρ as in Eq. (17.12) and using Lemma 17.2 we obtain $(\varphi_m^\dagger|\rho) = \rho_m$. On the other hand, by Eq. (17.13) we have that ρ_m is the m-th diagonal element of the matrix S_ρ: by definition of S_{φ_m} [Eq. (17.7)], this implies $\rho_m = Tr[S_{\varphi_m} S_\rho]$. Now, by construction we have $Tr[E_{\varphi_m^\dagger} S_\rho] = (\varphi_m^\dagger|\rho) = \rho_m = Tr[S_{\varphi_m} S_\rho]$ for every $\rho \in St_1(A)$. This implies that $Tr[E_{\varphi_m^\dagger} S_\rho] = Tr[S_{\varphi_m} S_\rho]$ for every $\rho \in St_\mathbb{R}(A)$. Finally, since the image of $St_\mathbb{R}(A)$ under the matrix representation S is the whole space of Hermitian matrices $Herm(\mathbb{C}^{d_A}) = \{S_\rho | \rho \in St_\mathbb{R}(A)\}$, we can conclude that $E_{\varphi_m^\dagger} = S_{\varphi_m}$. Similarly, for the generalized effects $\sigma_k^{mn\dagger}$, $k = x,y$, recalling Exercises 17.1, 17.2, and 17.3, and the expansion of ρ in Eq. (17.12) one has

$$Tr[E_{\sigma_k^{mn\dagger}} S_\rho] = (\sigma_k^{mn\dagger}|\rho)$$
$$= (\sigma_k^{mn\dagger}|\Pi_{\{m,n\}}|\rho)$$
$$= 2\rho_k^{mn}$$
$$= Tr[S_{\sigma_k^{mn}} S_\rho],$$

for every $\rho \in St_\mathbb{R}(A)$. Thus $E_{\sigma_k^{mn\dagger}} = S_{\sigma_k^{mn}}$. $\qquad \square$

This construction of the representation S along with the definition of the basis elements σ_k^{mn} in Eqs. (17.2) and (17.3) allows one to identify the blocks in S_ρ corresponding to the faces F_{mn} with states of a qubit. This result will play a crucial role in the forthcoming construction.

Lemma 17.5 *Let $\rho \in \mathsf{St}(A)$ be a state, and take $1 \leq m < n \leq d_A$. The matrix elements of the matrix representation of ρ given by $(S_\rho)_{mm}$, $(S_\rho)_{nn}$, $(S_\rho)_{mn}$, $(S_\rho)_{nm}$ coincide with the matrix representation of the qubit state $\mathcal{E}^{(mn)}\Pi_{\{m,n\}}\rho$ if $\lambda^{mn} = 1$, or with its transpose if $\lambda^{mn} = -1$.*

Proof By Eqs. (17.7), (17.8), and (17.9) and Lemma 17.4, we have that

$$(S_\rho)_{mm} = (\varphi_m^\dagger | \rho),$$

$$(S_\rho)_{mn} = \frac{1}{2}[(\sigma_x^{mn\dagger}|\rho) - i\lambda^{mn}(\sigma_y^{mn\dagger}|\rho)],$$

$$(S_\rho)_{nm} = \frac{1}{2}[(\sigma_x^{mn\dagger}|\rho) + i\lambda^{mn}(\sigma_y^{mn\dagger}|\rho)].$$

Moreover, by Lemma 17.2 one has $\Pi_{\{m,n\}}|\sigma_k^{pq}) = \delta_{mp}\delta_{nq}|\sigma_k^{pq})$. Thus, if we expand ρ as in Eq. (17.12), we have

$$\Pi_{\{m,n\}}|\rho) = (S_\rho)_{mm}|\varphi_m) + (S_\rho)_{nn}|\varphi_n)$$
$$+ [(S_\rho)_{mn} + (S_\rho)_{nm}]|\sigma_x^{mn}) + i\lambda^{mn}[(S_\rho)_{mn} - (S_\rho)_{nm}]|\sigma_y^{mn}).$$

Consequently

$$(S_{\Pi_{\{m,n\}}\rho})_{mm} = (S_\rho)_{mm}, \qquad\qquad (S_{\Pi_{\{m,n\}}\rho})_{nn} = (S_\rho)_{nn},$$
$$(S_{\Pi_{\{m,n\}}\rho})_{mn} = (S_\rho)_{mn}, \qquad\qquad (S_{\Pi_{\{m,n\}}\rho})_{nm} = (S_\rho)_{nm}.$$

Now, recalling the definition of σ_k^{mn} in Eqs. (17.2) and (17.3), one has the following decomposition for the qubit state $\mathcal{E}^{mn}\Pi_{\{m,n\}}\rho \in \mathsf{St}(Q)$

$$\mathcal{E}^{(m,n)}\Pi_{\{m,n\}}|\rho) = (S_\rho)_{mm}|\eta_0) + (S_{\rho_{nn}})|\eta_1)$$
$$+ [(S_\rho)_{mn} + (S_\rho)_{nm}]|\sigma_x^Q) + i\lambda^{mn}[(S_\rho)_{mn} - (S_\rho)_{nm}]|\sigma_y^Q).$$

Recalling that $(S_\nu)_{ii} = (\eta_i^\dagger|\nu)$, one has

$$(S_{\mathcal{E}^{(m,n)}\Pi_{\{m,n\}}\rho})_{00} = (S_\rho)_{mm}, \qquad (S_{\mathcal{E}^{(m,n)}\Pi_{\{m,n\}}\rho})_{11} = (S_\rho)_{nn}.$$

Finally, since $(S_\nu)_{01} = \frac{1}{2}[(\sigma_x^{Q\dagger}|\nu) - i(\sigma_y^{Q\dagger}|\nu)]$, we have that

$$(S_{\mathcal{E}^{(m,n)}\Pi_{\{m,n\}}\rho})_{01} = (S_\rho)_{mn}, \qquad (S_{\mathcal{E}^{(m,n)}\Pi_{\{m,n\}}\rho})_{10} = (S_\rho)_{nm},$$

if $\lambda^{mn} = 1$ and

$$(S_{\mathcal{E}^{(m,n)}\Pi_{\{m,n\}}\rho})_{01} = (S_\rho)_{nm}, \qquad (S_{\mathcal{E}^{(m,n)}\Pi_{\{m,n\}}\rho})_{10} = (S_\rho)_{mn},$$

if $\lambda^{mn} = -1$.   □

An analogous result can be proved for effects.

Lemma 17.6 *Let $a \in \mathsf{Eff}(A)$ be an effect, and take $m < n \leq d_A$. The matrix elements of the matrix representation of a, given by $(E_a)_{mm}$, $(E_a)_{nn}$, $(E_a)_{mn}$, $(E_a)_{nm}$ coincide with the matrix representation of the qubit effect $(a|\Pi_{\{m,n\}}\mathcal{D}^{(mn)}$ if $\lambda_{mn} = 1$, or with its transpose if $\lambda_{mn} = -1$.*

Proof By Exercise 17.2 one can expand the effect a and its representation E_a as

$$a = \sum_{j=1}^{d_A} a_j \varphi_j^\dagger + \sum_{j=1}^{d_A-1} \sum_{l=j+1}^{d_A} \sum_{k=x,y} a_k^{jl} \sigma_k^{jl\dagger},$$

$$E_a = \sum_{j=1}^{d_A} a_j E_{\varphi_j^\dagger} + \sum_{j=1}^{d_A-1} \sum_{l=j+1}^{d_A} \sum_{k=x,y} a_k^{jl} E_{\sigma_k^{jl\dagger}},$$

and by Lemma 17.4 we can write

$$E_a = \sum_{j=1}^{d_A} a_j S_{\varphi_j} + \sum_{j=1}^{d_A-1} \sum_{l=j+1}^{d_A} \sum_{k=x,y} a_k^{jl} S_{\sigma_k^{jl}}.$$

From Eqs. (17.7), (17.8), and (17.9), we then have

$$(E_a)_{mm} = a_m, \qquad\qquad\qquad (E_a)_{nn} = a_n,$$

$$(E_a)_{mn} = \frac{1}{2}(a_x^{mn} - i\lambda_{mn} a_y^{mn}) \qquad (E_a)_{nm} = \frac{1}{2}(a_x^{mn} + i\lambda_{mn} a_y^{mn}).$$

By the same Exercise 17.2, one also has $(\sigma_k^{pq\dagger}|\Pi_{\{m,n\}} = \delta_{mp}\delta_{nq}(\sigma_k^{pq\dagger}|$. Thus we have

$$(a|\Pi_{\{m,n\}} = a_m(\varphi_m^\dagger| + a_n(\varphi_n^\dagger| + \sum_{k=x,y} a_k^{mn}(\sigma_k^{mn\dagger}|,$$

and consequently, using Lemma 17.4

$$E_{(a|\Pi_{\{m,n\}}} = a_m E_{\varphi_m^\dagger} + a_n E_{\varphi_n^\dagger} + \sum_{k=x,y} a_k^{mn} E_{\sigma_k^{mn\dagger}}$$

$$= a_m S_{\varphi_m} + a_n S_{\varphi_n} + \sum_{k=x,y} a_k^{mn} S_{\sigma_k^{mn}}.$$

This implies

$$(E_{(a|\Pi_{\{m,n\}}})_{mm} = (E_a)_{mm}, \qquad\qquad (E_{(a|\Pi_{\{m,n\}}})_{nn} = (E_a)_{nn},$$

$$(E_{(a|\Pi_{\{m,n\}}})_{mn} = (E_a)_{mn}, \qquad\qquad (E_{(a|\Pi_{\{m,n\}}})_{nm} = (E_a)_{nm}.$$

Recalling now the definition of $\sigma_k^{mn\dagger}$ in Eq. (17.6), along with Eq. (14.6), one has the following representation for the qubit effect $(a|\Pi_{\{m,n\}}\mathcal{D}^{mn} \in \mathsf{Eff}(Q)$:

$$E_{(a|\Pi_{\{m,n\}}\mathcal{D}^{(mn)}} = a_m E_{\eta_0^\dagger} + a_n E_{\eta_1^\dagger} + \sum_{k=x,y} a_k^{mn} E_{\sigma_k^{Q\dagger}}$$

$$= (E_a)_{mm} S_{\eta_0} + (E_a)_{nn} S_{\eta_1} + [(E_a)_{mn} + (E_a)_{nm}] S_{\sigma_x^Q} + i\lambda_{mn}[(E_a)_{mn} - (E_a)_{nm}] S_{\sigma_y^Q}.$$

Thus we have

$$(E_{(a|\Pi_{\{m,n\}}\mathcal{D}^{(mn)}})_{00} = (E_a)_{mm}, \quad (E_{(a|\Pi_{\{m,n\}}\mathcal{D}^{(mn)}})_{11} = (E_a)_{nn},$$

as well as

$$(E_{(a|\Pi_{\{m,n\}}\mathcal{D}^{(mn)}})_{01} = (E_a)_{mn}, \quad (E_{(a|\Pi_{\{m,n\}}\mathcal{D}^{(mn)}})_{10} = (E_a)_{nm},$$

for $\lambda_{mn} = 1$, while

$$(E_{(a|\Pi_{\{m,n\}}\mathcal{D}^{(mn)})})_{01} = (E_a)_{nm}, \quad (E_{(a|\Pi_{\{m,n\}}\mathcal{D}^{(mn)})})_{10} = (E_a)_{mn},$$

for $\lambda_{mn} = -1$.   □

The remainder of the section is devoted to the proof that the set of matrices $\{S_\rho \mid \rho \in \mathsf{St}_1(A)\}$ is the whole set of positive Hermitian matrices with unit trace and that the set of matrices $\{E_a \mid a \in \mathsf{Eff}(A)\}$ is the set of positive Hermitian matrices bounded by the identity. Let us start from some simple facts.

A state whose representation is particularly simple is the invariant state χ_A:

Lemma 17.7   *The invariant state χ_A has matrix representation $S_{\chi_A} = \frac{I_{d_A}}{d_A}$, where I_{d_A} is the identity matrix in dimension d_A.*

Proof   Obvious from the expression $\chi_A = \frac{1}{d}\sum_m \varphi_m$ and from the matrix representation of the states $\{\varphi_m\}_{m=1}^{d_A}$ in Eq. (17.7).   □

Lemma 17.8   *The deterministic effect $e \in \mathsf{Eff}(A)$ has matrix representation $E_e = I_{d_A}$.*

Proof   This result follows straightforwardly from the expression $e = \sum_m \varphi_m^\dagger$, combined with Lemma 17.4 and Eq. (17.7).   □

Corollary 17.9   *For every state $\sigma \in \mathsf{St}(A)$ one has*

$$\mathrm{Tr}[S_\sigma] \leq 1$$

and for every state $\rho \in \mathsf{St}_1(A)$ one has

$$\mathrm{Tr}[S_\rho] = 1.$$

Proof   One has $\mathrm{Tr}[S_\sigma] = \mathrm{Tr}[E_e S_\sigma] = (e|\sigma)$. Thus, for $\sigma \in \mathsf{St}(A)$ one has $\mathrm{Tr}[S_\sigma] \leq 1$, while for $\sigma = \rho \in \mathsf{St}_1(A)$ one has $\mathrm{Tr}[S_\rho] = 1$.   □

Theorem 17.10   *The matrix elements of S_ψ for a pure state $\psi \in \mathsf{St}_1(A)$ are $(S_\psi)_{mn} = \sqrt{p_m p_n} e^{i\theta_{mn}(\psi)}$, with $\sum_{m=1}^{d_A} p_m = 1$, $\theta_{mn}(\psi) \in [0, 2\pi)$, $\theta_{mm}(\psi) = 0$, and $\theta_{mn}(\psi) = -\theta_{nm}(\psi)$.*

Proof   First of all, by Lemma 17.4 the diagonal elements of S_ψ are

$$(S_\psi)_{mm} = \mathrm{Tr}[E_{\varphi_m^\dagger} S_\psi]$$
$$= (\varphi_m^\dagger | \psi)$$
$$=: p_m.$$

We clearly have $\sum_{m=1}^{d_A} p_m = (e|\psi) = 1$. By Lemma 17.5 for every $1 \leq m < n \leq d_A$ the matrix elements $(S_\psi)_{mm}$, $(S_\psi)_{nn}$, $(S_\psi)_{mn}$, and $(S_\psi)_{nm}$ coincide with those of the pure qubit state $\mathcal{E}^{(mn)}\Pi_{\{m,n\}}\psi$, which is then represented by a non-negative rank-one matrix, whose diagonal elements are equal to $(S_{\mathcal{E}^{(mn)}\Pi_{\{m,n\}}\psi})_{00} = p_m$ and $(S_{\mathcal{E}^{(mn)}\Pi_{\{m,n\}}\psi})_{11} = p_n$.

Thus, the off-diagonal element $(S_{\mathcal{E}^{(m,n)}\Pi_{\{m,n\}}\psi})_{01}$ is equal to $\sqrt{p_m p_n}e^{i\theta_{mn}(\psi)}$, for some $\theta_{mn}(\psi) \in [0, 2\pi)$, and $(S_{\mathcal{E}^{(m,n)}\Pi_{\{m,n\}}\psi})_{10} = (S_{\mathcal{E}^{(m,n)}\Pi_{\{m,n\}}\psi})_{01}^*$. Finally, by Lemma 17.5 we have

$$(S_\psi)_{mn} = \sqrt{p_m p_n}e^{i\lambda^{mn}\theta_{mn}(\psi)}, \quad (S_\psi)_{nm} = (S_\psi)_{mn}^* \, .$$

Repeating the same argument for all choices of indices m, n, the thesis follows. $\quad\square$

Theorem 17.11 *For a pure state $\psi \in \mathrm{St}_1(A)$, the matrix representation $E_{\psi^\dagger}$ of the corresponding atomic effect $\psi^\dagger$ has the property that $E_{\psi^\dagger} = S_\psi$.*

Proof We already know that the statement holds for $d_A = 2$, as we proved in Eq. (14.6). Let us now consider a generic system A. If $\Pi_{\{m,n\}}|\psi) = 0$, then $(S_\psi)_{ij} = 0$ for $i, j \in \{m, n\}$. On the other hand, φ_m and φ_n must be perfectly discriminable from ψ, which implies $(\psi^\dagger|\varphi_m) = (\psi^\dagger|\varphi_n) = 0$, and thus $(\psi^\dagger|\Pi_{(m,n)} = 0$. This means that $(E_{\psi^\dagger})_{ij} = (S_\psi)_{ij} = 0$ for $i, j \in \{m, n\}$. Let us then consider those values $1 \leq m < n \leq d_A$ such that $\Pi_{\{m,n\}}\psi \neq 0$. Let $\psi_\perp^{(mn)}$ be the unique pure state in the face F_{mn} that is perfectly discriminable from $\Pi_{\{m,n\}}|\psi)$. Note that, since $\psi_\perp^{(mn)}$ belongs to the face F_{mn}, it is also perfectly discriminable from $\Pi_{\{1,...,d_A\}\setminus\{m,n\}}|\psi)$. Hence, by Lemma 15.36, $\psi_\perp^{(mn)}$ is perfectly discriminable from ψ and, in particular,

$$0 = (\psi^\dagger|\psi_\perp^{(mn)})$$
$$= (\psi^\dagger|\Pi_{\{m,n\}}|\psi_\perp^{(mn)})$$
$$= (\psi^\dagger|\Pi_{\{m,n\}}\mathcal{D}^{(m,n)}\mathcal{E}^{(m,n)}\Pi_{\{m,n\}}|\psi_\perp^{(mn)}).$$

This implies that the atomic effect $(\psi^\dagger|\Pi_{\{m,n\}}\mathcal{D}^{(m,n)}$ must be proportional to the effect $\xi^\dagger$, defined by

$$\xi := g_{mn}\mathcal{E}^{(m,n)}\Pi_{\{m,n\}}\psi$$
$$g_{mn} := (e|\mathcal{E}^{(m,n)}\Pi_{\{m,n\}}|\psi)^{-1}$$
$$= (e|\Pi_{\{m,n\}}|\psi)^{-1},$$

namely

$$(\psi^\dagger|\Pi_{\{m,n\}}\mathcal{D}^{(m,n)} = c_{mn}(\xi^\dagger|, \quad c_{mn} \geq 0.$$

Thus, using Eq. (14.6) in Corollary 14.5, we have

$$E_{(\psi^\dagger|\Pi_{\{m,n\}}\mathcal{D}^{(m,n)}} = c_{mn}E_{\xi^\dagger}$$
$$= c_{mn}S_\xi$$
$$= c_{mn}g_{mn}S_{\mathcal{E}^{(m,n)}\Pi_{\{m,n\}}|\psi)}.$$

Now, by Lemma 17.6, for every $m < n$, the matrix elements $(E_{\psi^\dagger})_{mm}$, $(E_{\psi^\dagger})_{nn}$, $(E_{\psi^\dagger})_{mn}$, $(E_{\psi^\dagger})_{nm}$ are proportional to those of $S_{\mathcal{E}^{(m,n)}\Pi_{\{m,n\}}|\psi)}$, which in turn coincide with those of S_ψ by virtue of Lemma 17.5. It is then clear that $c_{mn}g_{mn} = c_{mn'}g_{mn'} = c_{m'n}g_{m'n} =: c$, that is, $E_{\psi^\dagger} = cS_\psi$. Taking the trace on both sides we obtain $\mathrm{Tr}[E_{\psi^\dagger}] = c$. To prove that $c = 1$, we use the relation $\mathrm{Tr}[E_{\psi^\dagger}]/d_A = (\psi^\dagger|\chi_A) = 1/d_A$. $\quad\square$

We conclude with a simple corollary that will be used in the next subsection.

Corollary 17.12   *Let $\varphi \in \mathsf{St}_1(A)$ be a pure state and let $\{\gamma_i\}_{i=1}^r \subset \mathsf{St}_1(A)$ be a set of pure states. If the state φ can be written as*

$$|\varphi) = \sum_{i=1}^r x_i |\gamma_i)$$

for some real coefficients $\{x_i\}_{i=1}^r$, then the atomic effect $\varphi^\dagger$ is given by

$$(\varphi^\dagger| = \sum_{i=1}^r x_i (\gamma_i^\dagger|.$$

Proof   For every $\rho \in \mathsf{St}(A)$ by Theorem 17.11 one has

$$(\varphi^\dagger|\rho) = \mathrm{Tr}[E_{\varphi^\dagger} S_\rho] = \mathrm{Tr}[S_\varphi S_\rho] = \sum_i x_i \mathrm{Tr}[S_{\gamma_i} S_\rho]$$

$$= \sum_i x_i \mathrm{Tr}[E_{\gamma_i^\dagger} S_\rho] = \sum_i x_i (\gamma_i^\dagger|\rho),$$

thus implying the thesis.   □

17.3 Representation of Two-qubit Systems

If A and B are two systems with $d_A = d_B = 2$, then we can use two different types of matrix representations for the states of the composite system AB:

1. The first type of representation is the representation S_φ introduced through Lemma 17.3: here we will refer to it as the *standard representation*. Note that there are many different representations of this type because for every pair (m, n) there is freedom in choice of the x- and y-axis [cf. Eqs. (17.10) and (17.11)].
2. The second type of representation is the *tensor product representation T_φ*, defined in two steps: (i) first, it is defined on *product* states $\rho \otimes \sigma$ by the Kronecker tensor product of matrices, i.e. $T_{\rho \otimes \sigma} := S_\rho^A \otimes S_\sigma^B$, where S^A and S^B are the matrix representations for system A and B, respectively; (ii) then, it is extended by linearity to general states: for a state $|\rho) = \sum_{ij} \rho_{ij} |\alpha_i)|\beta_j)$, with $\alpha_i \in \mathsf{St}(A), \beta_j \in \mathsf{St}(B)$, we have

$$T_\rho := \sum_{i,j} \rho_{ij} S_{\alpha_i}^A \otimes S_{\beta_j}^B. \tag{17.15}$$

Here we have a freedom in the choice of the axes for the Bloch spheres of qubits A and B. Since A and B are operationally equivalent, we will denote the elements of the bases for $\mathsf{St}_\mathbb{R}(A)$ and $\mathsf{St}_\mathbb{R}(B)$ with the same letters: $\{\varphi_m\}_{m=1}^2$ for the two perfectly discriminable pure states and $\{\sigma_k\}_{k=x,y}$ for the remaining basis vectors.

We now show a few properties of the tensor representation. First of all, notice that the extension is well defined. Indeed, let $T_\rho = 0$, and expand $\rho = \sum_{i,j=1}^4 \rho_{ij} \alpha_i \otimes \beta_j$, where $\{\alpha_i\}_{i=1}^4$ and $\{\beta_j\}_{j=1}^4$ are two sets of pure states that span $\mathsf{St}_\mathbb{R}(A)$ and $\mathsf{St}_\mathbb{R}(B)$, respectively. Then

$$T_\rho = \sum_{i,j=1}^{4} \rho_{ij} S^A_{\alpha_i} \otimes S^B_{\beta_j}.$$

Since the matrix representations S^A and S^B are injective, the set $\{S^A_{\alpha_i} \otimes S^B_{\beta_j}\}$ is linearly independent, and thus the condition $T_\rho = 0$ implies $\rho_{ij} = 0$ for all $1 \leq i,j \leq 4$, namely $\rho = 0$.

Let now F_A denote the matrix corresponding to the effect $A \in \mathsf{Eff}(AB)$ in the tensor representation, that is, the matrix defined by

$$(A|\rho) := \mathrm{Tr}[F_A T_\rho] \qquad \forall \rho \in \mathsf{St}(AB). \tag{17.16}$$

It is easy to show that the matrix representation for effects must satisfy the analog of Eq. (17.15):

Lemma 17.13   *Let $A \in \mathsf{Eff}(AB)$ be a bipartite effect, written as $(A| = \sum_{i,j} A_{ij}(a_i|(b_j|$. Then one has*

$$F_A = \sum_{i,j} A_{ij} E^A_{a_i} \otimes E^B_{b_j},$$

where E^A and E^B are the standard matrix representations of effects on A and B, respectively.

Proof   For every bipartite state $|\rho) = \sum_{k,l} \rho_{kl} |\alpha_k)|\beta_l)$ one has

$$\mathrm{Tr}[F_A T_\rho] = (A|\rho)$$
$$= \sum_{i,j,k,l} A_{ij} \rho_{kl} (a_i|\alpha_k)(b_j|\beta_l)$$
$$= \sum_{i,j,k,l} A_{ij} \rho_{kl} \mathrm{Tr}[E^A_{a_i} S^A_{\alpha_k}] \mathrm{Tr}[E^B_{b_j} S^B_{\beta_l}]$$
$$= \sum_{i,j,k,l} A_{ij} \rho_{kl} \mathrm{Tr}[(E^A_{a_i} \otimes E^B_{b_j}) T_{\alpha_k \otimes \beta_l}]$$
$$= \sum_{i,j} A_{ij} \mathrm{Tr}[(E^A_{a_i} \otimes E^B_{b_j}) T_\rho]$$

which implies the thesis.   □

Corollary 17.14   *Let $\Psi \in \mathsf{St}_1(AB)$ be a pure state. Then one has $F_{\Psi^\dagger} = T_\Psi$.*

Proof   Let $\{\alpha_i\}_{i=1}^4$ and $\{\beta_j\}_{j=1}^4$ be two sets of pure states that span $\mathsf{St}_\mathbb{R}(A)$ and $\mathsf{St}_\mathbb{R}(B)$, respectively. Expand Ψ as $|\Psi) = \sum_{i,j} c_{ij} |\alpha_i)|\beta_j)$. Then, Corollary 17.12 yields $(\Psi^\dagger| = \sum_{i,j} c_{ij} (\alpha_i^\dagger|(\beta_j^\dagger|$. Therefore, we have

$$F_{\Psi^\dagger} = \sum_{i,j} c_{ij} E^A_{\alpha_i^\dagger} \otimes E^B_{\beta_j^\dagger} = \sum_{i,j} c_{ij} S^A_{\alpha_i} \otimes S^B_{\beta_j} = T_\Psi.$$

□

Corollary 17.15   *For every bipartite state $\rho \in \mathsf{St}_1(AB)$, $d_A = d_B = 2$ one has $\mathrm{Tr}[T_\rho] = 1$.*

Proof   For each qubit we have

$$E_{\varphi_1}^\dagger = \begin{pmatrix} 1 & 0 \\ 0 & 0 \end{pmatrix}, \ E_{\varphi_2}^\dagger = \begin{pmatrix} 0 & 0 \\ 0 & 1 \end{pmatrix}. \tag{17.17}$$

Hence, $E_{e_A}^A = E_{e_B}^B = I$, where I is the 2×2 identity matrix. By Lemma 17.13, we then have $F_{e_A \otimes e_B} = I \otimes I$ and, therefore, $\mathrm{Tr}[T_\rho] = \mathrm{Tr}[F_{e_A \otimes e_B} T_\rho] = (e_A \otimes e_B|\rho) = 1$. $\square$

Finally, an immediate consequence of local discriminability is the following.

Lemma 17.16   *Suppose that $\mathcal{U} \in \mathbf{G}_A$ and $\mathcal{V} \in \mathbf{G}_B$ are two reversible transformations for qubits A and B, respectively, and that $U, V \in \mathrm{SU}(2)$ are such that*

$$S_{\mathcal{U}\rho}^A = US_\rho^A U^\dagger \qquad \forall \rho \in \mathsf{St}_1(A)$$
$$S_{\mathcal{V}\sigma}^B = VS_\sigma^B V^\dagger \qquad \forall \sigma \in \mathsf{St}_1(B).$$

Then, we have $T_{(\mathcal{U} \otimes \mathcal{V})\tau} = (U \otimes V)T_\tau(U^\dagger \otimes V^\dagger)$ for every $\tau \in \mathsf{St}_1(AB)$.

Proof   Let us expand τ as $\tau = \sum_{i,j=1}^4 t_{ij}\alpha_i \otimes \beta_j$, where $\{\alpha_i\}_{i=1}^4$ and $\{\beta_j\}_{j=1}^4$ are bases for $\mathsf{St}_{\mathbb{R}}(A)$ and $\mathsf{St}_{\mathbb{R}}(B)$. Then, by the properties of parallel composition and by linearity of $\mathcal{U} \otimes \mathcal{V}$ we have

$$(\mathcal{U} \otimes \mathcal{V})\tau = \sum_{i,j=1}^4 t_{ij}(\mathcal{U}\alpha_i) \otimes (\mathcal{V}\beta_j).$$

Finally, since the representation T is linear, we have

$$T_{(\mathcal{U}\otimes\mathcal{V})\tau} = \sum_{i,j=1}^4 t_{ij}T_{(\mathcal{U}\alpha_i)\otimes(\mathcal{V}\beta_j)}$$

$$= \sum_{i,j=1}^4 t_{ij}S_{(\mathcal{U}\alpha_i)} \otimes S_{(\mathcal{V}\beta_j)}$$

$$= \sum_{i,j=1}^4 t_{ij}US_{\alpha_i}^A U^\dagger \otimes VS_{\beta_j}^B V^\dagger$$

$$= (U \otimes V)\left(\sum_{i,j=1}^4 t_{ij}S_{\alpha_i}^A \otimes S_{\beta_j}^B\right)(U^\dagger \otimes V^\dagger)$$

$$= (U \otimes V)T_\tau(U^\dagger \otimes V^\dagger).$$

$\square$

In the following, we will show that, with a suitable choice of matrix representation for system B, the standard representation coincides with the tensor representation, that is, $S_\rho = T_\rho$ for every $\rho \in \mathsf{St}(AB)$. This technical result is important because our derivation uses properties that are easily proved in the standard representation, along with properties that follow from the action of local reversible transformations, easily expressed in the tensor representation (Lemma 17.16). It is, however, essential to have a representation that enjoys simultaneously properties of both the standard and the tensor representation.

The four states $\{\varphi_m \otimes \varphi_n\}_{m,n=1}^2$ are clearly a maximal set of perfectly discriminable pure states in AB. In the following we will construct the standard representation starting from this set.

Let us start from the following result.

Lemma 17.17 *Let* $\Phi \in St_1(AB)$ *be a pure state such that*[3]

$$(\varphi_1^\dagger \otimes \varphi_1^\dagger | \Phi) = (\varphi_2^\dagger \otimes \varphi_2^\dagger | \Phi) = 1/2. \tag{17.18}$$

With a suitable choice of the matrix representation S^B, *the state* Φ *is represented by the matrix*

$$T_\Phi = \frac{1}{2} \begin{pmatrix} 1 & 0 & 0 & 1 \\ 0 & 0 & 0 & 0 \\ 0 & 0 & 0 & 0 \\ 1 & 0 & 0 & 1 \end{pmatrix}. \tag{17.19}$$

Moreover, one has

$$\Phi = \chi_A \otimes \chi_B + \frac{1}{4}(\sigma_x \otimes \sigma_x - \sigma_y \otimes \sigma_y + \sigma_z \otimes \sigma_z). \tag{17.20}$$

Proof Let us start with the proof of Eq. (17.19). First of all, we observe that by Eq. (17.18), one has

$$(\varphi_1^\dagger \otimes \varphi_2^\dagger | \Phi) = (\varphi_2^\dagger \otimes \varphi_1^\dagger | \Phi) = 0. \tag{17.21}$$

Equation (17.21) and (17.18) along with Exercise 15.2 imply that

$$(\varphi_i^\dagger |_B | \Phi) = \frac{1}{2} |\varphi_i)_A \tag{17.22}$$

and then $(e|_B|\phi) = |\chi)_A$. An analogous argument holds for the marginal on B. Thus, Φ is a minimal purification of χ_A. Now, for every reversible transformation $\mathcal{U} \in G_A$, let $\mathcal{U}^* \in G_B$ be the conjugate of $\mathcal{U}$, defined with respect to the state Φ. Since all 2×2 unitary irreducible representations of $G_A = SU(2)$ are unitarily equivalent, by a suitable choice of the standard representation S_ρ^B for system B, one has

$$S_{\mathcal{U}^*\rho}^B = U^* S_\rho^B U^T, \tag{17.23}$$

where U^* and U^T are the complex conjugate and the transpose, respectively, of the matrix $U \in SU(2)$ such that $S_{\mathcal{U}\rho}^A = U S_\rho^A U^\dagger$. Due to Eq. (17.23) and to Lemma 17.16, the state Φ must satisfy the condition $(U \otimes U^*) T_\Phi (U^\dagger \otimes U^T) = T_\Phi, \forall U \in SU(2)$. Now, the unitary representation $U \otimes U^*$ has two irreducible subspaces whose projections are given by the matrices

---

[3] The state in Eq. (17.18) exists thanks to the superposition principle.

$$P_0 = \frac{1}{2} \begin{pmatrix} 1 & 0 & 0 & 1 \\ 0 & 0 & 0 & 0 \\ 0 & 0 & 0 & 0 \\ 1 & 0 & 0 & 1 \end{pmatrix}$$

$$P_1 = \frac{1}{2} \begin{pmatrix} 1 & 0 & 0 & -1 \\ 0 & 2 & 0 & 0 \\ 0 & 0 & 2 & 0 \\ -1 & 0 & 0 & 1 \end{pmatrix} = I \otimes I - P_0,$$

where I is the 2×2 identity matrix. The most general form for T_Φ is then the following:

$$
\begin{aligned}
T_\Phi &= x_0 P_0 + x_1 P_1 \\
&= (x_0 - x_1) P_0 + x_1 I \otimes I \\
&= \begin{pmatrix} \alpha + \beta & 0 & 0 & \beta \\ 0 & \alpha & 0 & 0 \\ 0 & 0 & \alpha & 0 \\ \beta & 0 & 0 & \alpha + \beta \end{pmatrix},
\end{aligned}
$$

having defined $\alpha := x_1$ and $\beta := (x_0 - x_1)/2$. Now, by definition of the tensor representation, the conditional states $(\varphi_m^\dagger |_A| \Phi)_{AB}$ are described by the diagonal blocks of the matrix T_Φ, and by Eq. (17.22) one has

$$
S^B_{\frac{1}{2}\varphi_1} = \begin{pmatrix} \alpha + \beta & 0 \\ 0 & \alpha \end{pmatrix} \qquad S^B_{\frac{1}{2}\varphi_2} = \begin{pmatrix} \alpha & 0 \\ 0 & \alpha + \beta \end{pmatrix}. \tag{17.24}
$$

Since the states φ_1 and φ_2 are pure, the above matrices must be rank-one. Moreover, their trace must be equal to $(\varphi_m^\dagger \otimes e_B | \Phi) = 1/2(e_B | \varphi_m) = \frac{1}{2}$, $m = 1, 2$. Then we have two possibilities. Either (i) $\alpha = 0$ and $\beta = \frac{1}{2}$ or (ii) $\alpha = -\beta = \frac{1}{2}$. In the case (i), Eq. (17.19) holds. In the case (ii), to prove Eq. (17.19) we need to change our choice of matrix representation for the qubit B. Precisely, we make the following change:

$$
\begin{aligned}
S^B_{\sigma_x} &\mapsto \widetilde{S}^B_{\sigma_x} = -S^B_{\sigma_x}, \\
S^B_{\sigma_y} &\mapsto \widetilde{S}^B_{\sigma_y} = -S^B_{\sigma_y}, \\
S^B_{\sigma_z} &\mapsto \widetilde{S}^B_{\sigma_z} = -S^B_{\sigma_z},
\end{aligned} \tag{17.25}
$$

where $\sigma_z := \varphi_1 - \varphi_2$. Note that the inversion of the axes, sending σ_k to $-\sigma_k$ for every $k = x, y, z$ is not an allowed physical transformation, but this is not a problem here, because Eq. (17.25) is just a new choice of matrix representation, in which the set of states of system B is still represented by the Bloch sphere.

More concisely, the change of matrix representation $S^B \mapsto \widetilde{S}^B$ can be expressed as

$$
S^B_\rho \mapsto \widetilde{S}^B_\rho := Y \left[S^B_\rho \right]^T Y^\dagger \qquad Y := \begin{pmatrix} 0 & -1 \\ 1 & 0 \end{pmatrix}.
$$

Note that in the new representation $\widetilde{S}^B$ the physical transformation $\mathcal{U}^*$ is still represented as $\widetilde{S}^B_{\mathcal{U}\rho} = U^*\widetilde{S}^B_\rho U^T$: indeed we have

$$
\begin{aligned}
\widetilde{S}^B_{\mathcal{U}^*\rho} &= Y\left[S^B_{\mathcal{U}^*\rho}\right]^T Y^\dagger \\
&= Y(U^* S^B_\rho U^T)^T Y^\dagger \\
&= Y(U\left[S^B_\rho\right]^T U^\dagger) Y^\dagger \\
&= (YUY^\dagger)(Y\left[S^B_\rho\right]^T Y^\dagger)(YU^\dagger Y^\dagger) \\
&= U^*(Y\left[S^B_\rho\right]^T Y^\dagger)U^T \\
&= U^*\widetilde{S}^B_\rho U^T,
\end{aligned}
$$

having used the relations $Y^\dagger Y = I$ and $YUY^\dagger = U^*$, valid for every $U \in SU(2)$. Clearly, the change of standard representation $S \to \widetilde{S}$ for the qubit B induces a change of tensor representation $T \to \widetilde{T}$, where $\widetilde{T}$ is the tensor representation defined by $\widetilde{T}_{\rho\otimes\sigma} := S^A_\rho \otimes \widetilde{S}^B_\sigma$. With this change of representation, we have

$$
\widetilde{T}_\Phi = \frac{1}{2}\begin{pmatrix} 1 & 0 & 0 & 1 \\ 0 & 0 & 0 & 0 \\ 0 & 0 & 0 & 0 \\ 1 & 0 & 0 & 1 \end{pmatrix}.
$$

This concludes the proof of Eq. (17.19).

Let us now prove Eq. (17.20). Using the definition $T_{\rho\otimes\tau} := (S^A_\rho \otimes S^B_\tau)$ one can directly verify the relation

$$
T_\Phi = S^A_\chi \otimes S^B_\chi + \frac{1}{4}(S^A_{\sigma_x} \otimes S^B_{\sigma_x} - S^A_{\sigma_y} \otimes S^B_{\sigma_y} + S^A_{\sigma_z} \otimes S^B_{\sigma_z}).
$$

This is precisely the matrix version of Eq. (17.20). $\square$

We now show how to construct a standard and a tensor representation in such a way that $T = S$.

Lemma 17.18   *For a composite system AB with $d_A = d_B = 2$ one can choose the standard representation in such a way that the following equalities hold:*

$$
S_{\varphi_m\otimes\varphi_n} = T_{\varphi_m\otimes\varphi_n}, \tag{17.26}
$$

$$
S_{\varphi_m\otimes\sigma_k} = T_{\varphi_m\otimes\sigma_k}, \qquad k = x, y, \tag{17.27}
$$

$$
S_{\sigma_k\otimes\varphi_m} = T_{\sigma_k\otimes\varphi_m}, \qquad k = x, y. \tag{17.28}
$$

Proof   Let us choose single-qubit representations S^A and S^B, which by definition satisfy Eqs. (17.7), (17.8), and (17.9). On the other hand, choosing the states $\{\varphi_m \otimes \varphi_n\}^2_{m,n=1}$ in lexicographic order as the four discriminable states for the standard representation, we have

$$
[S_{\varphi_1\otimes\varphi_1}]_{rs} = \delta_{1r}\delta_{1s}, \qquad [S_{\varphi_1\otimes\varphi_2}]_{rs} = \delta_{2r}\delta_{2s},
$$

$$
[S_{\varphi_2\otimes\varphi_1}]_{rs} = \delta_{3r}\delta_{3s}, \qquad [S_{\varphi_2\otimes\varphi_2}]_{rs} = \delta_{4r}\delta_{4s}.
$$

With this choice, we get $S_{\varphi_m \otimes \varphi_n} = S^A_{\varphi_m} \otimes S^B_{\varphi_n} = T_{\varphi_m \otimes \varphi_n}$ for every $m, n = 1, 2$. This proves Eq. (17.26). Let us now prove Eqs. (17.27) and (17.28). Consider the two-dimensional face $F_{11,12}$, generated by the states $\varphi_1 \otimes \varphi_1$ and $\varphi_1 \otimes \varphi_2$. This face is the face identified by the state $\omega_{11,12} := \varphi_1 \otimes \chi_B$, and by Corollary 15.31 we have $F_{11,12} = \{\varphi_1\} \otimes \mathrm{St}_1(B)$. Therefore, by Exercise 8.7, we can choose the vectors $\sigma_k^{11,12}$, $k = x, y$ to satisfy the relation $\sigma_k^{11,12} := \varphi_1 \otimes \sigma_k$, $k = x, y$. Using Eqs. (17.8) and (17.9), in the standard representation we have

$$[S_{\sigma_x}^{11,12}]_{rs} = \delta_{r1}\delta_{s2} + \delta_{r2}\delta_{s1},$$
$$[S_{\sigma_y}^{11,12}]_{rs} = i\lambda^{11,12}(\delta_{r1}\delta_{s2} - \delta_{r2}\delta_{s1}).$$

If we choose $\lambda^{11,12} = \lambda_B^{12}$, this implies $S_{\sigma_k^{11,12}} = S^A_{\varphi_{11}} \otimes S^B_{\sigma_k} = T_{\varphi_{11} \otimes \sigma_k}$, for $k = x, y$. Repeating the same argument for the face $F_{22,21}$, $F_{11,21}$, and $F_{21,22}$ we obtain the proof of Eqs. (17.27) and (17.28). $\quad\square$

Note that the choice of S^B needed in Eq. (17.19) is compatible with the choice of S^B needed in Lemma 17.18. Indeed, by Eq. (17.20), to prove compatibility we only have to show that the representation S^B used in Eq. (17.19) has the property $[S^B_{\varphi_m}]_{rs} = \delta_{mr}\delta_{ms}$, $m = 1, 2$. This property is automatically guaranteed by the relation $(\varphi_m^\dagger|_A|\Phi)_{AB} = 1/2|\varphi_m)$, $m = 1, 2$ and by Eq. (17.24) with $\alpha = 0$ and $\beta = 1/2$.

We now define the reversible transformations $\mathcal{U}_{x,\pi}$ and $\mathcal{U}_{z,\frac{\pi}{2}}$ as follows:

$$S_{\mathcal{U}_{x,\pi}\rho} = XS_\rho X, \qquad X := \begin{pmatrix} 0 & 1 \\ 1 & 0 \end{pmatrix},$$
$$S_{\mathcal{U}_{z,\frac{\pi}{2}}\rho} = e^{-i\frac{\pi}{4}Z}S_\rho e^{i\frac{\pi}{4}Z}, \qquad Z := \begin{pmatrix} 1 & 0 \\ 0 & -1 \end{pmatrix}. \tag{17.29}$$

Also, we define the states Ψ, $\Phi_{z,\frac{\pi}{2}}$, and $\Psi_{z,\frac{\pi}{2}}$ as

$$|\Psi) := (\mathcal{U}_{x,\pi} \otimes \mathcal{I})|\Phi),$$
$$|\Phi_{z,\frac{\pi}{2}}) := (\mathcal{U}_{z,\frac{\pi}{2}} \otimes \mathcal{I})|\Phi),$$
$$|\Psi_{z,\frac{\pi}{2}}) := (\mathcal{U}_{z,\frac{\pi}{2}} \otimes \mathcal{I})|\Psi).$$

Lemma 17.19　*The states* Ψ, $\Phi_{z,\frac{\pi}{2}}$, *and* $\Psi_{z,\frac{\pi}{2}}$ *have the following tensor representation:*

$$T_\Psi = \frac{1}{2}\begin{pmatrix} 0 & 0 & 0 & 0 \\ 0 & 1 & 1 & 0 \\ 0 & 1 & 1 & 0 \\ 0 & 0 & 0 & 0 \end{pmatrix}, \quad T_{\Phi_{z,\frac{\pi}{2}}} = \frac{1}{2}\begin{pmatrix} 1 & 0 & 0 & -i \\ 0 & 0 & 0 & 0 \\ 0 & 0 & 0 & 0 \\ i & 0 & 0 & 1 \end{pmatrix}, \quad T_{\Psi_{z,\frac{\pi}{2}}} = \frac{1}{2}\begin{pmatrix} 0 & 0 & 0 & 0 \\ 0 & 1 & -i & 0 \\ 0 & i & 1 & 0 \\ 0 & 0 & 0 & 0 \end{pmatrix}.$$

$$\tag{17.30}$$

Moreover, one has

$$\Psi = \chi_A \otimes \chi_B + \frac{1}{4}(\sigma_x \otimes \sigma_x + \sigma_y \otimes \sigma_y - \sigma_z \otimes \sigma_z),$$
$$\Phi_{z,\frac{\pi}{2}} = \chi_A \otimes \chi_B + \frac{1}{4}(\sigma_y \otimes \sigma_x + \sigma_x \otimes \sigma_y + \sigma_z \otimes \sigma_z),$$
$$\Psi_{z,\frac{\pi}{2}} = \chi_A \otimes \chi_B + \frac{1}{4}(\sigma_y \otimes \sigma_x - \sigma_x \otimes \sigma_y - \sigma_z \otimes \sigma_z). \tag{17.31}$$

Proof Eq. (17.30) is obtained from Eq. (17.19) by explicit calculation using Lemma 17.16 and Eq. (17.29). Then, the validity of Eq. (17.31) is easily obtained from Eq. (17.20) using the relations

$$\mathcal{U}_{x,\pi}|\sigma_x) = |\sigma_x), \quad \mathcal{U}_{x,\pi}|\sigma_y) = -|\sigma_y), \quad \mathcal{U}_{x,\pi}|\sigma_z) = -|\sigma_z),$$

and

$$\mathcal{U}_{z,\pi/2}|\sigma_x) = |\sigma_y), \quad \mathcal{U}_{z,\pi/2}|\sigma_y) = -|\sigma_x), \quad \mathcal{U}_{z,\pi/2}|\sigma_z) = |\sigma_z).$$

$\square$

Lemma 17.20 *The states* Φ, $\Phi_{z,\frac{\pi}{2}}$, Ψ, *and* $\Psi_{z,\frac{\pi}{2}}$ *have a standard representation of the form*

$$S_\Phi = \frac{1}{2}\begin{pmatrix} 1 & 0 & 0 & e^{i\theta} \\ 0 & 0 & 0 & 0 \\ 0 & 0 & 0 & 0 \\ e^{-i\theta} & 0 & 0 & 1 \end{pmatrix}, \quad S_{\Phi,\frac{\pi}{2}} = \frac{1}{2}\begin{pmatrix} 1 & 0 & 0 & \lambda i e^{i\theta} \\ 0 & 0 & 0 & 0 \\ 0 & 0 & 0 & 0 \\ -\lambda i e^{-i\theta} & 0 & 0 & 1 \end{pmatrix},$$

$$S_\Psi = \frac{1}{2}\begin{pmatrix} 0 & 0 & 0 & 0 \\ 0 & 1 & e^{i\gamma} & 0 \\ 0 & e^{-i\gamma} & 1 & 0 \\ 0 & 0 & 0 & 0 \end{pmatrix}, \quad S_{\Psi,\frac{\pi}{2}} = \frac{1}{2}\begin{pmatrix} 0 & 0 & 0 & 0 \\ 0 & 1 & \mu i e^{i\gamma} & 0 \\ 0 & -\mu i e^{-i\gamma} & 1 & 0 \\ 0 & 0 & 0 & 0 \end{pmatrix}. \quad (17.32)$$

with $\theta, \gamma \in [0, 2\pi)$ *and* $\lambda, \mu \in \{-1, 1\}$.

Proof Let us start from Φ. By Lemma 17.18 and from the definition of Φ, we have

$$(\varphi_1^\dagger \otimes \varphi_1^\dagger|\Phi) = (\varphi_2^\dagger \otimes \varphi_2^\dagger|\Phi) = \frac{1}{2},$$

$$(\varphi_1^\dagger \otimes \varphi_2^\dagger|\Phi) = (\varphi_2^\dagger \otimes \varphi_1^\dagger|\Phi) = 0,$$

and thus the diagonal elements of S_Φ are $\frac{1}{2}, 0, 0, \frac{1}{2}$. Now, by Theorem 17.10, the off-diagonal elements of S_Φ must be as in equation (17.32). Consider now Ψ. First, from Eq. (17.31) it is immediate to obtain

$$(\varphi_1^\dagger \otimes \varphi_1^\dagger|\Psi) = (\varphi_2^\dagger \otimes \varphi_2^\dagger|\Psi) = 0,$$

$$(\varphi_1^\dagger \otimes \varphi_2^\dagger|\Psi) = (\varphi_2^\dagger \otimes \varphi_1^\dagger|\Psi) = 1/2.$$

This gives the diagonal elements of S_Ψ. Then, using Theorem 17.10 we obtain that S_Ψ must be as in Eq. (17.32), for some value of γ. Let us now consider $\Phi_{z,\frac{\pi}{2}}$. Again, the diagonal elements of the matrix $S_{\Phi_{z,\frac{\pi}{2}}}$ are obtained from Eq. (17.31), which in this case yields

$$(\varphi_1^\dagger \otimes \varphi_1^\dagger|\Phi_{z,\frac{\pi}{2}}) = (\varphi_2^\dagger \otimes \varphi_2^\dagger|\Phi_{z,\frac{\pi}{2}}) = 1/2,$$

$$(\varphi_1^\dagger \otimes \varphi_2^\dagger|\Phi_{z,\frac{\pi}{2}}) = (\varphi_2^\dagger \otimes \varphi_1^\dagger|\Phi_{z,\frac{\pi}{2}}) = 0.$$

Hence, by Theorem 17.10 one must have $(S_{\Phi_{z,\frac{\pi}{2}}})_{1,4} = \frac{1}{2}e^{i\theta'}$ for some value of $\theta' \in [0, 2\pi)$. However, the value of θ' is not arbitrary. Indeed, we have

$$(\Phi^\dagger|\Phi_{z,\frac{\pi}{2}}) = \text{Tr}[E_{\Phi^\dagger}S_{\Phi_{z,\frac{\pi}{2}}}] = \text{Tr}[S_\Phi S_{\Phi_{z,\frac{\pi}{2}}}] = \text{Tr}[F_{\Phi^\dagger}T_{\Phi_{z,\frac{\pi}{2}}}] = \text{Tr}[T_\Phi T_{\Phi_{z,\frac{\pi}{2}}}] = \frac{1}{2},$$

which implies that

$$\frac{1}{2}[1 + \cos(\theta - \theta')] = \mathrm{Tr}[S_\Phi S_{\Phi_{z,\pi/2}}] = \mathrm{Tr}[E_{\Phi^\dagger} + S_{\Phi_{z,\pi/2}}] = (\Phi^\dagger | \Phi_{z,\pi/2})$$

$$= \mathrm{Tr}[F_{\Phi^\dagger} T_{\Phi_{z,\pi/2}}] = \mathrm{Tr}[T_\Phi T_{\Phi_{z,\pi/2}}] = \frac{1}{2},$$

where the third equality follows from Theorem 17.11, the second to last equality follows from Eq. (17.30), and the last equality follows from Eqs. (17.19) and (17.32). Hence, we must have $\cos(\theta - \theta') = 0$, i.e. $\theta' = \theta \pm \frac{\pi}{2}$, as in Eq. (17.32). Finally, the same arguments can be used for $\Psi_{z,\frac{\pi}{2}}$: the diagonal elements of $S_{\Psi_{z,\frac{\pi}{2}}}$ are obtained from the relations

$$(\varphi_1^\dagger \otimes \varphi_1^\dagger | \Psi_{z,\frac{\pi}{2}}) = (\varphi_2^\dagger \otimes \varphi_2^\dagger | \Psi_{z,\frac{\pi}{2}}) = 0$$

$$(\varphi_1^\dagger \otimes \varphi_2^\dagger | \Psi_{z,\frac{\pi}{2}}) = (\varphi_2^\dagger \otimes \varphi_1^\dagger | \Psi_{z,\frac{\pi}{2}}) = 1/2,$$

which follow from Eq. (17.31). This implies that the matrix $S_{\Psi_{z,\frac{\pi}{2}}}$ has $(S_{\Psi_{z,\frac{\pi}{2}}})_{23} = \frac{1}{2} e^{i\gamma'}$ for some $\gamma' \in [0, 2\pi)$. The relation $\mathrm{Tr}[S_\Psi S_{\Psi_{z,\frac{\pi}{2}}}] = \mathrm{Tr}[T_\Psi T_{\Psi_{z,\frac{\pi}{2}}}] = 1/2$ then implies $\gamma' = \gamma \pm \frac{\pi}{2}$.   □

Let us now consider the four vectors $\Sigma_x^{(11,22)}, \Sigma_y^{(11,22)}, \Sigma_x^{(12,21)}, \Sigma_y^{(12,21)}$ defined as follows:

$$\Sigma_x^{(11,22)} = 2\left(\Phi - \chi_A \otimes \chi_B - \frac{1}{4}\sigma_z \otimes \sigma_z\right),$$

$$\Sigma_y^{(11,22)} = 2\left(\Phi_{z,\frac{\pi}{2}} - \chi_A \otimes \chi_B - \frac{1}{4}\sigma_z \otimes \sigma_z\right),$$

$$\Sigma_x^{(12,21)} = 2\left(\Psi - \chi_A \otimes \chi_B + \frac{1}{4}\sigma_z \otimes \sigma_z\right),$$

$$\Sigma_x^{(12,21)} = 2\left(\Psi_{z,\frac{\pi}{2}} - \chi_A \otimes \chi_B + \frac{1}{4}\sigma_z \otimes \sigma_z\right). \qquad (17.33)$$

By the previous results, it is immediate to obtain the matrix representations of these vectors. In the tensor representation, using Eqs. (17.19) and (17.30) we obtain

$$T_{\Sigma_x^{(11,22)}} = \begin{pmatrix} 0 & 0 & 0 & 1 \\ 0 & 0 & 0 & 0 \\ 0 & 0 & 0 & 0 \\ 1 & 0 & 0 & 0 \end{pmatrix}, \quad T_{\Sigma_y^{(11,22)}} = \begin{pmatrix} 0 & 0 & 0 & -i \\ 0 & 0 & 0 & 0 \\ 0 & 0 & 0 & 0 \\ i & 0 & 0 & 0 \end{pmatrix},$$

$$T_{\Sigma_x^{(12,21)}} = \begin{pmatrix} 0 & 0 & 0 & 0 \\ 0 & 0 & 1 & 0 \\ 0 & 1 & 0 & 0 \\ 0 & 0 & 0 & 0 \end{pmatrix}, \quad T_{\Sigma_y^{(12,21)}} = \begin{pmatrix} 0 & 0 & 0 & 0 \\ 0 & 0 & -i & 0 \\ 0 & i & 0 & 0 \\ 0 & 0 & 0 & 0 \end{pmatrix},$$

while in the standard representation, using Eq. (17.32), we obtain

$$S_{\Sigma_x^{(11,22)}} = \begin{pmatrix} 0 & 0 & 0 & e^{i\theta} \\ 0 & 0 & 0 & 0 \\ 0 & 0 & 0 & 0 \\ e^{-i\theta} & 0 & 0 & 0 \end{pmatrix} \qquad S_{\Sigma_y^{(11,22)}} = \begin{pmatrix} 0 & 0 & 0 & -\lambda i e^{i\theta} \\ 0 & 0 & 0 & 0 \\ 0 & 0 & 0 & 0 \\ \lambda i e^{-i\theta} & 0 & 0 & 0 \end{pmatrix}$$

$$S_{\Sigma_x^{(12,21)}} = \begin{pmatrix} 0 & 0 & 0 & 0 \\ 0 & 0 & e^{i\gamma} & 0 \\ 0 & e^{-i\gamma} & 0 & 0 \\ 0 & 0 & 0 & 0 \end{pmatrix} \qquad S_{\Sigma_x^{(11,22)}} = \begin{pmatrix} 0 & 0 & 0 & 0 \\ 0 & 0 & -\mu i e^{i\gamma} & 0 \\ 0 & \mu i e^{-\gamma} & 0 & 0 \\ 0 & 0 & 0 & 0 \end{pmatrix}.$$

Comparing the two matrix representations we are now in position to prove the desired result.

Lemma 17.21 *With a suitable choice of axes, one has $S_{\sigma_k \otimes \sigma_l} = T_{\sigma_k \otimes \sigma_l}$ for every $k, l = x, y$.*

Proof For the face $(11, 22)$, using the freedom coming from Eqs. (17.9) and (17.10), we redefine the x and y axes so that $\sigma_x^{(11,22)} = \Sigma_x^{(11,22)}$ and $\lambda \sigma_y^{(11,22)} = \Sigma_y^{(11,22)}$. In this way we have

$$S_{\Sigma_k^{(11,22)}} = T_{\Sigma_k^{(11,22)}} \qquad \forall k = x, y.$$

Likewise, for the face $(12, 21)$ we redefine the x and y axes so that $\sigma_x^{(12,21)} = \Sigma_x^{(12,21)}$ and $\mu \sigma_y^{(12,21)} = \Sigma_y^{(12,21)}$, so that we have

$$S_{\Sigma_k^{(12,21)}} = T_{\Sigma_k^{(12,21)}} \qquad \forall k = x, y.$$

The above definitions are allowed by Eqs. (17.10) and (17.11). Finally, using Eqs. (17.20), (17.31), and (17.33) we have the relations

$$\sigma_x \otimes \sigma_x = \Sigma_x^{(11,22)} + \Sigma_x^{(12,21)},$$
$$\sigma_y \otimes \sigma_y = \Sigma_x^{(11,22)} - \Sigma_x^{(12,21)},$$
$$\sigma_x \otimes \sigma_y = \Sigma_y^{(11,22)} - \Sigma_y^{(12,21)},$$
$$\sigma_y \otimes \sigma_x = \Sigma_y^{(11,22)} + \Sigma_y^{(12,21)}.$$

Since S and T coincide on the right-hand side of each equality, they must also coincide on the left-hand side. $\qquad \square$

Theorem 17.22 *With a suitable choice of axes, the standard representation coincides with the tensor representation, that is, $S_\rho = T_\rho$ for every $\rho \in \mathrm{St(AB)}$.*

Proof Combining Lemma 17.18 with Lemma 17.21 we obtain that S and T coincide on the tensor products basis $\mathcal{B} \otimes \mathcal{B}$, where $\mathcal{B} = \{\varphi_1, \varphi_2, \sigma_x, \sigma_y\}$. By linearity, S and T coincide on every state. $\qquad \square$

From now on, whenever we will consider a composite system AB where A and B are two-dimensional we will adopt the choice that guarantees that the standard representation coincides with the tensor representation.

17.4 Positive Matrices

In this section we show that the states in our theory can be represented by positive matrices. This amounts to prove that for every system A, the set of states $St_1(A)$ can be represented as a subset of the set of density matrices in dimension d_A. This result will be completed in Section 17.5, where we will see that, in fact, every density matrix in dimension d_A corresponds to some state of $St_1(A)$.

The starting point to prove positivity is the following lemma.

Lemma 17.23 *Let A and B be two-dimensional systems. Then, for every pure state $\Psi \in$ St(AB) one has $S_\Psi \geq 0$.*

Proof Take an arbitrary vector $Z \in \mathbb{C}^2 \otimes \mathbb{C}^2$, written in the Schmidt form as $|Z\rangle = \sum_{n=1}^{2} \sqrt{\lambda_n} |v_n\rangle |w_n\rangle$. Introducing the unitaries U, V such that $U|v_n\rangle = |n\rangle$ and $V|w_n\rangle = |n\rangle$ for every $n = 1, 2$, we have $|Z\rangle = (U^\dagger \otimes V^\dagger)|W\rangle$, where $|W\rangle = \sum_{n=1}^{2} \sqrt{\lambda_n} |n\rangle |n\rangle$. Therefore, we have

$$\langle Z|S_\Psi|Z\rangle = \langle W|S_{(\mathcal{U}\otimes\mathcal{V})\Psi}|W\rangle$$

where $\mathcal{U}$ and $\mathcal{V}$ are the reversible transformations defined by $S_{\mathcal{U}\rho} = US_\rho U^\dagger$ and $S_{\mathcal{V}\rho} = VS_\rho V^\dagger$, respectively ($\mathcal{U}$ and $\mathcal{V}$ are physical transformations by virtue of Corollary 14.6). Here we used the fact that the standard two-qubit representation coincides with the tensor representation and, therefore, $S_{(\mathcal{U}\otimes\mathcal{V})\Psi} = (U \otimes V)S_\Psi(U \otimes V)^\dagger$. Denoting the pure state $(\mathcal{U} \otimes \mathcal{V})|\Psi\rangle$ by $|\Psi'\rangle$ we then have

$$\langle Z|S_\Psi|Z\rangle = \lambda_1 [S_{\Psi'}]_{11,11} + \lambda_2 [S_{\Psi'}]_{22,22} + 2\sqrt{\lambda_1\lambda_2}\text{Re}\left([S_{\Psi'}]_{11,22}\right).$$

Since by Theorem 17.10 we have $[S_{\Psi'}]_{11,22} = \sqrt{[S_{\Psi'}]_{11,11}[S_{\Psi'}]_{22,22}}e^{i\theta}$, we conclude

$$\langle Z|S_\Psi|Z\rangle = \lambda_1 [S_{\Psi'}]_{11,11} + \lambda_2 [S_{\Psi'}]_{22,22} + 2\cos\theta \sqrt{\lambda_1\lambda_2[S_{\Psi'}]_{11,11}[S_{\Psi'}]_{22,22}}$$

$$\geq \left(\sqrt{\lambda_1[S_{\Psi'}]_{11,11}} - \sqrt{\lambda_2[S_{\Psi'}]_{22,22}}\right)^2 \geq 0.$$

Finally, since the vector $Z \in \mathbb{C}^2 \otimes \mathbb{C}^2$ is arbitrary, the matrix S_Ψ is positive. □

Corollary 17.24 *Let C be a system of dimension $d_C = 4$. Then, with a suitable choice of matrix representation the pure states of C are represented by positive matrices.*

Proof The system C is operationally equivalent to the composite system AB, where $d_A = d_B = 2$. Let $\mathcal{U} \in \text{Transf}(AB \to C)$ be the reversible transformation implementing the equivalence. Now, we know by Lemma 17.23 that the states of AB are represented by positive matrices. If we define the basis vectors for C by applying $\mathcal{U}$ to the basis for AB, we obtain that the states of C are represented by the same matrices as those representing the states of AB. □

Corollary 17.25 *Let* A *be a system with* $d_A = 3$. *With a suitable choice of matrix representation, the matrix* S_φ *is positive for every pure state* $\varphi \in \mathsf{St}(A)$.

Proof Let C be a system with $d_C = 4$. By Corollary 17.24 the states of C are represented by positive matrices. Consider an ideal compression scheme $(A, \mathcal{E}^{(123)}, \mathcal{D}^{(123)})$ for the face $F_{\{1,2,3\}}$. If we define the basis vectors for A by applying $\mathcal{E}$ to the basis vectors for the face $F_{\{1,2,3\}}$, then we obtain that the states of A are represented by the same matrices representing the states in the face $F_{\{1,2,3\}}$. Since these matrices are positive by Corollary 17.24, the thesis follows. $\qquad\square$

From now on, for every three-dimensional system A we will choose the matrix representation so that S_ρ is positive for every $\rho \in \mathsf{St}(A)$.

Corollary 17.26 *Let* $\varphi \in \mathsf{St}_1(A)$ *be a pure state of a system* A *with* $d_A = 3$. *Then, the corresponding matrix* S_φ, *given by*

$$
S_\varphi = \begin{pmatrix} p_1 & \sqrt{p_1 p_2}\, e^{i\theta_{12}(\varphi)} & \sqrt{p_1 p_3}\, e^{i\theta_{13}(\varphi)} \\ \sqrt{p_1 p_2}\, e^{-i\theta_{12}(\varphi)} & p_2 & \sqrt{p_2 p_3}\, e^{i\theta_{23}(\varphi)} \\ \sqrt{p_1 p_3}\, e^{-i\theta_{13}(\varphi)} & \sqrt{p_2 p_3}\, e^{-i\theta_{23}(\varphi)} & p_3 \end{pmatrix} \tag{17.34}
$$

satisfies the property

$$
e^{i\theta_{13}(\varphi)} = e^{i[\theta_{12}(\varphi)+\theta_{23}(\varphi)]}. \tag{17.35}
$$

Equivalently, $S_\varphi = |v\rangle\langle v|$, *where* $v \in \mathbb{C}^3$ *is the vector given by*

$$
|v\rangle := (\sqrt{p_1}, \sqrt{p_2}\, e^{-i\theta_{12}(\varphi)}, \sqrt{p_3}\, e^{-i\theta_{13}(\varphi)})^T. \tag{17.36}
$$

Proof The form (17.34) is proved by Theorem 17.10. The relation (17.35) among the phases $\theta_{ij}(\psi)$ can be trivially satisfied when $p_i = 0$ for some $i \in \{1,2,3\}$. Hence, let us assume $p_1, p_2, p_3 > 0$. Computing the determinant of S_φ one obtains $\det(S_\varphi) = 2p_1 p_2 p_3\{\cos[\theta_{12}(\varphi)+\theta_{23}(\varphi)-\theta_{13}(\varphi)]-1\}$. Since S_φ is positive, we must have $\det(S_\varphi) \geq 0$. If $p_1, p_2, p_3 > 0$, the only possibility is to have $\theta_{13}(\varphi) = \theta_{12}(\varphi)+\theta_{23}(\varphi) \mod 2\pi$. Finally, equivalence of (17.35) with $|v\rangle\langle v|$ for $|v\rangle$ given in (17.36) is given by a trivial check. $\quad\square$

Corollary 17.26 will now be extended to systems of arbitrary dimension. We recall that by Theorem 17.10, if $\psi \in \mathsf{PurSt}(A)$ is a pure state of a general system A, its standard representation is such that $(S_\psi)_{mn} = \sqrt{(S_\psi)_{mm}(S_\psi)_{nn}}\, e^{i\theta_{mn}(\psi)}$.

Consider then an arbitrary face $F_{\{p,q,r\}}$, with projection $\Pi_{\{p,q,r\}}$. Then $\Pi_{\{p,q,r\}}\psi$ is a pure state in $F_{\{p,q,r\}}$, and by virtue of Lemma 17.2, if we consider an ideal compression scheme $(Q, \mathcal{E}^{(pqr)}, \mathcal{D}^{(pqr)})$ with $d_Q = 3$, the matrix elements $(S_\psi)_{ij}$ for $i,j \in \{p,q,r\}$ coincide with those of the pure state $\Pi_{\{p,q,r\}}\psi$. In turn, the standard representation of states of Q can be taken such that the matrix elements of $S_{\mathcal{E}^{(pqr)}\Pi_{\{p,q,r\}}\varphi}$ coincide with those of $S_{\Pi_{\{p,q,r\}}\psi}$, i.e. $(S_{\mathcal{E}^{(pqr)}\Pi_{\{p,q,r\}}\psi})_{ij} = (S_\psi)_{ij}$ for $i,j \in \{p,q,r\}$.

Thus, if we choose the standard representation in which pure states of Q are positive, by Corollary 17.26 we have

$$
\theta_{pr}(\psi) = \theta_{pq}(\psi) + \theta_{qr}(\psi), \tag{17.37}
$$

for every pure state $\psi \in \mathsf{PurSt}(A)$.

The crucial step now is to prove that it is possible to choose a standard representation in such a way that the condition of Eq. (17.37) is satisfied for every three-dimensional face $F_{\{p,q,r\}}$ simultaneously, independently of the pure state ψ. To this end, it is useful to make the following remark.

Remark Let $\psi \in \mathsf{PurSt}(A)$ be a pure state of a general system A with $d_A \geq 3$. The standard representations $S : \mathsf{St}(A) \to \mathsf{Herm}(\mathbb{C}^{d_A})$ can only differ by an offset phase ϕ_{mn} and possibly a complex conjugation for every off-diagonal element $(S_\rho)_{mn}$. In other words, if S and S' are two different standard representations, one has either

$$(S'_\rho)_{mn} = (S_\rho)_{mn} e^{i\phi_{mn}}, \tag{17.38}$$

or

$$(S'_\rho)_{mn} = (S_\rho)^*_{mn} e^{i\phi_{mn}}, \tag{17.39}$$

where the choice of offset phases ϕ_{mn} as well as the choice between the form of Eq. (17.38) and that of (17.39) is made independently for every m, n. Notice also that this choice is unique, independently of ρ. For example, one could have $(S'_\rho)_{12} = (S_\rho)_{12} e^{i\phi_{12}}$ and $(S'_\rho)_{23} = (S_\rho)^*_{23} e^{i\phi_{23}}$. Indeed, these choices are accounted for by the parameters ϕ_{mn} and λ^{mn} in Eqs. (17.10) and (17.11), where $\lambda^{mn} \lambda'^{mn} = 1$ determines the choice (17.38) for the element mn, while $\lambda^{mn} \lambda'^{mn} = -1$ determines the choice (17.39). However, one cannot have $(S'_\rho)_{12} = (S_\rho)_{12}$ and $(S'_\sigma)_{12} = (S_\sigma)^*_{12} e^{i\phi'_{12}}$.

Lemma 17.27 *Let A be a system with $d_A \geq 3$. It is possible to choose the standard representation $S : \mathsf{St}(A) \to \mathsf{Herm}(\mathbb{C}^{d_A})$ in such a way that if $\psi \in \mathsf{PurSt}(A)$ is a pure state, then the phases $\theta_{mn}(\psi)$ satisfy equation (17.37) on every face $F_{\{p,q,r\}}$.*

Proof Let us first consider $p = 1$, $q = 2$, and $r = 3$. One can choose the standard representation S in such a way that one has $(S_\psi)_{ij} = (S_{\mathcal{E}^{(123)} \Pi_{\{1,2,3\}} \psi})_{ij}$ for $i, j \in \{1, 2, 3\}$. Since the matrix $S_{\mathcal{E}^{(123)} \Pi_{\{1,2,3\}} \psi}$ is non-negative, by Corollary 17.26 one has

$$e^{i\theta_{13}(\psi)} = e^{i[\theta_{12}(\psi) + \theta_{23}(\psi)]}.$$

Now, let us take $p = 1$, $q = 2$, and $3 < r \leq d_A$. In this case, the matrix elements $(S_\rho)_{11}$, $(S_\rho)_{12}$, $(S_\rho)_{21}$, and $(S_\rho)_{22}$ are already fixed by the choice in the first step. If we consider one standard representation where

$$e^{i\theta'_{1r}(\psi)} = e^{i[\theta'_{12}(\psi) + \theta'_{2r}(\psi)]},$$

remembering that $e^{i\theta_{12}} = e^{i[\theta'_{12}(\psi) + \phi_{12}]}$ is already fixed, we can choose e.g. $\phi_{1r} = \phi_{12}$, while $\phi_{2r} = 0$, thus obtaining

$$e^{i\theta_{1r}(\psi)} = e^{i[\theta'_{1r}(\psi) + \phi_{1r}]}$$
$$= e^{i[\theta'_{1r}(\psi) + \phi_{12}]}$$
$$= e^{i[\theta'_{12}(\psi) + \phi_{12} + \theta'_{2r}(\psi)]}$$
$$= e^{i[\theta_{12}(\psi) + \theta_{2r}(\psi)]}.$$

Now, let us consider triples with $p = 1, 2 < q < r \leq d_A$. At this stage, all the choices of $e^{i\theta_{1q}(\psi)} = e^{-i\theta_{q1}(\psi)}$ and $e^{i\theta_{1r}(\psi)} = e^{-i\theta_{r1}(\psi)}$ are fixed; however, the offset phase ϕ_{qr} is still free. This freedom is sufficient to complete the construction of the representation S. Indeed, let us fix the triple $1qr$, and consider a representation S' such that

$$e^{i\theta'_{1r}(\psi)} = e^{i[\theta'_{1q}(\psi)+\theta'_{qr}(\psi)]}.$$

Then, it is sufficient to choose $\phi_{qr} = \phi_{1r} - \phi_{1q}$. In this way we have

$$e^{i\theta_{1r}(\psi)} = e^{i[\theta'_{1r}(\psi)+\phi_{1r}]}$$
$$= e^{i[\theta'_{1q}(\psi)+\phi_{1q}+\theta'_{qr}(\psi)+\phi_{qr}]}$$
$$= e^{i[\theta_{1q}(\psi)+\theta_{qr}(\psi)]}.$$

At this point, we fixed all the possible choices and we concluded the construction of the representation S. It remains to check that the condition in Eq. (17.37) is satisfied also for triples pqr with $p > 1$. Indeed, by construction we have

$$e^{i\theta_{ij}(\psi)} = e^{i[-\theta_{1i}(\psi)+\theta_{1j}(\psi)]}$$
$$= e^{i[\theta_{i1}(\psi)+\theta_{1j}(\psi)]}.$$

Thus, we have

$$e^{i[\theta_{pq}(\psi)+\theta_{qr}(\psi)]} = e^{i[\theta_{p1}(\psi)+\theta_{1q}(\psi)+\theta_{q1}(\psi)+\theta_{1r}(\psi)]}$$
$$= e^{i[\theta_{p1}(\psi)+\theta_{1r}(\psi)]}$$
$$= e^{i\theta_{pr}(\psi)}.$$

$\square$

We can thus prove the following result.

Corollary 17.28 *One can choose a standard representation S_ρ in such a way that, if $\psi \in St_1(A)$ is a pure state and $d_A = N$, then $S_\psi = |v\rangle\langle v|$, where $v \in \mathbb{C}^N$ is the vector given by $v := (\sqrt{p_1}, \sqrt{p_2}e^{-i\alpha_2(\psi)}, \ldots, \sqrt{p_N}e^{-i\alpha_N(\psi)})^T$ with $\alpha_i(\psi) \in [0, 2\pi)$ $\forall i = 2, \ldots, N$.*

Proof By Lemma 17.26, for a pure state ψ we have $e^{i\theta_{pr}(\psi)} = e^{i[\theta_{pq}(\psi)+\theta_{qr}(\psi)]}$. Since this relation must hold for every choice of the triple $V = \{p, q, r\}$, if we define $\alpha_p(\psi) := \theta_{p1}(\psi)$, then we have $e^{i\theta_{pq}(\psi)} = e^{i[\theta_{p1}(\psi)+\theta_{1q}(\psi)]} = e^{i[\theta_{p1}(\psi)-\theta_{q1}(\psi)]} = e^{i[\alpha_p(\psi)-\alpha_q(\psi)]}$. It is now easy to verify that $S_\psi = |v\rangle\langle v|$, where $v = (\sqrt{p_1}, \sqrt{p_2}e^{-i\alpha_2(\psi)}, \ldots, \sqrt{p_N}e^{-i\alpha_N(\psi)})^T$.

$\square$

In conclusion, we proved the following:

Corollary 17.29 *For every system A, the state space $St_1(A)$ can be represented as a subset of the set of density matrices in dimension d_A.*

Proof For every state $\rho \in St_1(A)$ the matrix S_ρ is Hermitian by construction, with unit trace by Corollary 17.9, and positive since it is a convex mixture of positive matrices. $\square$

17.5 Quantum Theory in Finite Dimensions

Here we conclude our derivation of quantum theory by showing that every density matrix in dimension d_A corresponds to some state $\rho \in St_1(A)$. From now on it will always be implicitly assumed that we use a representation S such that S_ρ is non-negative.

We already know from the superposition principle (Lemma 16.1) that for every choice of probabilities $\{p_i\}_{i=1}^{d_A}$ there is a pure state $\varphi \in St_1(A)$ such that $\{p_i\}_{i=1}^{d_A}$ are the diagonal elements of S_φ. Thus, the set of density matrices corresponding to pure states contains at least one matrix of the form $S_\varphi = |v\rangle\langle v|$, with

$$|v\rangle = (\sqrt{p_1}, \sqrt{p_2}e^{-i\beta_2}, \dots, \sqrt{p_{d_A}}e^{-i\beta_{d_A}}).$$

It only remains to prove that every possible choice of phases $\beta_i \in [0, 2\pi)$ corresponds to some pure state.

We recall that for a face $F \subseteq St_1(A)$ we can define the group $\mathbf{G}_{F,F^\perp}$ to be the group of reversible transformations $\mathcal{U} \in \mathbf{G}_A$ such that $\mathcal{U} =_{\omega_F} \mathcal{I}_A$ and $\mathcal{U} =_{\omega_F^\perp} \mathcal{I}_A$. We then have the following lemma.

Lemma 17.30 *Consider a system A with $d_A = N$. Let $\{\varphi_i\}_{i=1}^N \subset St_1(A)$ be a maximal set of perfectly discriminable pure states, F be the face $F_{\{1,2,\dots,N-1\}}$, and $F^\perp$ its orthogonal face $F_{\{N\}}$. If $\mathcal{U}$ is a reversible transformation in $\mathbf{G}_{F,F^\perp}$, then the action of $\mathcal{U}$ is given by*

$$S_{\mathcal{U}\rho} = US_\rho U^\dagger \qquad U = \left(\begin{array}{ccc|c} & & & 0 \\ & I_{N-1} & & \vdots \\ & & & 0 \\ \hline 0 & \cdots & 0 & e^{-i\beta} \end{array} \right) \qquad (17.40)$$

where I_{N-1} is the $(N-1) \times (N-1)$ identity matrix and $\beta \in [0, 2\pi)$.

Proof Consider an arbitrary state $\rho \in St_1(A)$ and its matrix representation

$$S_\rho = \left(\begin{array}{c|c} S_{\Pi_F\rho} & \mathbf{f} \\ \hline \mathbf{f}^\dagger & S_{\Pi_F^\perp\rho} \end{array} \right),$$

where $\mathbf{f} \in \mathbb{C}^{N-1}$ is a suitable vector. Since $\mathcal{U} =_{\omega_F} \mathcal{I}_A$ and $\mathcal{U} =_{\omega_F^\perp} \mathcal{I}_A$, by Lemma 15.33 we have

$$\Pi_F\mathcal{U} = \Pi_F, \quad \Pi_{F^\perp}\mathcal{U} = \Pi_{F^\perp},$$

and consequently, $\Pi_F\mathcal{U}\rho = \Pi_F\rho$ and $\Pi_{F^\perp}\mathcal{U}\rho = \Pi_{F^\perp}\rho$, and thus

$$S_{\mathcal{U}\rho} = \left(\begin{array}{c|c} S_{\Pi_F\rho} & \mathbf{g} \\ \hline \mathbf{g}^\dagger & S_{\Pi_F^\perp\rho} \end{array} \right),$$

where $\mathbf{g} \in \mathbb{C}^{N-1}$ is a suitable vector. To prove Eq. (17.40), we will now prove that $\mathbf{g} = e^{i\beta}\mathbf{f}$ for some $\beta \in [0, 2\pi)$.

Let us start from the case $N = 3$. Since $\mathcal{U}|\varphi_i) = |\varphi_i)$ $\forall i = 1, 2, 3$, we have $(\varphi_i^\dagger|\mathcal{U} = (\varphi_i^\dagger|$ $\forall i = 1, 2, 3$ (Lemma 15.32). This implies that $\mathcal{U}$ sends states in the face F_{13} to states in the face F_{13}: indeed, for every $\rho \in F_{13}$ one has

$$(a_{13}|\mathcal{U}|\rho) = [(\varphi_1^\dagger| + (\varphi_3^\dagger|]\mathcal{U}|\rho) = [(\varphi_1^\dagger| + (\varphi_3^\dagger|]|\rho) = (a_{13}|\rho) = 1,$$

which implies $\mathcal{U}\rho \in F_{13}$ (Lemma 15.15). In other words, the restriction of $\mathcal{U}$ to the face F_{13} is a reversible qubit transformation. Therefore, the action of $\mathcal{U}$ on a state $\rho \in F_{13}$ must be given by

$$S_{\mathcal{U}\rho} = \begin{pmatrix} \rho_{11} & 0 & \rho_{13}e^{i\beta} \\ 0 & 0 & 0 \\ \rho_{31}e^{-i\beta} & 0 & \rho_{33,} \end{pmatrix}$$

for some $\beta \in [0, 2\pi)$. Similarly, we can see that $\mathcal{U}$ sends states in the face F_{23} to states in the face F_{23}. Hence, for every $\sigma \in F_{23}$ we have

$$S_{\mathcal{U}\sigma} = \begin{pmatrix} 0 & 0 & 0 \\ 0 & \sigma_{22} & \sigma_{23}e^{i\beta'} \\ 0 & \sigma_{32}e^{-i\beta'} & \sigma_{33} \end{pmatrix}$$

for some $\beta' \in [0, 2\pi)$. We now show that $e^{i\beta'} = e^{i\beta}$. To see that, consider a generic state $\varphi \in \mathsf{St}_1(A)$, with the property that $p_i = (a_i|\varphi) > 0$ for every $i = 1, 2, 3$ – such a state exists due to the superposition principle (Theorem 16.1). Writing S_φ as in Eq. (17.34) we then have

$$S_{\mathcal{U}\varphi} = \begin{pmatrix} p_1 & \sqrt{p_1p_2}e^{i\theta_{12}(\varphi)} & \sqrt{p_1p_3}e^{i[\theta_{13}(\varphi)+\beta]} \\ \sqrt{p_1p_2}e^{-i\theta_{12}(\varphi)} & p_2 & \sqrt{p_2p_3}e^{i[\theta_{23}(\varphi)+\beta']} \\ \sqrt{p_1p_3}e^{-i[\theta_{13}(\varphi)+\beta]} & \sqrt{p_2p_3}e^{-i[\theta_{23}(\varphi)+\beta']} & p_3 \end{pmatrix}.$$

Now, since φ and $\mathcal{U}\varphi$ are pure states, by Corollary 17.26 we must have

$$e^{i\theta_{13}(\varphi)} = e^{i[\theta_{12}(\varphi)+\theta_{23}(\varphi)]}$$

$$e^{i[\theta_{13}(\varphi)+\beta]} = e^{i[\theta_{12}(\varphi)+\theta_{23}(\varphi)+\beta']}.$$

By comparison we obtain $e^{i\beta} = e^{i\beta'}$. This proves Eq. (17.40) for $N = 3$. The proof for $N > 3$ is then immediate: for every three-dimensional face $F_{\{p,q,N\}}$ the action of $\mathcal{U}$ is given Eq. (17.40) for some β_{pq}. However, since $F_{\{p,q,N\}} \cap F_{\{p,q',N\}} = F_{pN}$, we must have $e^{i\beta_{pq}} = e^{i\beta_{pq'}}$. Similarly $e^{i\beta_{pq}} = e^{i\beta_{p'q}}$. We conclude that $e^{i\beta_{pq}} = e^{i\beta}$ for every p, q. This proves Eq. (17.40) in the general case. □

We now show that every possible phase shift in Eq. (17.40) corresponds to a physical transformation.

Lemma 17.31 *A transformation $\mathcal{U}$ of the form of Eq. (17.40) is a reversible transformation for every $\beta \in [0, 2\pi)$.*

Proof By Lemma 17.30, the group $\mathbf{G}_{F,F^\perp}$ is a subgroup of $U(1)$. Now, there are two possibilities: either $\mathbf{G}_{F,F^\perp}$ is a (finite) cyclic group or $\mathbf{G}_{F,F^\perp}$ coincides with $U(1)$. However, we know from Theorem 15.38 that $\mathbf{G}_{F,F^\perp}$ has a continuum of elements. Hence, $\mathbf{G}_{F,F^\perp} \simeq U(1)$ and β must then take every value in $[0, 2\pi)$. $\square$

An obvious corollary of the previous lemmas is the following.

Corollary 17.32 *The transformation $\mathcal{U}_{\boldsymbol{\beta}}$ defined by*

$$S_{\mathcal{U}_{\boldsymbol{\beta}} \rho} = U S_\rho U^\dagger, \tag{17.41}$$

where U is the diagonal matrix with diagonal elements $(1, e^{i\beta_1}, \ldots, e^{i\beta_{N-1}})$ is a reversible transformation for every vector $\boldsymbol{\beta} := (\beta_2, \ldots, \beta_N) \in [0, 2\pi) \times \cdots \times [0, 2\pi)$.

This leads directly to the conclusion of our derivation:

Theorem 17.33 *For every system A, the state space $\mathsf{St}_1(A)$ is the set of all density matrices on the Hilbert space $\mathbb{C}^{d_A}$.*

Proof Let us denote d_A as N. For every choice of probabilities $\mathbf{p} = (p_1, \ldots, p_N)$ there exists at least one pure state $\varphi_{\mathbf{p}}$ such that $p_k = (a_k | \varphi_{\mathbf{p}})$ for every $k = 1, \ldots, N$ (Lemma 16.1). This state is represented by the matrix $S_{\varphi_{\mathbf{p}}} = |v_{\mathbf{p}}\rangle\langle v_{\mathbf{p}}|$ with

$$|v_{\mathbf{p}}\rangle = (\sqrt{p_1}, \sqrt{p_2} e^{-i\alpha_2}, \ldots, \sqrt{p_N} e^{-i\alpha_N})^T,$$

(Lemma 17.28). Now, we can transform $\varphi_{\mathbf{p}}$ with every reversible transformation $\mathcal{U}_{\boldsymbol{\beta}}$ defined in Eq. (17.41), thus obtaining $S_{\mathcal{U}_{\boldsymbol{\beta}} \varphi_{\mathbf{p}}} = U_{\boldsymbol{\beta}} |v_{\mathbf{p}}\rangle\langle v_{\mathbf{p}}| U_{\boldsymbol{\beta}}^\dagger$, where $U_{\boldsymbol{\beta}} |v_{\mathbf{p}}\rangle = (\sqrt{p_1}, \sqrt{p_2} e^{-i(\alpha_2 + \beta_2)}, \ldots, \sqrt{p_N} e^{-i(\alpha_N + \beta_N)})^T$. Since $\mathbf{p}$ and $\boldsymbol{\beta}$ are arbitrary, this means that every rank-one density matrix corresponds to some pure state. Taking the possible convex mixtures we obtain that every $N \times N$ density matrix corresponds to some state of system A. $\square$

Choosing a suitable representation $\rho \mapsto S_\rho$, we proved that for every system A the set of normalized states $\mathsf{St}_1(A)$ is the whole set of density matrices in dimension d_A. Thanks to the purification postulate, this is enough to prove that all the effects $\mathsf{Eff}(A)$ and all the transformations $\mathsf{Transf}(A \to B)$ allowed in our theory are exactly the effects and the transformations allowed in quantum theory. Precisely we have the following:

Corollary 17.34 *For every couple of systems A and B the set of physical transformations $\mathsf{Transf}(A \to B)$ coincides with the set of all completely positive trace-non-increasing maps from $M_{d_A}(\mathbb{C})$ to $M_{d_B}(\mathbb{C})$.*

Proof We proved that our theory has the same normalized states as quantum theory. On the other hand, quantum theory is a theory with purification and local discriminability. The thesis then follows from the fact that two theories with purification and local discriminability that have the same set of normalized states are necessarily the same (Proposition 7.10). $\square$

17.6 Summary

We showed that our set of axioms uniquely identifies quantum theory. In particular, we proved that:

1. the set of states for a system A of dimension d_A is the set of density matrices on the Hilbert space $\mathbb{C}^{d_A}$;
2. the set of effects is the set of positive matrices bounded by the identity;
3. the pairing between a state and an effect is given by the trace of the product of the corresponding matrices.

Using the fact that two theories with purification and local discriminability having the same set of states for every system must coincide (see Section 7.12), we then obtain that all the physical transformations in our theory are exactly the physical transformations allowed in quantum mechanics. This concludes our derivation of quantum theory.

Solutions to Selected Problems and Exercises

Exercise 17.1

It is sufficient to write

$$
\begin{aligned}
(\sigma_k^{mn\dagger}|\varphi_m) &= (\sigma_k^{mn\dagger}|\mathcal{D}^{(mn)}\mathcal{E}^{(mn)}|\varphi_m) \\
&= (\sigma_k^{Q\dagger}|\eta_{z,-}) \\
&= [(\eta_{k,+}^\dagger| - (\eta_{k,-}^\dagger|]|\eta_{z,-}) \\
&= 0,
\end{aligned}
$$

where we use the result of Exercise 14.2. The calculation is completely analogous for φ_n, with $\mathcal{E}^{(mn)}|\varphi_n) = |\eta_{z,+})$.

Exercise 17.2

We start observing that, due to Corollary 14.5, one has the following matrix representation of $(\varphi_m^\dagger|\mathcal{D}^{(mn)}, (\varphi_n^\dagger|\mathcal{D}^{(mn)}$ and $(\sigma_k^{mn\dagger}|\mathcal{D}^{(mn)}$:

$$
E_{(\varphi_m^\dagger|\mathcal{D}^{(mn)}} = S_{\mathcal{E}^{(mn)}|\varphi_m)}, \qquad\qquad E_{(\varphi_n^\dagger|\mathcal{D}^{(mn)}} = S_{\mathcal{E}^{(mn)}|\varphi_n)},
$$

$$
E_{(\sigma_k^{mn\dagger}|\mathcal{D}^{(mn)}} = S_{\mathcal{E}^{(mn)}|\sigma_k^{mn})}.
$$

Thus, since the map $\mathcal{E}^{(mn)}$ acts injectively on effects $\mathsf{Eff}_\mathbb{R}(Q)$, and the inverse representation map E^{-1} acts injectively from $\mathsf{Herm}(\mathbb{C}^2)$ to $\mathsf{Eff}_\mathbb{R}(Q)$, linear independence of the four elements $\varphi_m^\dagger, \varphi_n^\dagger, \sigma_x^{mn\dagger}, \sigma_y^{mn\dagger}$ is a consequence of linear independence of the four matrices $S_{\mathcal{E}^{(mn)}|\varphi_m)}, S_{\mathcal{E}^{(mn)}|\varphi_n)}, S_{\mathcal{E}^{(mn)}|\sigma_x^{mn})}, S_{\mathcal{E}^{(mn)}|\sigma_y^{mn})}$. Now, by Eq. (17.5), we have that for a set V such that $m \notin V$ or $n \notin V$, $(\sigma_k^{mn\dagger}|\Pi_V = 0$: if both $m, n \notin V$, it is clear that F_{mn} is contained in the face orthogonal to the one on which Π_V projects, and thus

$$(\sigma_{k,\pm}^{mn\dagger}|\Pi_V = [(\varphi_{k,+}^{mn\dagger}| - (\varphi_{k,-}^{mn\dagger}|]\Pi_V$$
$$= 0,$$

while for $n \notin V$ one has

$$(\sigma_k^{mn\dagger}|\Pi_V = (\sigma_{k,}^{mn\dagger}|\Pi_{\{m,n\}}\Pi_V$$
$$= (\sigma_k^{mn\dagger}|\Pi_{\{m\}}$$
$$= (\sigma_k^{mn\dagger}|\varphi_m)(\varphi_m^\dagger|$$
$$= 0,$$

where we used Corollary 15.30 and Exercise 17.1. Since the cardinality of the set $\{\varphi_n^\dagger\}_{m=1}^{d_A} \cup \{\sigma_k^{mn\dagger}\}_{n>m=1,\dots,d_A\ k=x,y}$ is precisely $d_A^2 = \dim \mathsf{Eff}_\mathbb{R}(A)$, it is sufficient to prove their linear independence. Suppose then that

$$\sum_m c_m \varphi_m^\dagger + \sum_{n>m,\ k=x,y} c_k^{mn} \sigma_k^{mn\dagger} = 0.$$

This implies that

$$0 = \sum_m c_m (\varphi_m^\dagger|\Pi_{\{m,n\}} + \sum_{n>m,\ k=x,y} c_k^{mn}(\sigma_k^{mn\dagger}|\Pi_{\{m,n\}}$$
$$= c_m(\varphi_m^\dagger| + c_n(\varphi_n^\dagger| + \sum_{k=x,y} c_k^{mn}(\sigma_k^{mn\dagger}|.$$

Finally, by linear independence of $\varphi_m^\dagger, \varphi_n^\dagger, \sigma_x^{mn\dagger}, \sigma_y^{mn\dagger}$ it must be $c_m = c_n = c_k^{mn} = 0$ for all m, n, k, and this proves linear independence of the set $\{\varphi_n^\dagger\}_{m=1}^{d_A} \cup \{\sigma_k^{mn\dagger}\}_{n>m=1,\dots,d_A\ k=x,y}$.

Exercise 17.3

One can write

$$(\sigma_k^{mn\dagger}|\sigma_{k'}^{mn}) = (\sigma_k^{mn\dagger}|\mathcal{D}^{(mn)}\mathcal{E}^{(mn)}|\sigma_{k'}^{mn})$$
$$= (\sigma_k^{Q\dagger}|\sigma_{k'}^{Q})$$
$$= \sum_{s,s'=\pm} ss'(\eta_{k,s}^\dagger|\eta_{k',s'}).$$

Now, by the Bloch representation of qubit states and effects, one has $\sum_{s,s'=\pm} ss'(\eta_{k,s}^\dagger|\eta_{k',s'}) = 2\delta_{kk'}$.

References

Aharonov, D., Kitaev, A., and Nisan, N. 1998. Quantum circuits with mixed states. In: *Proceedings of the 30th Annual ACM Symposium on Theory of Computing (STOC)*. New York, NY: ACM.

Aharonov, Y., Anandan, J., and Vaidman, L. 1993. Meaning of the wave function. *Phys. Rev. A*, **47**, 4616–4626.

Aharonov, Y. and Vaidman, L. 1993. Measurement of the Schrödinger wave of a single particle. *Phys. Lett. A*, **178**, 38–42.

Alfsen, E. M. and Shultz, F. W. 2001. *State Spaces of Operator Algebras: Basic Theory, Orientations, and C*-products*. Boston, MA: Birkhäuser.

Alfsen, E. M. and Shultz, F. W. 2003. *Geometry of State Spaces of Operator Algebras*. Boston, MA: Birkhäuser.

Alicki, R. and Lendi, K. 1987. *Quantum Dynamical Semigroups and Applications*. Lecture Notes in Physics, vol. 286. Berlin: Springer.

Alter, O. and Yamamoto, Y. 1995. Inhibition of the measurement of the wave function of a single quantum system in repeated weak quantum nondemolition measurements. *Phys. Rev. Lett.*, **74**, 4106–4109.

Araki, H. 1980. On a characterization of the state space of quantum mechanics. *Comm. Math. Phys.*, **75**(1), 1–24.

Barnum, H., Nielsen, M. A., and Schumacher, B. 1998. Information transmission through a noisy quantum channel. *Phys. Rev. A*, **57**, 4153–4175.

Barnum, H., Gaebler, C. P., and Wilce, A. 2009. Ensemble steering, weak self-duality, and the structure of probabilistic theories. *arXiv*, quant-ph.

Barnum, H., Mueller, M. P., and Ududec, C. 2014. Higher-order interference and single-system postulates characterizing quantum theory. *arXiv:1403.4147*.

Barrett, J. 2007. Information processing in generalized probabilistic theories. *Phys. Rev. A*, **75**(3), 032304.

Barvinok, A. 2002. *A Course in Convexity*. Graduate Studies in Mathematics. Providence, RI: American Mathematical Society.

Baumeler, Ä. and Wolf, S. 2015. Device-independent test of causal order and relations to fixed-points. *arXiv.org*.

Belavkin, V. P. and Staszewski, P. 1986. A Radon Nikodym theorem for completely positive maps. *Rep. Math. Phys.*, **24**, 49–53.

Belavkin, V. P., D'Ariano, G. M., and Raginsky, M. 2005. Operational distance and fidelity for quantum channels. *J. Math. Phys.*, **46**(6), 062106.

Béllissard, J. and Iochum, B. 1978. Homogeneous self-dual cones, versus Jordan algebras: the theory revisited. *Ann. Inst. Fourier (Grenoble)*, **28**(1), v, 27–67.

Beltrametti, E. G., Cassinelli, G., Rota, G.-C., and Carruthers, P. A. 2010. *The Logic of Quantum Mechanics*. Vol. 15. Cambridge: Cambridge University Press.

Bennett, C. H. and Wiesner, S. J. 1992. Communication via one-and two-particle operators on Einstein-Podolsky-Rosen states. *Phys. Rev. lett.*, **69**(20), 2881.

Bennett, C. H., Brassard, G., *et al.* 1984. Quantum cryptography: public key distribution and coin tossing. In: *Proceedings of IEEE International Conference on Computers, Systems and Signal Processing*, vol. 175. New York: IEEE.

Bennett, C. H., Brassard, G., Crépeau, C., Jozsa, R., Peres, A., and Wootters, W. K. 1993. Teleporting an unknown quantum state via dual classical and Einstein-Podolsky-Rosen channels. *Phys. Rev. Lett.*, **70**(13), 1895.

Bennett, C. H., Bernstein, H. J., Popescu, S., and Schumacher, B. 1996. Concentrating partial entanglement by local operations. *Phys. Rev. A*, **53**, 2046.

Bernstein, E. and Vazirani, U. 1993. Quantum complexity theory. Pages 11–20 of: *Proceedings of the Twenty-fifth Annual ACM Symposium on Theory of Computing*. New York, NY: ACM.

Bhatia, R. 1997. *Matrix Analysis*. New York: Springer-Verlag.

Birkhoff, G. 1984. Lattice theory. In: Dilworth, R. P. (ed.), *Proceedings of the Second Symposium in Pure Mathematics of the American Mathematical Society April 1959*, vol. 175. Providence, RI: American Mathematical Society.

Birkhoff, G. and von Neumann, J. 1936. The logic of quantum mechanics. *Math. Annal.*, **37**, 823.

Bisio, A., Chiribella, G., D'Ariano, G. M., Facchini, S., and Perinotti, P. 2009a. Optimal quantum tomography. *IEEE J. Select. Topics Quantum Elec.*, **15**(6), 1646.

Bisio, A., Chiribella, G., D'Ariano, G. M., Facchini, S., and Perinotti, P. 2009b. Optimal quantum tomography of states, measurements, and transformations. *Phys. Rev. Lett.*, **102**(1), 010404.

Bisio, A., Chiribella, G., D'Ariano, G. M., and Perinotti, P. 2012. Quantum networks: general theory and applications. *Acta Phys. Slovaca*, **61**(1), 273–390.

Brandenburger, A. and Yanofsky, N. 2008. A classification of hidden-variable properties. *J. Phys. A*, **41**(42), 425302.

Brassard, G. 2005. Is information the key? *Nature Phys.*, **1**(1), 2–4.

Brukner, C. 2014a. Bounding quantum correlations with indefinite causal order. *arXiv.org*.

Brukner, C. 2014b. Quantum causality. *Nature Phys.*, **10**(4), 259–263.

Bruss, D., D'Ariano, G. M., Macchiavello, C., and Sacchi, M. F. 2000. Approximate quantum cloning and the impossibility of superluminal information transfer. *Phys. Rev. A*, **62**(6), 062302.

Buscemi, F., D'Ariano, G. M., and Perinotti, P. 2004. There exist nonorthogonal quantum measurements that are perfectly repeatable. *Phys. Rev. Lett.*, **92**, 070403.

Buscemi, F., Chiribella, G., and D'Ariano, G. M. 2005. Inverting quantum decoherence by classical feedback from the environment. *Phys. Rev. Lett.*, **95**, 090501.

Busch, P., Lahti, P. J., and Mittelstaedt, P. 1991. *The Quantum Theory of Measurement*. Berlin: Springer.

Childs, A. M., Chuang, I. L., and Leung, D. W. 2001. Realization of quantum process tomography in NMR. *Phys. Rev. A*, **64**(Jun), 012314.

Chiribella, G. and Spekkens, R. (eds). 2015. *Quantum Theory: Informational Foundations and Foils*. Japan: Springer Verlag.

Chiribella, G., D'Ariano, G. M., and Schlingemann, D. 2007. How continuous quantum measurements in finite dimensions are actually discrete. *Phys. Rev. Lett.*, **98**(19), 190403.

Chiribella, G., D'Ariano, G. M., and Perinotti, P. 2008a. Quantum circuit architecture. *Phys. Rev. Lett.*, **101**(6), 060401.

Chiribella, G., D'Ariano, G. M., and Perinotti, P. 2008b. Optimal cloning of unitary transformation. *Phys. Rev. Lett.*, **101**, 180504.

Chiribella, G., D'Ariano, G. M., and Perinotti, P. 2008c. Quantum circuit architecture. *Phys. Rev. Lett.*, **101**, 060401.

Chiribella, G., D'Ariano, G. M., and Perinotti, P. 2009. Theoretical framework for quantum networks. *Phys. Rev. A*, **80**(2), 022339.

Chiribella, G., D'Ariano, G. M., and Perinotti, P. 2010a. Probabilistic theories with purification. *Phys. Rev. A*, **81**(6), 062348.

Chiribella, G., D'Ariano, G. M., and Schlingemann, D. 2010b. Barycentric decomposition of quantum measurements in finite dimensions. *J. Math. Phys.*, **51**(2), 022111.

Chiribella, G., D'Ariano, G. M., and Perinotti, P. 2010c. Probabilistic theories with purification. *Phys. Rev. A*, **81**(6), 062348.

Chiribella, G., D'Ariano, G. M., and Perinotti, P. 2011. Informational derivation of quantum theory. *Phys. Rev. A*, **84**(1), 012311.

Chiribella, G., D'Ariano, G. M., Perinotti, P., and Valiron, B. 2013a. Quantum computations without definite causal structure. *Phys. Rev. A*, **88**(Aug), 022318.

Chiribella, G., D'Ariano, G. M., and Perinotti, P. 2013b. A short impossibility proof of quantum bit commitment. *Phys. Lett. A*, **377**, 1076–1087.

Chiribella, G., D'Ariano, G. M., and Perinotti, P. 2014. Non-causal theories with purification. *arxiv*.

Chiribella, G., D'Ariano, G. M., and Perinotti, P. 2015. Non-causal theories with purification. *in preparation*.

Choi, M.-D. 1972. Positive linear linear maps on C^* algebras. *Canad. J. Math.*, **XXIV**(3), 520–529.

Choi, M.-D. 1975. Completely positive linear maps on complex matrices. *Linear Algebra Appl.*, **10**, 285–290.

Chuang, I. L. and Nielsen, M. A. 2000. *Quantum Information and Quantum Computation*. Cambridge: Cambridge University Press.

Clifton, R., Bub, J., and Halvorson, H. 2003. Characterizing quantum theory in terms of information-theoretic constraints. *Foundations of Phys.*, **33**(11), 1561–1591.

Coecke, B. 2008. Introducing categories to the practicing physicist. *Adv. Stud. Math. Logic*, **30**, 45.

Coecke, B. 2006. Introducing categories to the practicing physicist. Pages 289–355 of: *What is Category Theory?* Advanced Studies in Mathematics and Logic, vol. 30. Milan, Italy: Polimetrica Publishing.

Coecke, B., Moore, D., and Wilce, A. 2000. *Current research in operational quantum logic: algebras, categories, languages*. Vol. 111. New York: Springer Science & Business Media.

Cox, R. T. 1961. *The Algebra of Probable Inference*. Baltimore, OH: Johns Hopkins University.

Dakic, B. and Brukner, C. 2011. Quantum theory and beyond: is entanglement special? Pages 365–392 of: Halvorson, H. (ed.), *Deep Beauty: Understanding the Quantum World through Mathematical Innovation*. Cambridge University Press.

D'Ariano, G. M. 1997. Homodyning as universal detection. Pages 365–392 of: Hirota, O., Holevo, A. S., and Caves, C. M. (eds), *Quantum Communication, Computing and Measurement*. New York and London: Plenum.

D'Ariano, G. M. 2005. Homodyning as universal detection. Pages 494–508 of: Hayashi, M. (ed.), *Asymptotic Theory of Quantum Statistical Inference, Selected Papers*. Singapore: World Scientific.

D'Ariano, G. M. 2006a. How to derive the Hilbert-space formulation of quantum mechanics from purely operational axioms. In: Bassi, A., Duerr, D., Weber, T., and Zanghi, N. (eds), *Quantum Mechanics*, vol. 844. USA: American Institute of Physics.

D'Ariano, G. M. 2006b. On the missing axiom of Quantum Mechanics. Pages 114–130 of: *AIP Conference Proceedings*, vol. 810.

D'Ariano, G. M. 2007a. Operational axioms for a C^*,-algebraic formulation for Quantum Mechanics. Page 191 of: Hirota, O., Shapiro, J. H., and Sasaki, M. (eds), *Proceedings of the 8th Int. Conf. on Quantum Communication, Measurement and Computing*. Japan: NICT press.

D'Ariano, G. M. 2007b. Operational axioms for quantum mechanics. Page 79 of: Adenier, G., Fuchs, C. A., and Khrennikov, A. Yu. (eds), *AIP Conference Proceedings*, vol. 889. USA: American Institute of Physics.

D'Ariano, G. M. 2010. Probabilistic theories: what is special about quantum mechanics? Chap. 5 of: Bokulich, A. and Jaeger, G. (eds), *Philosophy of Quantum Information and Entanglement*. Cambridge: Cambridge University Press.

D'Ariano, G. M. and Lo Presti, P. 2001. Quantum tomography for measuring experimentally the matrix elements of an arbitrary quantum operation. *Phys. Rev. Lett.*, **86**(May), 4195–4198.

D'Ariano, G. M. and Lo Presti, P. 2003. Imprinting a complete information about a quantum channel on its output state. *Phys. Rev. Lett.*, **91**, 047902.

D'Ariano, G. M. and Perinotti, P. 2005. Efficient universal programmable quantum measurements. *Phys. Rev. lett.*, **94**(9), 090401.

D'Ariano, G. M. and Perinotti, P. 2007. Optimal data processing for quantum measurements. *Phys. Rev. Lett.*, **98**, 020403.

D'Ariano, G. M. and Tosini, A. 2010. Testing axioms for quantum theory on probabilistic toy-theories. *Quantum Inf. Proc.*, **9**, 95–141.

D'Ariano, G. M. and Tosini, A. 2013. Emergence of space-time from topologically homogeneous causal networks. *Studies in History and Philosophy of Modern Physics*.

D'Ariano, G. M. and Yuen, H. P. 1996. Impossibility of measuring the wave function of a single quantum system. *Phys. Rev. Lett.*, **76**(Apr), 2832–2835.

D'Ariano, G. M., Macchiavello, C., and Paris, M. G. A. 1994. Detection of the density matrix through optical homodyne tomography without filtered back projection. *Phys. Rev. A*, **50**(Nov), 4298–4302.

D'Ariano, G. M., Lo Presti, P., and Sacchi, M. 2000. Bell measurements and observables. *Phys. Lett. A*, **272**, 32.

D'Ariano, G. M., Lo Presti, P., and Perinotti, P. 2005. Classical randomness in quantum measurements. *J. Phys. A: Math. Gen.*, **38**(26), 5979–5991.

D'Ariano, G. M., Giovannetti, V., and Perinotti, P. 2006. Optimal estimation of quantum observables. *J. of Math. Phys.*, **47**, 022102–1.

D'Ariano, G. M., Kretschmann, D., Schlingemann, D., and Werner, R. F. 2007. Reexamination of quantum bit commitment: the possible and the impossible. *Phys. Rev. A*, **76**, 032328.

D'Ariano, G. M, Manessi, F., and Perinotti, P. 2014a. Determinism without causality. *Physica Scripta*, **T163**, 014013.

D'Ariano, G. M., Manessi, F., Perinotti, P., and Tosini, A. 2014b. Fermionic computation is non-local tomographic and violates monogamy of entanglement. *EPL (Europhysics Letters)*, **107**(2), 20009.

Davies, E. B. 1977. Quantum dynamical semigroups and the neutron diffusion equation. *Rep. Math. Phys.*, **11**, 169.

Dieks, D. G. B. J. 1982. Communication by EPR devices. *Phys. Lett. A*, **92**(6), 271–272.

Dowe, P. 2007. *Physical Causation*. Cambridge Studies in Probability, Induction and Decision Theory. Cambridge: Cambridge University Press.

Eberhard, P. H. 1978. Bell's theorem and the different concepts of locality. *Il Nuovo Cimento B Series 11*, **46**(2), 392–419.

Einstein, A., Podolsky, B., and Rosen, N. 1935. Can quantum-mechanical description of physical reality be considered complete? *Phys. Rev.*, **47**(10), 777–780.

Ekert, A. K. 1991. Quantum cryptography based on Bell's theorem. *Phys. Rev. Lett.*, **67**(6), 661.

Ellis, G. F. R. 2008. On the flow of time. *arXiv preprint arXiv:0812.0240.*

Feynman, R. 1965. *The Character of Physical Law.* London: BBC Books.

Foulis, D. J. and Randall, C. H. 1984. A note on misunderstandings of Piron's axioms for quantum mechanics. *Found. Phys.*, **14**, 65–88.

Foulis, D. J., Piron, C., and Randall, C. H. 1983. Realism, operationalism, and quantum mechanics. *Found. Phys.*, **13**, 813–841.

Fuchs, C. A. 2002. Quantum mechanics as quantum information (and only a little more). *arXiv preprint quant-ph/0205039.*

Fuchs, C. A. 2003. Quantum mechanics as quantum information, mostly. *J. of Modern Optics*, **50**(6-7), 987–1023.

Fuchs, C. A. and Schack, R. 2013. Quantum-Bayesian coherence. *Rev. Mod. Phys.*, **85**, 1693–1715.

Fuchs, C. A., *et al.* 2001. Quantum foundations in the light of quantum information. *NATO Science Series Sub Series III Comp. Syst. Sci.*, **182**, 38–82.

Ghirardi, G.-C., Rimini, A., and Weber, T. 1980. A general argument against superluminal transmission through the quantum mechanical measurement process. *Lettere Al Nuovo Cimento (1971–1985)*, **27**(10), 293–298.

Gillies, D. 2000. *Philosophical Theories of Probability*. Move Psychology Press.

Gordon, J. P. and Louisell, W. H. 1966. Simultaneous measurements of noncommuting observables. *Phys. of Quantum Elec.*, **1**, 833–840.

Gorini, V., Kossakowski, A., and Sudarshan, E. C. G. 1976. Completely positive dynamical semigroups of N-level systems. *J. Math. Phys.*, **17**, 821.

Goyal, P., Knuth, K. H., and Skilling, J. 2010. Origin of complex quantum amplitudes and Feynman's rules. *Phys. Rev. A*, **81**(Feb), 022109.

Gregoratti, M. and Werner, R. F. 2003. Quantum lost and found. *Journal of Modern Optics*, **50**(6-7), 915–933.

Gross, D., Müller, M., Colbeck, R., and Dahlsten, O. C. O. 2010. All reversible dynamics in maximally nonlocal theories are trivial. *Phys. Rev. Lett.*, **104**(Feb), 080402.

Grover, L. K. 1996. A fast quantum mechanical algorithm for database search. Pages 212–219 of: *Proceedings of the Twenty-eighth Annual ACM Symposium on Theory of Computing*. STOC '96. New York, NY: ACM.

Haag, R. 1993. *Local Quantum Physics*. Berlin: Springer-Verlag.

Haag, R. and Haag, R. 1996. *Local Quantum Physics: Fields, Particles, Algebras*, Vol. 2. Berlin: Springer.

Hardy, L. 2001. Quantum theory from five reasonable axioms. *arXiv:quant-ph/0101012*.

Hardy, L. 2011. Reformulating and reconstructing quantum theory. *quant-ph*, Apr.

Hardy, L. and Wootters, W. K. 2012. Limited holism and real-vector-space quantum theory. *Found. Phys.*, **42**(3), 454–473.

Heisenberg, W. 1930. *The Physical Principles of the Quantum Theory*. Trans. C. Eckart and F. C. Hoyt. Chicago, IL: Chicago University Press.

Helstrom, C. W. 1976. *Quantum Detection and Estimation Theory*. Mathematics in Science and Engineering, Vol. 123. New York NY: Academic Press.

Herbert, N. 1982. FLASH – A superluminal communicator based upon a new kind of quantum measurement. *Found. Phys.*, **12**, 1171.

Hilbert, D., von Neumann, J., and Nordheim, L. 1928. Über die Grundlagen der Quantenmechanik. *Mathematische Annalen*, **98**(1), 1–30.

Holevo, A. S. 1982. *Probabilistic and Statistical Aspects of Quantum Theory*. Series in Statistics and Probability, vol. 1. Amsterdam: North-Holland.

Holevo, A. S. 2011. The Choi–Jamiolkowski forms of quantum Gaussian channels. *J. Math. Phys.*, **52**(4), 042202.

Horodecki, M., Horodecki, P., and Horodecki, R. 1996. Separability of mixed states: necessary and sufficient conditions. *Phys. Lett. A*, **223**, 1–8.

Imamoglu, A. 1993. Logical reversibility in quantum-nondemolition measurements. *Phys. Rev. A*, **47**, R4577–R4580.

Jamiolkowski, A. 1972. Linear transformations which preserve trace and positive semidefiteness of operators. *Rep. Math. Phys.*, **3**, 275.

Jauch, J. M. and Piron, C. 1963. Can hidden variables be excluded in quantum mechanics? *Helv. Phys. Acta*, **36**, 827.

Jaynes, E. T. 2003. *Probability Theory: The Logic of Science.* Cambridge: Cambridge University Press.

Johnson, D. and Feige, U. (eds) 2007. *Toward a General Theory of Quantum Games.* Vol. STOC '07 Symposium on Theory of Computing Conference. New York, NY: ACM.

Jordan, P., von Neumann, J., and Wigner, E. 1934. On an algebraic generalization of the quantum mechanical formalism. *Ann. Math.*, **35**, 29.

Joyal, A. and Street, R. 1991. The geometry of tensor calculus, I. *Adv. Math.*, **88**(1), 55–112.

Kaiser, D. 2011. *How the Hippies Saved Physics: Science, Counterculture, and the Quantum Revival.* New York: WW Norton & Company.

Kato, T. 1980. *Perturbation Theory for Linear Operators.* New York, NY: Springer.

Keynes, J. M. 2004. *A Treatise on Probability.* Dover Books on Mathematics Series. Mineola, NY: Dover Publications.

Knill, E. and Laflamme, R. 1997. Theory of quantum error-correcting codes. *Phys. Rev. A*, **55**(Feb), 900–911.

Kretschmann, D. and Werner, R. F. 2005. Quantum channels with memory. *Phys. Rev. A*, **72**, 062323.

Leung, D. 2001. *Towards Robust Quantum Computation.* PhD thesis, Stanford University.

Lindblad, G. 1976. On the generators of quantum dynamical semigroups. *Comm. Math. Phys.*, **48**, 199.

Lindblad, G. 1999. A general no-cloning theorem. *Lett. Math. Phys.*, **47**(2), 189–196.

Lo, H.-K. and Chau, H. F. 1997. Is quantum bit commitment really possible? *Phys. Rev. Lett.*, **78**, 3410.

Ludwig, G. 1983. *Foundations of Quantum Mechanics.* New York, NY: Springer-Verlag.

Mac Lane, S. 1978. *Categories for the Working Mathematician*, Vol. 5. Berlin: Springer Science & Business Media.

Mackey, G. W. 1963. *The Mathematical Foundations of Quantum Mechanics*, Vol. 1. Mineola, NY: Dover Publications.

Masanes, L. and Müller, M. P. 2011. A derivation of quantum theory from physical requirements. *New J. Phys.*, **13**(6), 063001.

Masanes, L., Müller, M. P, Augusiak, R., and Pérez-García, D. 2013. Existence of an information unit as a postulate of quantum theory. *Proc. Nat. Acad. Sci.*, **110**(41), 16373–16377.

Mayers, D. 1997. Unconditionally secure quantum bit commitment is impossible. *Phys. Rev. Lett.*, **78**, 3414.

Nielsen, M. A. and Chuang, I. L. 1997. Programmable quantum gate arrays. *Phys. Rev. Lett.*, **79**(2), 321–324.

Oreshkov, O., Costa, F., and Brukner, C. 2012. Quantum correlations with no causal order. *Nature Commun.*, **3**, 1092.

Ozawa, M. 1984. Quantum measuring processes of continuous observables. *J. Math. Phys.*, **25**, 79.

Ozawa, M. 1997. Quantum state reduction and the quantum bayes principle. Pages 233–241 of: Hirota, O., Holevo, A. S., and Caves, C. M. (eds), *Quantum Communication, Computing and Measurement.* New York, NY: Plenum.

Paris, M. and Rehacek, J. 2004. *Quantum State Estimation*, Vol. 649. Berlin: Springer.

Parthasarathy, K. R. 1999. Extremal decision rules in quantum hypothesis testing. *Infin. Dimens. Anal. Quantum. Probab. Relat. Top.*, **02**(04), 557–568.

Paulsen, V. I. 1986. *Completely Bounded Maps and Dilations*. Harlow, UK: Longman Scientific and Technical.

Pearl, J. 2012. *Causality: Models, Reasoning, and Inference*. Cambridge: Cambridge University Press.

Penrose, R. 1971. Applications of negative dimensional tensors. Pages 221–244 of: Welsh, D. J. A. (ed.), *Combinatorial Mathematics and its Applications*. New York, NY: Academic Press.

Piron, C. 1964. Axiomatique quantique. *Helvetica Phys. Acta*, **37**(4-5), 439.

Piron, C. 1976. *Foundations of Quantum Physics*. Mathematical Physics Monograph Series. Benjamin-Cummings Publishing Company.

Planck, M. 1941. *Der Kausalbegriff in der Physik*. Verlag von S. Hirzel.

Plenio, M. B. and Virmani, S. 2007. An introduction to entanglement measures. *Quantum Inf. Comput.*, **7**, 1.

Procopio, L. M, Moqanaki, A., Araújo, M., Costa, F., Alonso Calafell, I., Dowd, E. G., *et al.* 2015. Experimental superposition of orders of quantum gates. *Nature Commun.*, **6**(Aug.), 7913.

Rambo, T., Altepeter, J., D'Ariano, G. M., and Kumar, P. 2012. Functional quantum computing: an optical approach. *arXiv:1211.1257*.

Rockafellar, R. T. 2015. *Convex Analysis*. Princeton, NJ: Princeton University Press.

Royer, A. 1994. Reversible quantum measurements on a spin 1/2 and measuring the state of a single system. *Phys. Rev. Lett.*, **73**, 913–917.

Royer, A. 1995. Reversible quantum measurements on a spin 1/2 and measuring the state of a single system (erratum). *Phys. Rev. Lett.*, **74**, 1040–1040.

Russel, B. 1912. On the notion of cause. *Proc. Aristotelian Soc.*, **13**, 1–26.

Salmon, W. 1967. *The Foundations Of Scientific Inference*. Pittsburgh, PA: University of Pittsburgh Press.

Salmon, W. C. 1998. *Causality and Explanation: Wesley C. Salmon*. Oxford Scholarship online. Oxford University Press.

Schrödinger, E. 1935a. Die gegenwärtige Situation in der Quantenmechanik. *Naturwissenschaften*, **23**, 807–812; 823–828; 844–849.

Schrödinger, E. 1935b. Probability relations between separated systems. Pages 555–563 of: *Proc. Camb. Phil. Soc*, vol. 31. Cambridge Univ Press.

Schumacher, B. 1996. Sending entanglement through noisy quantum channels. *Phys. Rev. A*, **54**, 2614–2628.

Schumacher, B. and Nielsen, M. A. 1996. Quantum data processing and error correction. *Phys. Rev. A*, **54**, 2629–2635.

Schumacher, B. and Westmoreland, M. D. 2012. Modal quantum theory. *Found. Phys.*, **42**(7), 918–925.

Scott, A. J. and Grassl, M. 2009. SIC-POVMs: a new computer study. *arXiv preprint arXiv:0910.5784*.

Selinger, P. *A Survey of Graphical Languages for Monoidal Categories*. Available at www.mathstat.dal.ca/selinger/papers/graphical.pdf.

Selinger, P. 2011. A survey of graphical languages for monoidal categories. Pages 289–355 of: *New Structures for Physics*. Berlin: Springer.

Shor, P. W. 1997. Polynomial-time algorithms for prime factorization and discrete logarithms on a quantum computer. *SIAM J. Comp.*, **26**(5), 1484–1509.

Smithey, D. T., Beck, M., Raymer, M. G., and Faridani, A. 1993. Measurement of the Wigner distribution and the density matrix of a light mode using optical homodyne tomography: Application to squeezed states and the vacuum. *Phys. Rev. Lett.*, **70**(Mar), 1244–1247.

Solovay, R. M. 1970. A model of set theory in which every set of reals is Lebesgue measurable. *Annals of Mathematics*, **92**, 1–56.

Ueda, M. and Kitagawa, M. 1992. Reversibility in quantum measurement processes. *Phys. Rev. Lett.*, **68**, 3424–3427.

Uhlmann, A. 1977. Relative entropy and the Wigner–Yanase–Dyson–Lieb concavity in an interpolation theory. *Commun. Math. Phys.*, **54**(1), 21–32.

Varadarajan, V. S. 1962. Probability in physics and a theorem on simultaneous observability. *Comm. Pure Appl. Math.*, **15**, 189.

Vogel, K. and Risken, H. 1989. Determination of quasiprobability distributions in terms of probability distributions for the rotated quadrature phase. *Phys. Rev. A*, **40**(Sep), 2847–2849.

von Neumann, J. 1932. *Mathematische Grundlagen der Quantenmechanik*. Berlin: Springer-Verlag. Translated as *Mathematical Foundations of Quantum Mechanics*, Princeton University Press, 1955. Chap. 4.

von Neumann, J. 1996. *Mathematical Foundations of Quantum Mechanics*, Vol. 2. Princeton, NJ: Princeton university press.

Werner, R. F. 1989. Quantum states with Einstein–Podolsky–Rosen correlations admitting a hidden-variable model. *Phys. Rev. A*, **40**, 4277–4281.

Wheeler, J. A. 1990. Complexity, entropy and the physics of information. In: Zurek, W. H. (ed.), *Santa Fe Institute Studies in the Sciences of Complexity, Proceedings of the 1988 Workshop on Complexity, Entropy and the Physics of Information, Santa Fe, New Mexico, May–June, 1989*. Redwood City, CA: Addison-Wesley.

Wiesner, S. 1983. Conjugate coding. *Sigact News*, **15**, 78.

Wilce, A. 2010. Formalism and interpretation in quantum theory. *Foundations of Physics*, **40**(4), 434–462.

Wilce, A. 2012. Conjugates, correlation and quantum mechanics. *arXiv:1206.2897*.

Wootters, W. K. and Zurek, W. H. 1982. A single quantum cannot be cloned. *Nature*, **299**(5886), 802–803.

Yao, A. C.-C. 1993. Quantum circuit complexity. Pages 352–361 of: *Proceedings of Thirty-fourth IEEE Symposium on Foundations of Computer Science (FOCS1993)*.

Yuen, H. P. 1986. Amplification of quantum states and noiseless photon amplifiers. *Phys. Lett. A*, **113**, 405–407.

Yuen, H. P. 2012. An unconditionally secure quantum bit commitment protocol. *arXiv preprint arXiv:arXiv:1212.0938v1*.

Index

adjoint
  representation, 82
  map, 35
ancilla-assisted tomography, 54
anyons, 113
arrow of time, 142
atomic
  effect, 24
  instrument, 46
  point in a convex cone, 23
  state, 24
  transformation, 24, 43
atomicity of composition, 130

$B_s(\mathcal{H})$, 14
Banach spaces
  of operators, 74
base of convex cone, 146
Bell measurement, 33, 41
Bhatia dilation of effect, 83
Bloch
  ball, 18
  sphere, 18

Carathéodory's theorem, 173
category theory, 5
  strictly symmetric monoidal, 5
causal chain, 142
causality, 20, 135
  axiom, 141
  causal theories, 141
  in philosophy, 6
  no signaling from the future, 142
  philosophy, 139
CB norm, 123, 128
central extension, 100
channel, 119
  correctable upon input, 66
  correlation-erasing, 66
  purification-preserving, 66
Choi–Jamiołkowski
  isomorphism, 41
  operator, 41
circuit, 112
classical theory, 49

closed circuits, 115, 125
  event, 119
  test, 119
coarse-graining, 13, 22, 119
cocycle, 100
commuting POVM, 59
complementary channel, 67
complete OPT, 149
complete positivity, 36
completely bounded map, 128
completely bounded norm, 123
completely mixed state, 34, 54, 132
completely positive map, 36
composite system, 113
compression
  efficient, 134
  ideal, 34, 134
  lossless, 134
conditional preparation, 145
conditional state, 19, 38, 43
conditioned test, 142
conditioning-equivalent transformations, 149
conic combinations, 117
conservation of information, 170
contractive operator, 44
convex
  OPT, 149
  refinement, 23
  refinement set, 23
CP, 42
CP map, 36
  iff condition, 84

dagger correspondence, 229
decoding, 134
deletion channel, 66
density matrix, 16
density operator, 16
deterministic OPT, 148
diamond norm, 123, 128
directed acyclic graph, 4
discrimination, 121
dissipative evolution, 107
double-covering, 102
double-ket notation, 39
dual set, 28

Printed in the United States
By Bookmasters